Invertebrate Conservation and Agricultural Ecosystems

Invertebrate Conservation and Agricultural Ecosystems is both an introduction to invertebrate conservation biology for agriculturists and an introduction to crop protection for conservation biologists, demonstrating how these two disparate fields may draw on each other for greater collective benefit. It makes use of recent literature to show how invertebrate conservation in highly altered landscapes may be promoted and enhanced.

The book deals with problems of, and approaches to, invertebrate conservation in highly managed agricultural ecosystems, and examines how biodiversity may be promoted without compromising agricultural production. It draws attention to the importance of invertebrates in agricultural systems and their role in ecosystem functions.

Dr Tim New is Reader and Associate Professor in Zoology at La Trobe University, Melbourne, Australia. He has broad interests in insect conservation, systematics and ecology, and he has published extensively on these topics, with over 350 research papers and more than 20 books. In 2003 he was awarded the Marsh Christian Trust Award for insect conservation by the Royal Entomological Society. Dr New is currently Editor-in-Chief of the *Journal of Insect Conservation*.

ECOLOGY, BIODIVERSITY, AND CONSERVATION

The world's biological diversity faces unprecedented threats. The urgent challenge facing the concerned biologist is to understand ecological processes well enough to maintain their functioning in the face of the pressures resulting from human population growth. Those concerned with the conservation of biodiversity and with restoration also need to be acquainted with the political, social, historical, economic and legal frameworks within which ecological and conservation practice must be developed. This series will present balanced, comprehensive, up-to-date and critical reviews of selected topics within the sciences of ecology and conservation biology, both botanical and zoological, and both 'pure' and 'applied'. It is aimed at advanced final-year undergraduates, graduate students, researchers, and university teachers, as well as ecologists and conservationists in industry, government and the voluntary sectors. The series encompasses a wide range of approaches and scales (spatial, temporal, and taxonomic), including quantitative, theoretical, population, community, ecosystem, landscape, historical, experimental, behavioural, and evolutionary studies. The emphasis is on science related to the real world of plants and animals, rather than on purely theoretical abstractions and mathematical models. Books in this series will, wherever possible, consider issues from a broad perspective. Some books will challenge existing paradigms and present new ecological concepts, empirical or theoretical models, and testable hypotheses. Other books will explore new approaches and present syntheses on topics of ecological importance.

Ecology and Control of Introduced Plants
Judith H. Myers and Dawn R. Bazely

Invertebrate Conservation and Agricultural Ecosystems

T. R. NEW

La Trobe University, Melbourne, Australia

CAMBRIDGE UNIVERSITY PRESS
Cambridge, New York, Melbourne, Madrid, Cape Town, Singapore, São Paulo

Cambridge University Press
The Edinburgh Building, Cambridge CB2 2RU, UK

www.cambridge.org
Information on this title: www.cambridge.org/9780521825032

First published 2005

Printed in the United Kingdom at the University Press, Cambridge

A catalogue record for this book is available from the British Library

ISBN-13 978-0-521-82503-2 hardback
ISBN-10 0-521-82503-2 hardback

ISBN-13 978-0-521-53201-9 paperback
ISBN-10 0-521-53201-9 paperback

Contents

Preface

Increasing commitment of land to agriculture and the intensification of agricultural practices in many parts of the world have been associated with widespread and substantial loss of biodiversity, much of it undocumented and unheralded. As the most diverse larger components of that biodiversity, insects and other invertebrate animals play major roles in ecosystem function and sustainability in both natural and modified ecosystems, including agricultural ecosystems. However, agriculturists and conservationists have a long history of differing attitudes towards these animals. The former are concerned with crop and other commodity protection, in which depredations by invertebrate pests cause major economic losses, and suppression of their effects (commonly by killing them) is a major management necessity. The latter are concerned over the replacement of complex natural habitats with relatively simple 'agroecosystems', which reputedly support only very limited numbers of species, and with the wider environmental ramifications of crop-protection measures, such as use of pesticides and exotic species on non-target areas and species.

This book is both an introduction to the principles of agricultural pest management for invertebrate conservation biologists and an introduction to invertebrate importance and conservation for applied entomologists and pest managers. My major aim is for each party to appreciate more fully the perspectives of the other and to realise the amount of common ground that their viewpoints may encompass. This appreciation is vital in order to promote more holistic management for the benefits of invertebrates and other organisms, and for wider ecological sustainability in the mosaic of modified and more natural areas that comprise modern landscapes. Whereas the principles are of much wider relevance than to invertebrates alone, the relatively fine ecological scales on which many invertebrates operate provide wide lessons for increased understanding of their responses to environmental changes and of the ways in which their wider wellbeing can be promoted. Agroecosystems are a critical focus in any such endeavour.

Many of the topics discussed here have been treated extensively in a series of papers, symposia and books in the past decade or so, and part of the aim of this book is to provide a balanced synthesis of these diverse and scattered publications. The major focus is reducing damage to environments and enhancing conservation prospects through refining agricultural practices and harmonising these with wider conservation needs, and demonstrating the sincere and substantial progress that has been made in this endeavour. Many of these recent publications are cited in the text, and the great variety of scientific journals that encompass relevant issues are evident from the references provided; some, indeed, cover little but these topics. And, as with much other scientific advance, the amount of pertinent material on the World Wide Web is both daunting and of very mixed relevance and accuracy.

The first chapter introduces the changes to natural ecosystems wrought by agricultural development, the characteristics of agroecosystems, and the importance of incorporating these modified ecosystems into a wider conservation perspective to complement the conventional 'conservation estate' founded on protected areas. The broader ecological importance and diversity of invertebrates in agroecosystems, together with background on the main groups involved, are introduced in Chapter 2, and this is followed by a broad appraisal of the threats they face and their values in wider assessments of environmental change and quality (Chapter 3). A practical agricultural perspective of invertebrates as pests is given in Chapter 4, together with a summary of the development of the science of integrated pest management. The consequences and concerns of one of the three main early strands of crop protection, application of pesticides, are also discussed. The wider concerns of other aspects of pest management on crops are discussed in the next two chapters, which deal respectively with biological and cultural control measures. Expansion to the wider environment and considerations of landscape ecology to harmonise pest management and conservation ideals comprise Chapters 7 and 8, with the latter emphasising the consequences of habitat fragmentation and the need to promote wider connectivity in landscapes. Conservation concerns in pasture management are treated briefly in Chapter 9. Finally, Chapter 10 evaluates and discusses ways in which agroecosystem management and more conventional conservation management may both benefit from lessons from terrestrial invertebrate ecology, and how these may be integrated to promote more holistic conservation for invertebrates without unduly compromising the needs for agricultural commodity protection and assured economic sustainability.

Acknowledgements

The following publishers have kindly given permission for me to use or modify diagrams or tabular material from books to journals to which they hold copyright. Individual references to sources are given in the relevant legends throughout the book: Association of Applied Biologists, Wellesbourne, UK; Birkhauser-Verlag AG, Basel, Switzerland; Blackwell Publishing, Malden, USA and Oxford, UK; British Crop Protection Enterprises Ltd, Alton, UK; CABI Publishing, Wallingford, UK; Cambridge University Press, Cambridge, UK; CRC Press, Boca Raton, FL, USA; Ecological Society of America; Elsevier Science, Oxford, UK (also for Academic Press); Intercept Ltd, Andover, UK; John Wiley & Sons Ltd, Chichester, UK and Hoboken, USA; Kluwer Academic Publishers, Dordrecht, the Netherlands and New York, USA; Netherlands Entomological Society, Amsterdam, the Netherlands; Springer-Verlag GmbH & Co KG, Heidelberg, Germany; University of California Press, Berkeley, CA, USA; and Urban & Fischer Verlag, Jena.

I appreciate greatly the enthusiastic support for this project given by Ward Cooper at Cambridge University Press, together with the support of Jayne Aldhouse and Carol Miller (as production editors) and the patient copy-editing of Colette Holden.

1 · *Introduction: agricultural ecosystems and conservation*

Development of agriculture has had profound effects on natural communities and has involved changes in land-use patterns over much of the world. In general, the changes have involved replacement of complex natural communities by much simpler and less diverse systems, and loss or reduced abundance of numerous species of the fauna (of which invertebrates are predominant components) and flora previously present. Intensification of agriculture has escalated these changes and led to the development of so-called 'agroecosystems' as substantial and highly managed terrestrial ecosystems, in which commodity production (rather than conservation) is the major aim and outcome. Agroecosystems are an important arena in which to enhance conservation, including that of invertebrates, for major benefits to biodiversity.

Introduction

Agriculture has been defined as 'the art, practice, or science of crop and livestock production on organised farm units' (Pesek, 1993) and is the world's most extensive industry, one upon which humankind depends. Indeed, the future of human populations is linked fundamentally with sustaining agricultural production. Agriculture is thus the single largest component of global land use, with some 36% of the world's land surface devoted to providing the primary produce needed to sustain people (Gerard, 1995). Clearing of natural vegetation to increase this proportion is continuing in many parts of the world. Agriculture is also amongst the most varied suite of activities that may compromise the integrity of natural environments, and the variety of scales and activities associated with a very broad range of products renders generalisations about its wider effects difficult. However, agricultural activity, and its enhancement to cater for the needs of growing human populations, has led to remodelling of natural landscapes in many countries, to the extent that agricultural landscapes encompassing pasture and crops are often regarded as largely 'natural', particularly in parts of the northern hemisphere, where centuries of change have occurred. In parts of Europe, the transformation of natural ecosystems has rendered it impossible to find existing woodland habitats that can be considered realistic analogues of forests characteristic of the climatic climax vegetation of only some 5000–7000 years ago (Mannion, 1995). In much of the tropics and the southern hemisphere, such massive changes have occurred over a shorter period – in Australia, for example, only over about the past 200 years – and can be considered realistically in relation to the properties of the more pristine parental environments that they have replaced.

The history of agricultural development, though, shows parallels in many places, with changes to increase productivity, efficiency and economic wellbeing imposing the need to oppose factors (such as pests that attack crops and livestock) that may compromise those aims. These trends have also been associated with changing scales of operation. Agriculture requires environmental modification, but the extent of this varies and is reflected in the enormous variety of global agricultural systems. At one end of the spectrum of effects, practices such as nomadic pastoralism may necessitate little change, other than for transitory or seasonal effects, and the host ecosystems may remain largely unaltered. At the other extreme, intensive crop production involves massive changes, with severe implications for environmental wellbeing, and that may be regarded as 'permanent'. In principle, systems such as nomadic pastoralism and small-scale shifting agriculture reflect the 'low-input' end of the technological gradients (see below) and represent the beginnings from which more intensive systems have been developed. Neither practice involves use of fossil fuels, or other than very minimal chemical inputs, and both involve temporary use of land areas. Shifting agriculture is generally limited by availability of nutrients and commonly involves clearing of natural vegetation that is collected, dried and burned before cultivation of crops. Most occurs in the tropics, and the major variants in practice reflect local cultural practices and environmental conditions.

More permanent 'settled agriculture' is based largely on arable systems, i.e. those that historically require ploughing, tillage or other ground preparation before crop cultivation and that comprise the most diverse category of agricultural production systems. Croplands occupy some 11% of land area, not including the somewhat special case of agroforestry (see p. 296). In contrast, settled pastoral systems (extended to include the world's rangelands) occupy more land (with estimates of up to 50% of land area) (Mannion, 1995) but lesser variety, reflecting the regimes in which the relatively few species of domestic livestock are produced. Both these major categories have diversified in relation to (1) what is grown and (2) how it is grown, and the purposes for which the commodities are intended. Both occur in all except the most inhospitable parts of the world, in which productivity is limited by climate extremes such as lack of rainfall (although irrigation may, in places, counter this) and nutrient/food availability. Mixed farming systems involve integrating livestock and crop production, in many cases involving fodder crops for the livestock and human commodity crops. Collectively, these practices have affected all major terrestrial ecosystems directly, and aquatic systems more indirectly, with Hart *et al.* (1994) claiming that around 75% of Earth's habitable land has been disturbed to some extent by agriculture-related activities.

These activities thus range from local, largely subsistence agriculture to intensive operations over large land areas to satisfy export needs as major components of national economies. Very broadly, this variety allows agricultural activities and systems to be divided into three major categories, representing points in a cultural and economic continuum as progressive increases in complexity and ecological impact:

1 No-input agriculture. This is the simplest form of agriculture, involving very few inputs for management, and is involved predominantly with controlling amounts taken from a system and thereby leaving it sustainable. Such harvest-only systems are exemplified by gathering wild crops (rather than planting crops under 'improved' conditions) and grazing stock on open rangelands rather than on improved pasture.
2 Low-input agriculture. This level includes more intensive management than the above and is exemplified by many kinds of subsistence agriculture in which desirable crop species are planted and protected by removal of competitors, such as weeds and pests, to assure and increase yields.
3 High-input agriculture. This is the most intensive category of agriculture, involving substantial human management of production systems, and is typified by development of many forms of agroecosystem (see p. 7). Substantial amounts of energy and chemicals may be needed to maintain, promote and protect species cultivated in large areas of monoculture, and intensive management is usually a necessary cost in the production system. This category therefore typifies much agricultural production in industrialised countries and may have severe effects on local environments and biota (Gerard, 1995).

Agricultural intensification increases in the above sequence. Intensification has three major axes, each associated with changes of farming practice and each affecting the parental environment and the biota of surrounding areas. Intensification reflects, amongst other things, that (1) land available is finite, so better use must be made of land already in production, and (2) even if further land is available, concerns over continued intrusion into natural ecosystems through additional land conversion for agriculture commonly render this undesirable or controversial. These axes are:

1 Space. Spatial intensification involves a greater proportion of available land being used intensively. Increased clearing of natural vegetation may eventuate, sometimes to create larger cultivable units. Loss of native biota may be relevant to producers through loss of refuges or habitat for natural enemies of crop pests, and the major obvious (visual) environmental effect is loss of remaining natural habitat fragments or other residual vegetation such as that around field margins.
2 Time. Temporal intensification involves higher land use made possible by practices such as irrigation, increased fertiliser inputs, and development of fast-growing plant varieties. Increased inputs of fertilisers and pesticides for crop promotion and protection have, in some cases, obviated the need for fallow periods or crop rotations, as losses of soil quality and continuity of pest populations in the area are countered by these other means.
3 Technology. Technological intensification, touched on above, is reflected in the reliance on massive amounts of chemical use as a staple way to maintain and improve crop yields. This has led, amongst other things, to needs for intensive management, for example for resistance to insecticides, and innovative measures such as genetically modified plants to increase uniformity, predictability of yield and protection against pests.

In each of these aspects, intensification practices modify the characteristics of the system. Spatial intensification is associated directly with declines in native species that previously could be sustained in the mosaic of habitats in the landscape. Temporal intensification may omit rotation of recuperative crops in favour of further sequences of profitable extractive crops, so that maintenance of soil quality depends increasingly on greater use of fertilisers. Technological intensification leads to reduced variability in the systems, facilitating efficient pest management, harvesting procedures and predictability of phenology and production. Very broadly, each may reduce independent sustainability of the systems, and greater levels of intensification demand progressively greater levels of external management to assure productivity.

With this realisation that continued high-input intensification is associated with increasing levels of non-sustainability, considerable emphasis is now being placed on modification of agricultural practices to increase and assure sustainable agriculture. Broadly, the term 'sustainable' implies regenerative processes that preferentially use locally available resources and natural processes, such as nutrient recycling, and limit the use of external inputs of agrochemicals and non-renewable energy. 'Regenerative agriculture' requires that any such external inputs are used efficiently, so that by-products can be recycled and absorbed. The term 'sustainable agriculture' first appeared in the literature in 1978 (Kogan, 1998). However, the more than 80 proposed definitions of 'sustainability' advanced in the past decade or so emphasise a considerable variety of different values, goals and postulates. It may imply, for example, persistence and capacity to continue; or it may primarily imply not damaging or losing natural resources. In addition, it has wider connotations of environmental conservation and product safety. The pragmatic aims thereby converge with environmentalists' ideals in encompassing benefits allied with 'ecologically clean', 'low input', 'organic', 'alternative' and the like in emphasising reduced technological and chemical inputs to the agricultural systems.

The major trends in management (after NRC, 1989) are:

1 Diversification rather than continuous planting of fields to single or few annual crops.
2 Biological and other pest controls or management, including integrated pest management (IPM) (Chapter 4); measures that reduce pesticide applications.
3 Disease prevention in livestock, rather than routine prophylactic antibiotic doses.
4 Genetic improvements in crops for purposes such as (1) pest and disease resistance, (2) drought tolerance and (3) increased efficiency of nutrient use.

These principles have been enunciated in various, overlapping ways. Thus, Oades & Walters (1994) noted that sustainable agriculture should:

1 maintain or improve the production of 'clean' foods;
2 maintain or improve the quality of landscapes, which include soils, water, biota and aesthetic attributes;

Table 1.1. *Objectives of sustainable agriculture (after Pretty, 1998)*

1. Incorporation of natural processes, such as nutrient recycling, nitrogen fixation and pest–predator relationships into agricultural production processes, so ensuring profitable and efficient food production.
2. Reduction in use of those external and non-renewable inputs with greatest potential to damage the environment or harm health of farmers and consumers. More targeted use of remaining inputs, with a view to minimising costs.
3. Full participation of farmers and rural people in all processes of problem analyses, technology development, adaptation and extension.
4. More equitable access to productive resources and opportunities, and progress towards more socially just forms of agriculture.
5. Greater productive use of local knowledge and practices, including innovative processes.
6. Increase in self-reliance amongst farmers and local people.
7. Improved matching between cropping patterns and productive potential and environmental constraints of the climate and landscape to ensure long-term sustainability of current production levels.

3 have minimal impacts on the wider environment;
4 be economically viable.
5 be acceptable to society.

Pretty (1998) considered a wider range of farming objectives (Table 1.1), recognising the philosophical viewpoint that 'sustainable agriculture seeks the integrated use of a wide range of pest, nutrient, and soil and water technologies' with community level involvement to facilitate increasing substitution of natural processes for external inputs 'so the impact on the environment is reduced'.

Management of agricultural systems when the emphasis is on enhancing basic biological systems differs greatly from that based on continued, accelerated use of external inputs. Much of this book draws on one major context of massive relevance in sustainability – that of pest management in agricultural systems and the accompanying concerns for the most diverse components of animal diversity, the non-target invertebrates. The milieux of concern are so-called 'agroecosystems', a term that can be applied either to areas used primarily for cultivation or (on much larger scales) to areas such as natural catchments in which the effects of agricultural practices may ramify.

Table 1.2. *Structural and functional differences between natural ecosystems and agroecosystems (from Gliessman, 1997, after Odum, 1969)*

	Natural ecosystems	Agroecosystems
Net productivity	Medium	High
Trophic interactions	Complex, webs	Simple, linear
Species diversity	High	Low
Genetic diversity	High	Low
Nutrient cycles	Closed	Open
Resilience	High	Low
Duration/permanence	Long	Short
Habitat heterogeneity	Complex	Simple
Human control	Independent	Dependent

Agroecosystems and agroecology

The development of the concept of 'agroecosystems' has helped ecologists to focus on the peculiar features of agricultural systems and the many ways in which they may differ from natural ecosystems. The field of agroecology has grown in parallel to reflect those differences (Gliessman, 1990, 1997).

Most fundamentally, an agroecosystem is an agricultural production unit (such as a farm, field or orchard) understood as an ecosystem (Gliessman, 1997) and thus conventionally has imposed physical boundaries. However, changes within those boundaries transcend them to penetrate much of the surrounding area. Agroecosystems differ from natural ecosystems in some important ways (Table 1.2), with the extent of differences reflecting the intensity of agricultural practice. They are subject to a similar variety of constraints and ecological rules.

Their major features (after Pedigo, 1996) are:

- Agroecosystems often have very limited duration because the life of crops can be very short. Each crop may be removed completely (many field crops, in particular) at harvest, so any equivalent to long-term successional change is then absent.
- Agroecosystems commonly undergo massive changes from external management, such as tilling, ploughing, chemical applications, and changes in microclimate and soil quality.

- Agroecosystems commonly are dominated by exotic species. If native species are used, then these may have been modified substantially from their ancestral forms through long histories of artificial selection or imposed genetic uniformity.
- Most agroecosystems are assumed to have very low species diversity (see p. 12); many are monocultures, areas of a single plant species or variety with little intraspecific variability and with measures taken to prevent increase of diversity during crop life.
- Many agroecosystems also consist of plants of imposed uniform age and size, so development from germination to harvest is uniform, with phases such as flowering or seeding occurring simultaneously in all individuals. Density of the plants may be much higher than in natural communities.
- Agroecosystems are commonly 'enriched' by addition of fertilisers, rendering the plants nutritious and attractive to many herbivores. Rapid growth may prolong the presence of tender, palatable tissues.

To a large extent, agroecosystems are 'purpose-built', and the maintenance of natural ecosystem functions has been regarded as low-priority, in relation to increased efficiency of production. Much of this trend has been driven by simple economic need; for example, labour costs in Germany rose by 300% between 1950 and 1970, whereas prices for agricultural products rose by only 25% during this period. Survival of farmers necessitated increased efficiency and productivity, obtained through increased specialisation, reduced machinery requirements, concentration on a few cash crops, and increasing fertiliser and pesticide inputs (Kuhbauch, 1998). This example is paralleled in many other places and indicates a major driver for intensification and accelerated ecological change associated largely with decreased agricultural diversity (Table 1.3).

Each of these changes links with perceived decline of biodiversity. Thus, for example, the larger the number of crop species, the larger the overall pool of resident species of animals and plants is likely to be, because each crop provides different resources and different environmental conditions for coexisting species. Each crop species may have its characteristic complement of plant weeds as well as its own consumers and their predators and parasitoids (see p. 142). Many of these taxa are the direct targets for pesticide use or other management, so that higher diversity of crops may be accompanied by more varied pesticide applications.

Table 1.3. *Indicators of diversity in agriculture (Kuhbauch, 1998)*

1. Number of crop species used in single farms or regions.
2. Subdividing arable fields into individual smaller fields.
3. Size of individual fields.
4. Percentage area and distribution of land not used by crops (i.e. with hedges, trees, etc.).
5. Intensity and frequency of farm-management inputs (fertilisation, pesticide applications, harvesting, etc.).
6. Number and species used in animal husbandry, per farm, per area unit, per region.
7. Diversity of non-agricultural vegetation, including weeds.
8. Diversity of companion plants.
9. Number of individual farms in region.

Accompanying organisms of no interest to agriculturists are likely to be displaced by management.

Establishment and maintenance of agroecosystems thereby encompasses several levels of change. The initial establishment may entail removal of long-lived native vegetation and its replacement by a few transient exotic species, so that loss of natural habitats is the major initial perturbation. This is followed by persistent attempts to block any natural tendency for diversity to increase again, and the large areas treated are associated with landscape-level effects such as fragmentation and progressive isolation of remaining natural habitats. Indeed, in some places, patches of natural habitat remaining in largely agricultural landscapes are mostly fortuitous. The Western Australian 'wheat belt', for example, has been almost wholly cleared, other than for small patches that were too rocky or too steep for easy cultivation. Many such patches are now valued as remnant habitats for endemic biota largely extirpated from the otherwise altered landscape.

The pattern painted for Europe (Kuhbauch, 1998) is of much wider relevance. Four likely regimes resulting from agricultural intensification were perceived there, namely:

- areas with intensive agricultural production and low biodiversity;
- nearly bare regions, some of them reforested, with medium biodiversity;
- a small proportion of areas and farms in which sustainable farming methods lead to relatively high biodiversity;
- some uncultivated 'islands' of natural resources with oligotrophic soils that preserve high biodiversity.

Agroecosystems and natural ecosystems

The axes of change from natural ecosystems to agroecosystems encompass the most significant ecological concerns for conservationists, many of them centering on diversity, stability (van Emden & Williams, 1974) and levels of disturbance, each reflected in the resilience of the systems – that is, their ability to buffer internally against imposed changes. 'Diversity' is often interpreted simply as species richness, i.e. the number of species present. However, ecological diversity may be assessed also at the genetic level (this approach is relatively rare in practice) or at various structural levels, such as the number of horizontal layers (more complicated 'architecture') within the vegetation present, or patchiness of distribution within an area, in addition to functional complexity (such as analysis of food webs to indicate the complexity of interactions between the species present). High diversity is characteristic of many natural ecosystems and is seen widely as being correlated with resilience or stability. Diversity increases with time, particularly with time since disturbance, so that the later stages in an ecological succession typically support more species, more complex interactions and more structural complexity than earlier successional stages. Agroecosystems only rarely proceed far along successional gradients, and the regular and intensive disturbance is linked not only with loss of biodiversity but also with the inability of the systems to accumulate diversity – which, in any case, is often undesirable to growers as constituting pests (herbivores on crops) or competitors (plant weeds competing within the same nutrient pool as the crop). When an ecosystem is disturbed, structural changes may be apparent as loss of species and of architecture, but these accompany functional simplification, with their extent often reflecting the intensity and frequency of disturbance. In essence, such changes stop succession and cause it to recommence from the new beginning imposed. 'Secondary succession' is succession on disturbed sites, in contrast to 'primary succession' on previously uninhabited sites, such as volcanic ash and lava flows.

Disturbance to ecosystems varies in (1) scale – the spatial extent; (2) intensity – the extent of change by loss of biomass or species; and (3) frequency – with lower frequency implying longer recovery intervals between successive disturbances. These are not always easy to distinguish clearly. Recovery involves invasion and establishment of species, leading to re-establishment of ecological function. The initial species present, the pioneer species or '*r*-selected species', tend to be characterised by their ability to exploit low-diversity environments. They are typically good colonisers with high fecundity (hence the name '*r*-selected', meaning that their intrinsic rate of increase, *r*, is high), so that populations increase rapidly. They are typified by many pest arthropods on crops, and by plant weeds, which can have high impacts on agroecosystems because of the rapidity and intensity of their interactions with them, hence the need for suppression of such species for crop protection.

In nature, many pioneer species cannot cope with more complex environments, either because conditions change or because of interference from other, later-arriving species. As secondary succession proceeds, *r*-species are eliminated or disperse elsewhere and are replaced by a different suite of species with longer-term investment in staying in their habitat. These are typified by 'K-species', i.e. those that are competitively able, can coexist with numerous other species in the same environments, but have lower fecundity. The term 'K' reflects the carrying capacity of the environment, in which the species persist in relatively low numbers. Many K-species are ecological specialists, whereas many *r*-species are regarded as ecological generalists, but these constructs are simply the extremes of a continuous spectrum of adaptations amongst the numerous species constituting a natural community. The main point is that many agroecosystems, by the nature of their temporary existence, are dominated by *r*-species and never furnish the conditions under which the diverse suites of taxa, predominant in climax and mature natural ecosystems, thrive. Partial exceptions occur. Orchard crops and agroforestry crops may persist for at least several decades, during which time substantial accompanying communities can develop and persist. However, many of these crops are still subject to considerable external management, such as control of key pests, so that only rather tenuous parallels occur with succession in natural ecosystems.

Some of the major ecological differences related to duration of natural ecosystems and agroecosystems are summarised in Table 1.4. In climax and near-climax natural ecosystems, stability is implied in that the characteristics listed facilitate internal regulation and balancing of the basic ecological processes. Agroecosystems tend to be 'forced' to retain early successional characteristics, and the scale over which this occurs differentiates them from the syndrome termed 'intermediate disturbance' in natural ecosystems, in which infrequent disturbances may allow continuity of high productivity and high diversity. Localised disturbances of this kind, such as patchy incidence of fires in a landscape, can produce spatial mosaics of different successional stages within the same basic ecosystem, be it grassland, shrubland or forest, and

Table 1.4. *Some changes in ecosystems during secondary succession, each viewed as a continuum*

	Characteristic form in:	
	Early-stage succession	Late-stage succession
Species diversity	Low	High
Species turnover	Rapid replacement	Slow to little change
Species characteristics	Generalists	Specialists
Biomass	Low, rapid rate of increase	High, slow rate of increase
Food chains/food webs	Simple	Complex
Species interactions	Simple	Complex
Population features	*r*-Selected	K-selected
Longevity	Short-lived	Longer-lived

some workers have used this as an analogy to patchy agricultural landscapes. Pickett & White (1985) documented numerous examples of such parallels as relevant to management of agricultural systems. Natural ecosystems tend to be self-sustaining; agroecosystems do not, and much modern management of the latter seeks to enhance within them those features of natural ecosystems that increase sustainability, whilst still producing harvests for human benefit. Thus, as Gliessman (1997) summarised, 'sustainable agroecosystems' and 'conventional agroecosystems' also constitute a functional transition between different ecological states, as in Table 1.5. Sustainable agroecosystems thereby mimic natural ecosystems in reducing dependence on external inputs, but they may be subject to variations in – for example – crop yields in different seasons through reflecting natural variations to a greater extent because of this.

The changes wrought in establishing, expanding and intensifying production within the agricultural estate over several thousand years, but particularly the intensification during the past half century, have thus been accompanied by loss of habitat and loss of independent ecological capability over large areas of the world. 'Intensification has lowered heterogeneity, element diversity and contact between neighbouring habitats' (Fry, 1991).

The most obvious manifestation has been ecological simplification in each of a number of parameters: species diversity, resilience, vegetation structure and composition, and so on. Such practices have been heralded widely as inimicable with conservation of natural biota and ecosystems, in that the goals of productive agriculture and of practical conservation

Table 1.5. *Some characteristics of conventional and sustainable agroecosystems (Gliessman, 1997)*

	Conventional	Sustainable
Diversity	Low	Medium
Resilience	Low	Medium
Flexibility	Low	Medium
Reliance on external inputs	High	Medium
Human displacement of ecological processes	High	Medium
Autonomy	Low	High
Output stability	High	Low/medium

have been deemed incompatible (Gillespie & New, 1998) (see Chapter 9).

Nevertheless, the importance of the agricultural estate has become appreciated increasingly as vital in seeking practicable conservation at many levels (van Hook, 1994), and landscape-wide conservation scenarios are gradually replacing the polarised attitudes between conservation and agriculture reflected in the 'reserves for nature versus other land for exploitation' attitude formerly so prevalent in many sectors. Warnings such as 'if we can't save nature outside protected areas, not much will survive inside' (Western, 1989) and 'the struggle to maintain biodiversity is going to be won or lost in agricultural ecosystems' (McIntyre *et al.*, 1992) have ominous forebodings and more than a little truth. They imply the urgency for wider, holistic considerations for land use and care. As Fry & Main (1993) (see also Fry, 1991) put it, 'Sorting out conservation priorities and agricultural practicalities are a priority even though these are different tasks.' The conservation priorities have accelerated recently, not least because of concerns over overproduction in parts of Europe, and monitoring environmental problems that can be attributed, at least in part, directly to the effects of intensive agriculture. They also extend into other areas – the effects of salinity in south-east Australia, of soil erosion in many places, acid rains and chemical pollution of waterways, continued pressures on marginal habitats and remnant habitats, long-lasting soil contamination, and so on. Despite the vastly different details of agricultural systems in various parts of the world, Fry & Main (1993) were able to formulate some general principles for restoration of agricultural lands that seemed to transcend the differences (see Chapter 7). These principles seek to improve the hospitality of the agricultural matrix and to bring its characteristics closer to those of natural ecosystems for the benefit of the wide array of biota that are disadvantaged by agricultural development. Matrix hospitality is a key theme in such conservation and is taken up later in this book (Chapter 10) in the context of integrating landscape-wide conservation effort for invertebrates.

The protected area system as the conservation estate

Reliance on protected areas, such as national parks, as the 'conservation estate' presupposes that the reserve system is adequate for this task. It is not, from a variety of viewpoints. Consider, for example, the following factors:

- Many protected areas have been based on selective acquisition of land in the past, rather than seeking to comprehensively represent all major ecosystems and habitats, so that much biodiversity has been excluded fortuitously from the reserve system.
- Many habitat patches (fragments) remain primarily for logistical accident or other reasons, such as difficulty of access for cultivation, rather than for proven biological importance or relevance. Their very existence is fortuitous rather than planned on any grounds related to conservation.
- Many important protected areas (such as many national parks) have multiple purposes rather than being specifically for conservation, although this is implied. The World Conservation Union (IUCN) (1994) recognised six categories of protected area, all with conservation aims but with some allowing substantial degrees of human interference for purposes such as tourism and recreation, education and maintenance of cultural attributes, as well as for ecologically sustainable uses (which may include extractive industry). Protected areas are therefore not necessarily immune from many of the threats to biodiversity that operate in the wider environment, and the categories (Table 1.6) show a gradation of possible human interference from low (category 1) to relatively high (category 6).
- Many species of animals and plants occur only outside protected areas, and no populations are known in high-quality reserves.
- For many other species, sustainability in reserves cannot be taken for granted. Many reserves are not sufficiently large, natural or well insulated from outside effects (see p. 299) to guarantee the survival of resident species without intensive (and, often, expensive) management.
- Many species utilise a much wider landscape than reserves alone, so that their wellbeing may depend heavily on maintenance of resources and habitat over a much broader scale.

Table 1.6. *The World Conservation Union categories of protected areas and their conservation roles (IUCN, 1993)*

Category
1 Protected area managed for (1a) science or (1b) wilderness protection: Strict Nature Reserves or Wilderness Areas.
2 Protected area managed for ecosystem protection and recreation: National Park.
3 Protected area managed for conservation of specific natural features: National Monuments.
4 Protected areas managed for conservation through management intervention: Habitat and Species Management Area.
5 Protected area managed for landscape/seascape conservation and recreation: Protected Landscape or Seascape.
6 Protected area managed for sustainable use of natural ecosystems: Managed Resources Protected Area.

- Most reserves are 'isolated' and lack continuity between them. They may simply maintain isolated populations as 'spots' within the broader range of the species, whilst allowing them to become extinct elsewhere in the landscape.

Such points were advocated by Bennett (1999) in support of an integrated landscape-wide approach to conservation. Within this, the agricultural estate is of particular importance:

- Many of the species occurring only outside protected areas are found on lands managed for agriculture and/or forestry (Gerard, 1995).
- Agricultural practices, if planned properly, can enhance the conservation values of isolated reserves (such as by providing buffer zones (see p. 254) or facilitating connectivity between them (see p. 253)).
- Some species of conservation concern are adapted for survival in particular agricultural landscapes, of kinds not generally considered suitable for reserves.
- Agricultural areas can constitute or provide refugia for many taxa, including threatened species.

Augmenting the current reserves system for greater conservation values is increasingly difficult in an era of multiple demands for land use, despite the increasing groundswell of support for such designations to occur. Collectively, reserves are unlikely to exceed an average of around 10% of the land area of each country; even this is a major advance since

McNeely (1988) noted that around 4% of the world's land surface had been set aside for biological conservation. In addition, practices such as agriculture essentially limit options for augmenting reserve systems – as Fry (1991) noted, 'damage' is one of the factors limiting the list of potential reserves in Europe.

In some places, the agricultural and conservation estates overlap because of programmes that acknowledge the needs for conservation on private lands. Thus, classic categories of agricultural landscapes in Europe have sometimes been designated as protected areas because of their biological, cultural or landscape interest, through (for example) the Environmentally Sensitive Areas designation of the then European Economic Community (EEC) or recognition as Sites of Special Scientific Interest (SSSI) (UK). More recent trends towards abandonment of agricultural land have allowed this to revert partially to 'natural' status, with two main drivers for this in Europe leading to a decline of several million hectares of productive agricultural land in recent decades. These drivers have been (1) the low viability of marginal farming land in relation to the costs of using it productively and (2) production of food surpluses, with the result that productive land has been removed from production cycles.

Such 'set-aside' schemes in the 1980s had the stated aim of taking up to ten million hectares of EEC land out of production by the end of the twentieth century, enabling considerable opportunity for reversion towards the conservation estate. In addition, production of food surpluses has helped greatly to contain calls for increased intensification of agriculture in such places. Long-term set-aside may help to re-establish succession and lead to re-establishment of more natural communities (Corbet, 1995). Decisions on long-term or short-term set-aside thereby influence the future age profile of ecological communities on now-uncultivated land.

More generally, several national attempts have been made to promote effective protected area systems in various parts of the world. In New Zealand, development of the Protected Natural Areas Programme in the mid 1980s sought to address under-representation of natural habitats in the conservation estate (Kelly & Park, 1986), while Australia's aim to establish a Comprehensive, Adequate and Representative (CAR) reserve system (largely in conjunction with forestry production) accelerated through the 1990s. The Australian system is based on a series of 'bioregions' (Brunkhorst *et al.*, 1998). Selection of sites for future reserves includes the steps of (1) identifying gaps in the current system and setting priorities for redressing lacunae; (2) identifying key sites as representatives to fill those gaps; (3) selecting potential candidate reserve areas; (4) assessing the feasibility of using these; and (5) establishing reserves and implementing management. Algorithms and other approaches to optimising reserve selection have proliferated in recent years. The need for these reflects the limited resources available for purposes such as land purchase or other change of tenure, and the need for any new reserves to be 'the best available' if such costs are involved. The most rational choices are likely to incorporate a range of biological values, such as seeking complementarity (additional values or species) to increase the overall biodiversity in reserves.

However, even massive (and potentially unattainable) increase of the world's reserve systems does not obviate need for wider conservation considerations. For example, anticipated or projected climate changes over the next century or so are likely to result in many of the current reserve areas changing substantially in character and not fulfilling their current roles as sanctuaries or aids to practical conservation. The distributions of very many species will change in response to global warming and other climatic factors, and only harmonising the qualities of reserves with features of the wider landscape is likely to provide the ecological gradients that such species may be able to track in order to persist, as the present environment becomes unsuitable for them. Many currently isolated 'hotspots' of biodiversity in the landscape are likely to become imperilled as conditions change, so that planning for the future rather than just for the present is a vital component of conservation activity for the next few decades. Agricultural landscapes, as the major component of the present inter-reserve matrix, are of major importance in this endeavour. What happens in agroecosystems, and the ways in which they are managed, affects the environment and biota in much of the rest of the landscape.

2 · *Agriculture and biodiversity: the place of invertebrates*

The vast taxonomic and functional diversity of terrestrial invertebrates gives them massive importance in sustaining ecosystem functions, so that their conservation has substantial practical importance. This chapter introduces the importance of invertebrate biodiversity in agroecosystems, particularly arable systems, and how this importance may be appraised. The taxonomic variety of invertebrates is summarised, and brief comments on some major taxonomic groups illustrate their functions and roles as a basic rationale for conservation need.

Introduction

Agriculture is recognised widely as one of the largest contributors to worldwide loss of biodiversity. As McLaughlin & Mineau (1995) noted, 'if we are serious about our commitment to conserve biodiversity, we will need to fully consider the effects of common agricultural practices, such as tillage, drainage, intercropping, rotation, grazing, pesticide and fertiliser use on wild flora and fauna'. Robinson & Sutherland (2002) noted that whilst agricultural intensification has had a wide range of impacts on biodiversity, data on many species are insufficient to enable any detailed assessment to be made of the factors involved. Writing particularly of Britain, they commented that initially (in the 1950s and 1960s), reduction of habitat diversity was the important factor, whilst more recently, reduction in habitat quality is probably more important. Not surprisingly, declines have been most marked in habitat specialists, so that many of the species at present common on farmland are relative generalists, able to withstand considerable disturbance.

Although they encompass a range broader than invertebrates alone, the figures for quantities of soil biota listed by Saunders (2000) are impressive. Writing on Australia, for a hectare of soil in temperate regions, Saunders claimed that there are about:

- 20 000 kg of microscopic organisms, such as bacteria and fungi;
- 50 kg of microfauna, i.e. organisms less than 2 mm in length, such as protozoans and small nematodes;
- 20 kg of slightly larger (2–10 mm) organisms, such as microarthropods; and
- 900 kg of larger organisms, such as earthworms and termites.

This biomass considerably exceeds that of most agricultural products grown on the same land area, and its contribution to ecological services (see p. 28) is clearly enormous. However, because this vast array of organisms is largely unknown and unnoticed, any detailed or purposeful moves to protect them and to sustain their ecological contributions is commonly absent in agricultural management, where intensification (such as fertiliser applications) may be used to substitute for loss of capability resulting from their loss.

Invertebrates are the most diverse macroscopic components of this biodiversity – in addition to that above ground, which tends to attract more attention – in terms of their numbers of species and higher taxonomic groups, biomass, and the variety of their ecological roles and functions. They are also the most important components in terms of facilitating and sustaining many ecological processes and ecosystem services. Invertebrates are also generally the least acknowledged components of faunal biodiversity. Much appraisal of organismal diversity emphasises vertebrates and vascular plants, groups that are much easier to identify, count and evaluate than most invertebrates (and their botanical equivalents, the non-vascular plants). Somewhat ironically, the impacts of invertebrates are acknowledged strongly in agricultural systems, with the relatively few pest species being conduits for much wider adverse effects on their less heralded relatives through measures taken for crop and other commodity protection. Efforts to sustain or enhance wider invertebrate conservation in agroecosystems have focused on the relatively small cohorts of immediate tangible benefit to agriculturists, namely pollinators and the natural enemies of pest species, and have largely ignored others. The various influences of these priorities are addressed in Chapters 3–5. For many of these 'other groups', changes in farming practices may influence populations in ways that are not always obvious, through changing the balance of resources present at any given time. Thus, reducing the availability of suitable grass hosts on arable farms in Britain can reduce sawfly abundance (Barker *et al.*, 1999), as also may increasing the proportion of land cultivated after harvest or increasing grazing pressures on grassland.

Changes in patterns of crop rotation have been implicated strongly as the main cause of this decline, because of removal of undersown grass fields from many rotation cycles. Overwintering disturbance of sawfly prepupae and pupae in soil by cultivation was a major cause of mortality of *Dolerus* and *Pachynematus* sawflies in such environments.

In this chapter, the diversity and importance of invertebrates in terrestrial ecosystems are introduced to help formulate a case for their conservation rather than their continued neglect. Many invertebrates are not conspicuous. For example, Hansen (2000) claimed that at practically any terrestrial site, the 'vast majority' of animal species are invertebrate members of the decomposer community, mainly inhabiting the top few centimetres of soil and unknown to most people. Even restricting this spectrum to a single phylum, Arthropoda, allowed Seastedt (2000) to designate these as 'webmasters of ecosystems', reflecting their central roles in influencing breakdown and release of nutrients from litter and influencing (or regulating) nutrient uptake by plants. Nevertheless, quantitative data on such effects are relatively scarce. For earthworms, Groffman & Jones (2000) noted that despite the massive attention they have received in studies on soil biology, 'surprisingly few studies' have addressed their importance to fundamental ecosystem functions through acting as decomposers and recycling agents.

Further suites of invertebrates play key roles in terrestrial food webs as herbivores or higher-level consumers. A relatively few pest herbivore species found on crops have been estimated to consume an average of 30% or more of primary production, with total destruction reported in many individual cases, simply attests to their effects within agroecosystems, admittedly under conditions that may be especially favourable to herbivore attack, in providing a superabundance of highly nutritious food under near-ideal conditions for this to be exploited with minimal effort. The roles of invertebrates in natural ecosystems are at least equally significant, with modern views of their centrality in ecosystem maintenance encapsulated in Wilson's (1987) designation of invertebrates as 'the little things that run the world'.

The broad taxonomic spectrum of animals relevant to this book is listed by major category in Table 2.1, in which all invertebrate groups with terrestrial representation are noted. Some of these have only tangential relevance here. For example, Acanthocephala are almost all parasites of vertebrates and intrude little into considerations of conventional agriculture, other than as sporadic pests of livestock. Others, predominantly Nematoda, Mollusca, Annelida and many of the numerous groups

Table 2.1. *Terrestrial invertebrate groups*

Phylum	Subphylum/ class	Comments
Platyhelminthes	Turbellaria	Flatworms. Few species of land planarians, most common in forested habitats, not usual in agroecosystems. One New Zealand species is an exotic pest elsewhere. Tapeworms (Cestoda) are parasites of vertebrates.
Nematoda		Roundworms. Ubiquitous in soil, with many also being significant pests of plants or parasites of animals.
Nematomorpha	Gordiodea	Bootlace worms. Parasites of terrestrial arthropods, including Orthoptera and Coleoptera.
Acanthocephala		Hookworms. Internal parasites, mainly of vertebrates.
Mollusca	Gastropoda	Snails and slugs. Include some important pests of plants, and predatory species variously seen as dangerous predators of native taxa or useful biological control agents.
Annelida	Oligochaeta	Earthworms. Most feed on organic matter in soil or on the ground surface.
Onychophora		Velvetworms, Peripatus. Mainly found in cryptic habitats such as rotting logs and leaf litter in natural ecosystems.
Tardigrada		'Water bears'. Minute invertebrates, rarely more than 1 mm long, some living in wet soil and on the damp surfaces of low-growing and non-vascular plants.
Arthropoda		
Hexapoda	Ellipura	Proturans and springtails. Small, soil- and litter-dwelling 'near-insects'.
	Diplura	Diplurans. Leaf-litter dwellers.
	Insecta	Insects. Abundant and diverse in all except the most extreme terrestrial environments.
Myriapoda	Chilopoda	Centipedes. Most in wet habitats near the ground, in litter and rotting wood, under stones and similar environments.
	Diplopoda	Millipedes. As for centipedes.
	Symphyla	Symphylans. Small, pale myriapods found mainly in upper soil layers.
Crustacea	Malacostraca	Crustaceans. Few species of crabs (coconut crab, red crab) and yabbies (crayfish) terrestrial; greater diversity of Isopoda (slaters) and Amphipoda (hoppers) in leaf litter and similar environments.
Chelicerata	Arachnida	Spiders, mites, scorpions, pseudoscorpions, opilionids. As for insects, abundant and diverse in most terrestrial environments.

of Arthropoda, dominate both agricultural systems and natural ecosystems. Still others, such as Onychophora, occur predominantly in natural ecosystems and have not played any major role (as far as is documented) in agroecosystems. They are still of concern, however, as displaced or likely to be threatened by agricultural activities.

However, this very broad categorisation hides innumerable variations, and most of these large groups are more diverse and better represented in natural ecosystems than in human-modified environments. Some of them have attracted considerable attention as functionally significant groups or simply because they are conspicuous and occur in very large numbers. Others gain attention because of needs for their conservation wrought by externally imposed changes to their environments and perceptions of declining distribution and numbers, and still others are notorious for their harmful or beneficial properties that accord them direct value in relation to agricultural wellbeing.

In addition to such tangible notice and benefits (Qualset *et al.*, 1995), invertebrates are important as indicators of environmental change and health (see p. 70). However, whatever their functional and practical importance, invertebrates command attention for their ecological and taxonomic diversity, these features rendering them by far the most numerous and varied recognisable animals with which people share terrestrial environments and resources. Their interactions with human needs merit close scrutiny in relation to sustainability of both agroecosystems and natural ecosystems. Invertebrates must be major constituents and considerations in any plans to document and understand 'biodiversity', and to sustain it.

Biodiversity in agroecosystems

Biodiversity, a term that has become very widespread (and has achieved the status of a political catch-cry) in only about 20 years, has a range of different meanings. These encompass the major levels of biological diversity, namely genetic variety within populations and species, the species (or other taxon) variety, and ecosystem variety, with different workers emphasising these in different ways. Each level can be considered at scales from very local, bounded by physical or ecological limits, to global. Thus, 'biodiversity' can be the totality of biological and ecological diversity or variety present on a site or in a particular habitat or political entity (country, state, region), or it can be focused on a single level, taxonomic group or species as smaller and more tangible components.

Biodiversity is linked with both stability and the processes involved in sustaining ecosystem functions (ecosystem services, p. 28). Thus, the presumptions listed below underpin much of broad agricultural management as well as wider conservation paradigms:

- The greater the richness of species present, the greater the diversity of functional groups (see p. 77) present and the more stable the system will be.
- High diversity of functional groups equates to high levels of 'goods and services'.
- Lowered diversity of functional groups implies needs to introduce substitutes, with cost implications for ecosystem management and, perhaps, introducing alien elements or taxa.
- Because the most common interactions between species are trophic, systems with high diversity have more complex food webs, which, in turn, allow for greater flexibility in how food resources may be changed or allocated.
- More options in food resources leads to sustainability of more species, equating to more functional groups and fulfilling more ecological roles.

In the words of Colloff *et al.* (2003), 'biodiversity drives ecosystem processes', and structural complexity facilitates biodiversity. For example, in the soil, (1) soil structure is influenced by earthworms, termites, ants and others; (2) decomposition and nutrient cycling are influenced by these and a variety of other decomposers; (3) predators affect various food-web characteristics and may affect agricultural pests; and so on.

Biodiversity has thus become a very general and basic conservation parameter, with the widespread rationale that 'high biodiversity is good' and, conversely, 'lowered biodiversity is bad'. The fact that species are the most tangible and most easily evaluated units of biodiversity has led to species richness being the most frequently cited aspect for discussions of biodiversity enumeration and importance. Within agroecosystems, different workers have attempted to clarify biodiversity in several ways, reflecting the values that agriculturists recognise and attribute to organisms, in addition to less tangible features.

For example, Vandermeer & Perfecto (1995) recognised the dichotomy of 'planned biodiversity' and 'associated biodiversity' in agroecosystems. The first category is the biodiversity associated with agricultural crops and/or livestock used by the farmers, and that will vary according to the local environment, the scale of operation, and the degree of management imposed. The second category is the remaining biodiversity that either

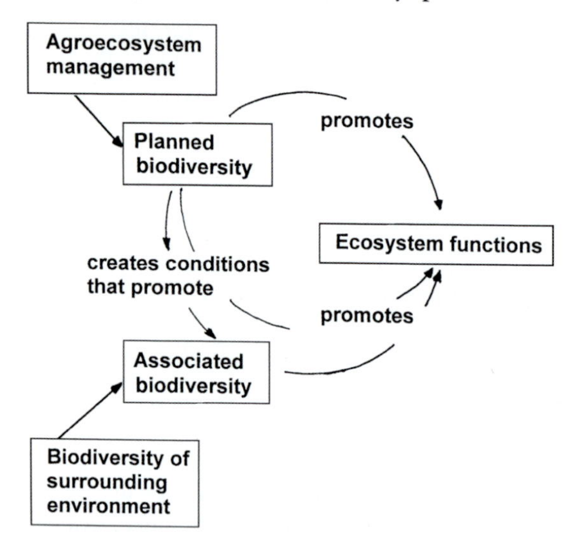

Figure 2.1. Categories of biodiversity: planned biodiversity (determined by the farmer) is linked with associated biodiversity (colonises agroecosystems following their establishment) to promote ecosystem function (Vandermeer & Perfecto, 1985).

(1) colonises agroecosystems from surrounding environments or (2) is present, is not disrupted by agricultural activity, and continues to thrive in the managed environments, the extent of this reflecting the intensity of management and variety of structure present. Associated biodiversity can thereby include considerable ecological variety, such as soil biota, and consumers at all trophic levels. These categories, and the relationships between them, are shown in Fig. 2.1.

This figure emphasises that the two categories of biodiversity are interdependent and together can help to harmonise aspects of wider biodiversity to promote ecosystem function. As an example from Vandermeer & Perfecto (1995) of how associated biodiversity has functions mediated through planned biodiversity, trees in an agroforestry system may create shade (planned function), which might favour nearby cropping based on sun-intolerant crops; however, the same trees may help to foster natural enemies of pests, for example by providing nectar sources for parasitoid wasps and predators. Those natural enemies, part of the 'associated biodiversity', are thus the consequence of an indirect function of the trees, which thereby contribute additionally to functional biodiversity.

Table 2.2. *Ways to think about species in considering interactions between agricultural practices and organisms living on exploited lands (Gall & Orians, 1992)*

Exploited species (target species). Those species for which the system is primarily managed.

Species incompatible with human needs. Species directly dangerous to people (e.g. lions) or whose presence is incompatible with agriculture (e.g. elephants); necessarily managed by segregating them completely from agricultural systems.

Species partially compatible with human needs. Species damaging at particular times and in particular places but compatible with the system if they are controlled or excluded at particular times.

Beneficial species. Species whose presence in agroecosystems is desirable: pollinators, natural enemies, species important in ecosystem functions. Practices that encourage these are preferred over practices that do not.

Harmful species (pests). Consumers of, or competitors with, target species, and whose presence is truly undesirable.

Rare and endangered species. Species of particular conservation concern or significance and that should not be placed at risk; sometimes under legal obligation.

Neutral species. Species not known to be beneficial, harmful or of conservation significance. These tend to be ignored in agricultural management, and some may become threatened.

Species harmful elsewhere. Species not favoured or harmed by agricultural practices but that are disruptive elsewhere, such as by spreading diseases, damaging crops or competing with desirable species.

More emphasis on the roles of biodiversity in cropping systems leads to other broad groupings, such as the following (Swift & Anderson, 1993):

- Productive biota – the crops, trees and animals selected by farmers and that play a determining role in the diversity and complexity of the agroecosystem.
- Resource biota – organisms that contribute to productivity through processes such as pollination, biological control, decomposition and nutrient recycling.
- Destructive biota – weeds, arthropods and other pests and pathogens that farmers seek to reduce or suppress as a means of crop (or other productive commodity) protection.

This categorisation by perceived roles is a useful approach by which farmers and others can consider the impacts of different species and come to appreciate the variety of roles they may play. Table 2.2 illustrates a

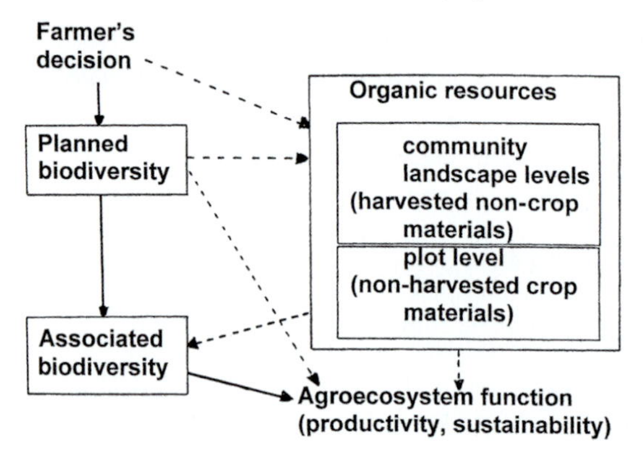

Figure 2.2. Relationships between categories of biodiversity (Vandermeer *et al.*, 1998): the basic transformation processes in an agroecosystem to benefit farming.

pragmatic view of 'agricultural biodiversity'. Indeed, this term is sometimes employed to differentiate the diversity of 'applied significance' from that less obvious or more tangential to agroecosystem management for production.

In general, biodiversity in agroecosystems depends on four broad characteristics of the systems (Southwood & Way, 1970):

- The diversity of vegetation within and around the agroecosystem.
- The permanence or longevity of the various crops within the agroecosystem.
- The intensity (extent, frequency, variety) of management imposed.
- The degree of isolation of the agroecosystem from natural vegetation.

These parameters guide many of the developments by which biodiversity can be retained or enhanced in agroecosystems. Many of the approaches depend on manipulating invertebrate biodiversity at the landscape level, involving practices that can be optimised only by developing a reasonably strong framework of biological knowledge of the most important invertebrate groups (higher taxa) involved with the major biomes of soil, crops, and local more natural vegetation (including, where possible, that typical of the regional precursors of change, such as native grassland or forest), so that their diversity and ecological roles can be evaluated and changes appraised in relation to continuing management decisions.

'Associated biodiversity' is related positively to 'planned biodiversity' in many groups of organisms (Vandermeer *et al.*, 1998, and included references), but the precise nature of the relationships are less certain,

Table 2.3. *Non-harvested components of plant biodiversity in agroecosystems, and some of their ecological values (Vandermeer* et al., *1998)*

Component	Agroecosystem functions
Trees	Shade; aesthetics; habitats and resources for associated biodiversity; wind breaks; litter values (as for crop residues, below); influences on hydrology; maintenance of soil fertility.
Weeds and cover legumes	Capture of plant nutrient excess (e.g. in fallow periods); protection of soil surface from erosion; habitats and resources for associated biodiversity; resources for soil organisms mediating soil processes; nitrogen fixation (legumes); microclimatic change; enhancement of sediments; dung from stock increases plant nutrient availability.
Crop residues	Mulching effects on moisture; protection of soil from erosion; habitats and resources for associated biodiversity; nutrient carry-over between crops; microbial products promoting aggregates stabilisation; maintenance of soil organic matter; dung from domestic stock.
Roots	Binding soil structure; creation of macropores and water conduits; contribution to soil organic matter; nutrient carry-over between crops; resources for associated biodiversity, especially soil organisms.

as emphasised in Fig. 2.2, which shows several possible non-linear interrelationships at different levels of agricultural intensification. Additional associated species are likely to enter the system with the addition of each planned species, but combinations of planned species are likely to create additional habitats and resources for associated biodiversity, probably changing the intensity of interactions amongst them and facilitating coexistence of greater numbers of species. The kind of agroecosystem and the particular taxa involved are likely to be important determinants of the outcome, so that there is no simple predictive model of universal application in predicting levels of associated biodiversity. Vandermeer *et al.* (1998) noted that associated biodiversity is often overlooked in considerations of planning choices available to farmers, but may be very large (perhaps particularly in less intensive agroecosystems), capable of being enhanced by management changes, and with its functions largely undocumented in any detail. The non-harvestable components of biodiversity within agroecosystems (Table 2.3) can be important influences on ecosystem function and facilitators of increased biodiversity, and many parameters of these are open to management and manipulation to further enhance the benefits they provide.

Despite the earlier assertion on lowered invertebrate species diversity within agroecosystems, many agricultural systems are 'surprisingly rich in species' (Sunderland *et al.*, 1997), with Nentwig (1995) estimating that a typical temperate agroecosystem contains at least 1500–3000 species. Other workers (e.g. Horne & Edward, 1997) have also shown that species richness in agroecosystems can, in many instances, approximate that of more natural systems. Data are most complete for predators and parasitoids, which, in contrast to many other groups, are of positive interest to growers (see p. 142); around 400 predatory species have been reported from cereals in the UK (Sunderland & Chambers, 1983), and about 600–1000 have been reported in US cotton and soybean crops (Whitcomb & Bell, 1964; Gross, 1987). Other examples are cited by Sunderland *et al.* (1997), who also noted that richness of spiders and carabid beetles, as common 'focal groups' for enumeration, can be particularly high. Thus, more than 400 species of spiders have been estimated to occur in US cotton fields (Young & Edwards, 1990) and more than 100 species of Carabidae in other, named crops (Horne & Edward, 1997). As the latter authors noted, some such estimates are difficult to evaluate because of lack of comparable information from adjacent natural habitats (if these exist) but are evidence of considerable species richness per se in arable systems.

Ecosystem services

A major putative function of biodiversity is to enhance or maintain what ecologists have termed 'ecosystem services', the multitude of largely unseen and unheralded processes, commonly involving interactions between the biota and abiotic components of the environment that assure environmental health and resilience. Should such processes diminish, the entire system may suffer and – in agroecosystems and other anthropogenic environments – such services may need replenishment by deliberate intervention or management. Thus, services such as pollination, maintenance of soil quality, nutrient recycling and many others (Costanza *et al.*, 1998) are significant concerns in agriculture but are commonly taken for granted in more natural environments. Invertebrates play significant roles in many ecosystem services, but quantifying and defining those roles is often difficult and usually has not been attempted.

However, attempts to do so are important in demonstrating practical values of invertebrates and pragmatic need for them in managed ecosystems. Thus, as noted earlier, the more diverse and complex the interactions between species in an ecosystem, the greater may be its resilience, and, in general, the greater the complexity of the plant communities, the greater will be the spectrum of habitats available to invertebrates and other animals.

An unusual study on *Citrus* in South Australia (Colloff *et al.*, 2003) investigated two key ecosystem services – pest control and nutrient cycling – emphasising the roles of various components of soil biodiversity in delivering these benefits. One innovative step was to ascribe dollar values to total soil invertebrate populations, based on the dollar value of the nitrogen/hectare they contained: this index provided insight into the relative soil biomass on different farm properties and its capacity to provide nitrogen to the soil after death and decomposition through natural turnover. Strong positive relationships between soil invertebrate biomass and soil carbon, and weaker correlations with soil nitrogen and phosphorus levels, were found. However, there was no relationship between soil invertebrate biomass and generic management practices of the properties. Specific management for enhanced levels of pest control and nutrient cycling included steps such as:

- maintenance of ground cover of perennial grasses and herbs to provide habitat and refuges for natural enemies (Chapter 5);
- use of ground cover as a nutrient source and to conserve soil moisture, through slashing and mulching;
- use of fertiliser regimes and nutrient management that depend on decomposition of organic carbon and nitrogen, rather than relying on direct inputs of inorganic materials;
- relatively low inputs of organic fertilisers, pesticides and heavy metals.

The exercise was exemplified by studies of the interactions between a major pest (Kelly's citrus thrips, *Pezothrips kellyanus*) and a complex of predatory mites as significant natural enemies, and the development of holistic management for IPM (Chapter 4). The study is important in demonstrating that soil biodiversity per se can be a significant and positive resource for farmers (Colloff *et al.*, 2003).

Ecological redundancy

The relationships between maintenance of ecological function and maintenance of high biodiversity as a prerequisite for this are by no means clear. Simplified ecosystems (including many agroecosystems) are characterised by relatively low biodiversity, with the common presumption that this

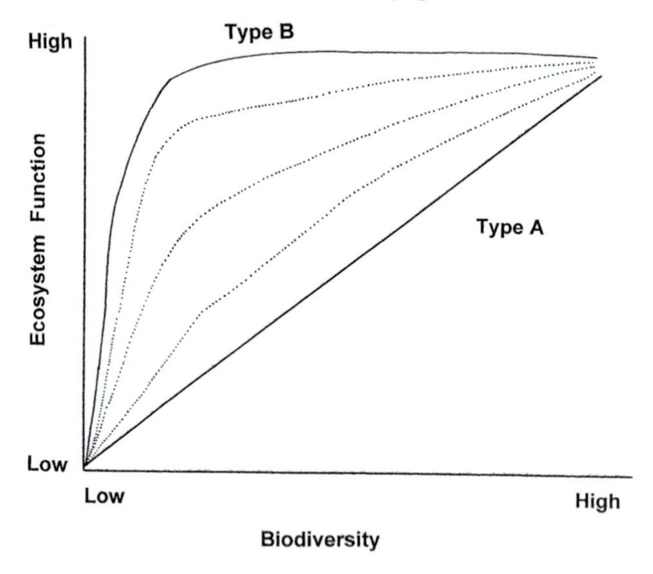

Figure 2.3. Relationships between biodiversity and ecosystem functions (Schwartz *et al.*, 2000). Line A represents the basic assumption of many conservationists, that ecosystem function depends directly on the 'amount' of biodiversity present; line B represents a case where low levels of biodiversity (most commonly equated to a few species) still enable high levels of ecosystem function; the intermediate lines represent intermediate cases.

lessens ecological functions, which then have to be compensated for by human inputs. It is also held widely that a large variety of invertebrates are critical components of this 'functional diversity', reflecting their massive diversity, ecological variety and biomass. However, a strict correlation between high level of ecological integrity and high biodiversity has been questioned.

Consider the graph lines shown in Fig. 2.3 (Schwartz *et al.*, 2000). The above relationship, considered widely to be true by many conservationists as approaching an ideal for conservation ambition of 'saving all biodiversity', is indicated by the straight line A. In contrast, the upper line B illustrates a system in which few species (or other biodiversity units) contribute to ecosystem function, which thereby reaches a high level if those organisms alone are present. The other species, probably the great majority of those present, contribute little or nothing to ecosystem function. The other lines demonstrate possible intermediate scenarios, in each of which some proportion of the representative total biodiversity is sufficient to sustain the functioning ecosystem.

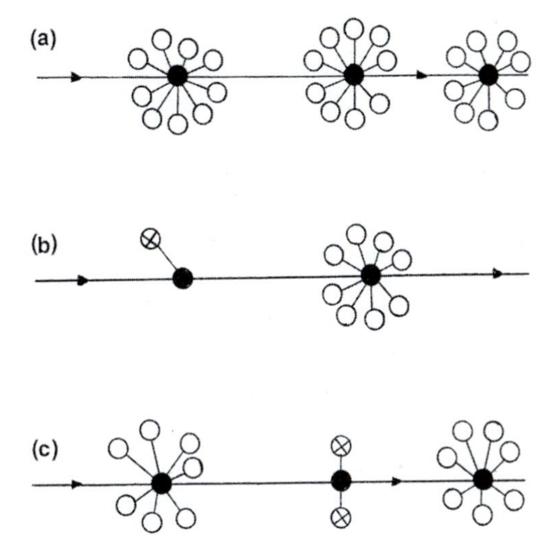

Figure 2.4. Biodiversity and functional guilds. In an ecological process such as a food chain, various 'nodes' occur, where changes in energy flow or nutrient allocation eventuate (solid points). At each point, taxa with similar functions (each species denoted by an open circle, and collectively a 'guild') mediate such changes. If numerous species occur in a guild, then loss of one or some may be functionally buffered by others and function may be maintained; if only one or a few species are present in a guild, then such compensation may not possible ('X' in the figure), and the system may collapse. (a) All nodes supported by many species; (b) first node with only a single species (keystone), whose loss affects all species at subsequent nodes; (c) central node vulnerable if one or other supporting species lost and buffering capacity reduced.

This inference of 'ecological redundancy' continues to be debated. One of the strongest pragmatic arguments for conserving biodiversity is to conserve the ability for species within a system to be able to compensate for, or buffer against, loss of function brought about by loss of taxa. Thus, Walker (1992) suggested that categorisation of biota by functional guilds might help in assessing conservation needs and focus. Fig. 2.4 schematises several food chains within communities, with each group of taxa (open circles) constituting a trophic level or node (solid circle), a point at which energy is partitioned for discarding from or retention within the ecosystem. If one species of a group of trophically similar taxa at the same node is lost, then its role has been supposed to be taken up by another member of the same guild; this has been extrapolated to a major feature by which diversity and stability have been functionally linked within ecosystems. In this way, even if a particular species participates normally in only a minor way in the functioning of the system, then its supposed capacity to increase that role if necessary to do so ensures that it is not permanently redundant. If there is only a single species at a particular linkage point (Fig. 2.4b), then loss of that species may be critical because no functional equivalent is available to replace it and the system may or will collapse. Such species are indeed functional keystones whose removal or loss has far-reaching effects and, if they can be detected, whose conservation may be critical in sustaining that system.

However, Schwartz *et al.*'s diagram (line B in Fig. 2.3) could imply that many species are truly of little or no ecological importance and are passengers in ecosystems driven by a low proportion of the species present. If this is so, then a strong argument for their conservation on functional importance becomes considerably weaker. Our present levels of knowledge of invertebrate functions, however, are inadequate to condone this implied lessening of importance, but values beyond 'ecological importance' assume relatively greater roles in conservation advocacy.

Invertebrates: knowledge of functional diversity

The invertebrate groups listed in Table 2.1 include massive ecological variety. They also reflect the difficulties of summarising basic knowledge of their systematic diversity and ecological roles. Some large invertebrate groups are acknowledged universally as ecologically important in agriculture and natural systems; others are familiar, even in general terms, to only a handful of specialists.

Within any large group of invertebrates, ecological variety at both the major trophic interaction level and in more subtle ecological strategies is likely to be high. Closely related species, even in the same habitat and with similar feeding habits, may differ considerably in facets of their abundance, reproductive rate, adaptations to particular microclimate regimes, duration of development and other life-history parameters. These differences mirror and extend the intermediate continuum between *r*-strategists and K-strategists (see Chapter 1) but emphasise that seeking generalisations about invertebrate functions can be both difficult and misleading if used for management recommendations or similar purpose without corroboration. A close parallel occurs in agroecosystems, where congeneric (or other closely related) species may differ considerably in their basic biology, so that failure to identify the species reliably, or to recognise such interspecific differences, may lead to considerable wasted effort in management. The two polyphagous species of *Helicoverpa* moths (Noctuidae) in Australia (*H. armigera*, *H. punctigera*), for example, are severe pests of many field crops, but their basic biology is very different. These differences lead to serious consequences should management target the 'wrong' species.

Furthermore, any widely distributed species is likely to show biological variation across its distributional range, over which it may encounter different climatic regimes, different complexes of natural enemies, seasonal opportunities, and relative abundance and availability of foods. Even very detailed studies of a pest species' biology in one area will generally need to be augmented in planning for its management elsewhere. One of the consequences of the development of integrated pest management (Chapter 4) founded on sound ecological knowledge is that understanding such variations has assumed central importance in comparison with – for example – earlier practices such as simply using a broad-spectrum chemical pesticide to kill the pests wherever they occurred and caused (or seemed likely to cause) damage.

Classical 'species-level' taxonomy has predominantly involved evaluating differences based on consistently different morphological character states. However, progressive elucidation of the biological species concept has, rather, emphasised that suites of biologically distinctive sibling or cryptic species may be difficult or impossible to separate on simply diagnosed structural features but may be abundantly distinct on biological features. Detailed studies of many pest groups and complexes, and of their arthropod natural enemies, have revealed consistently different entities, which may, for example, differ in their acceptance of different host plants, prey or host species or may 'perform' very differently on each of several such interacting species. Many such entities are referred to as 'biotypes', but other terms such as 'race' overlap with this.

As Claridge (1989) noted, careful studies of biotypes of pests have often shown them to be either (1) distinct clones in parthenogenetic taxa, such as many aphids, or (2) genetically distinct, reproductively isolated sibling species or genetic variants within panmictic populations of sexually reproducing species. Previously unrecognised sibling groups continue to be discovered in pest invertebrates, often by application of electrophoretic or other genetically related analyses to help surmount the difficulties of lack of apparent morphological differences. Behavioural differences, such as the different 'song patterns' of sibling green lacewings (*Chrysoperla carnea* s.l.) (Henry *et al.*, 2001, 2002), have also been used to separate biologically distinct forms of these economically important predators.

Once such differences are detected, they may prompt searches for small morphological differences previously considered to be insignificant. Electrophoretic techniques (see Loxdale & Den Hollander, 1989) have been instrumental in increasing the precision of invertebrate taxonomy and in helping to determine continuing evolutionary change and diversification, including traits of management relevance such as development of insecticide resistance.

The existence of numerous biotypes and similar entities reflects that insects, as the most intensively studied group, but also other invertebrates, manifest very subtle ecological specialisations (Berlocher, 1989), so that their diversity considered at this level of distinctive functional entities is often far greater than at the 'morphospecies' level alone. Such biotypes may be reproductively isolated from each other or may be compatible in various ways and to different extents. Berlocher gave the hypothetical example of a population of a pest whose immature stages (representing different genotypes) are restricted to one or other of a variety of host plant taxa but whose adults mate randomly, should they meet. 'Host races', often recognised as a distinct infraspecific category of phytophagous insects, generally are defined as 'populations of a species partially reproductively isolated from other conspecific populations as a direct consequence of adaptation to a specific host' (Diehl & Bush, 1984).

Recognition that many pest insects include such biological complexes has been a major stimulus to advances in recognition and taxonomic methodology, not least because many historical management programmes have been thwarted by failure to recognise differences at these levels. Such studies as those important in elucidating biotypes in pest complexes in agriculture, together with their predators and parasitoids, have never been undertaken for most innocuous insects or other invertebrates. The existence of such biological variation and the values of incorporating biological information into taxonomic interpretation are indeed widespread – for example, recognition of the specific or population-specific features of the songs of many grasshoppers and cicadas – but many subtle biological differences are unlikely ever to be elucidated without applied incentive, and enumeration of invertebrate biodiversity, both in agroecosystems and in natural ecosystems, relies heavily on more limited morphological interpretation.

For most groups of invertebrates, only 'guesstimates' of their total species richness exist, and many large groups are very poorly documented and evaluated. Whereas we know that some 600 distinct species of insects are major or significant pests in agricultural crops, we do not know whether the world's total insect species richness is ten million or several times this figure (see Stork (1998) for a discussion of approaches to this evaluation). Insects are considered generally to be the most speciose of all invertebrate groups, although increasing knowledge of nematode worms and mites suggests that these might be serious rivals. Erwin's (1982, 1983) early estimates (see also Erwin, 1998) of 30–50 million insect species rendered 'this pervasive terrestrial arthropod group 97% of global biodiversity' (Erwin, 1998). Most of these species are small and elusive and have never been examined closely or even collected. The total of named (formally described) species is around 1.4 million, of which some 400 000 are beetles (Hammond, 1992). Stork (1998) suggested that a 'true' figure for global insect diversity is likely to be in the range of 5–15 million species, although this does not account for biological variant categories of 'cryptic species' such as those noted above. Stork also noted that we know virtually nothing about almost all of these species. Many, indeed, are known only by the single specimens from which they were diagnosed and described.

Despite the intrinsic interest of documenting invertebrate biodiversity at the species level, for most groups listed in Table 2.1 global totals of any reality are still far off. Many biologists have expressed concern over our lack of such basic inventory of life on Earth, which they see as the foundation for understanding biodiversity, its ecological patterns and significance, and its conservation. Many have commented also on the likelihood that such definitive inventory will never become available and have thus queried its central importance in conservation assessment and planning, which has to proceed without those data. Tighter focus, such as the information derived from only a select suite of taxonomic or ecological groups, may be obtainable more easily for use in management.

Evaluating invertebrate biodiversity

The foregoing discussion leads to consideration of three major facets of documenting invertebrate biodiversity in terrestrial ecosystems:

- Overcoming, or compensating for, lack of taxonomic knowledge – particularly that a high proportion of the species present in samples will be undescribed, many others difficult to recognise and differentiate, and many described species determinable only by a specialist in that particular group.

- Evaluating and interpreting the functional and ecological diversity in any invertebrate group, and its relationships to sustainability and management.
- Sampling invertebrates sufficiently to provide comparative information useful in (1) basic inventory, (2) detecting changes in diversity, species composition and relative abundance and (3) monitoring imposed changes in space and time.

Each of these topics has received considerable attention in recent years. They are introduced here as a foundation for themes developed in later chapters. A fourth important theme for invertebrate studies, that of selecting which particular invertebrates to appraise, underlies all of these and is discussed in Chapter 3.

Taxonomic analysis

'Species' are inherently attractive to scientists as fundamental, tangible biological units, and many evaluations of 'biodiversity' involve enumeration of species within particular groups of organisms. The 'taxonomic impediment' (Taylor, 1983) for invertebrates poses both practical problems of interpreting biodiversity in this way and wider problems of communicating the results effectively. Any substantial sample of invertebrates from soil, vegetation or other biome is likely to contain numerous species. Many of these may be undescribed, and even those that are known may be difficult to recognise, diagnose and name correctly.

Recognition of the major groups of terrestrial invertebrates, commonly to order level, is relatively straightforward, but sorting and categorising them to finer taxonomic levels can commonly be a laborious and demanding process, with the results probably needing to be checked carefully for consistency and accuracy by a specialist. Once invertebrates of a given group have been sorted to putative species, finding names (with the rare exceptions of well-known flagship groups, such as butterflies, or those groups for which up-to-date comprehensive regional taxonomic handbooks have been produced) is often very difficult. Without specialist knowledge of the relevant scattered specialised literature and/or access to a substantial reference collection, names are often impossible to apply reliably. Without such support, serious errors are almost inevitable, and these may have far-reaching consequences, as exemplified earlier for agricultural pests. Correct names are keys to finding information; incorrect names, conversely, lead to false or misleading information.

The very extent of invertebrate alpha diversity (species richness) ensures that no scientist can 'handle' such enormous groups as insects or spiders in their entireity, or to the equivalent extent that a single person may be able to recognise and name, for example, birds or mammals. Many invertebrate systematists may spend their entire careers studying one order or family of insects, and expecting such a person to be able to name to species level members of even another group closely related to his or her own speciality is, in principle, equivalent to asking a specialist ornithologist to name fish or frogs. Indeed, because of the availability of well-illustrated guidebooks to vertebrates in many parts of the world, the latter request is likely to be far more rewarding. Some insect families contain more species than the world totals for birds and mammals together.

In short, there is a lack of expertise to classify and identify invertebrates and, at a time when calls to enumerate and define 'biodiversity' have never been louder or more urgent, taxonomic expertise in the world's museums and universities is declining rapidly (Lee, 2000; New, 2000a). Becoming familiar with the systematics of any large invertebrate group is a daunting task. As Moldenke *et al.* (1991) put it, writing specifically on soil fauna but with their central message of much wider relevance, interest of investigators 'often wanes when . . . faced with the investment in time and effort required to obtain the taxonomic skills necessary to expand their research progress. . . [they] frequently view learning the taxonomy of soil fauna as an insurmountable task.'

Taxonomic literature is frequently dificult for non-specialists to assimilate, not least because most specialists have traditionally tended to write for other specialists, using considerable specialised technical vocabulary to encapsulate information in forms largely unintelligible to other people. Even when 'user-friendly' keys and other identification guides are available, they often induce or force users to make mistakes. For example, a key may give the impression that all species of a particular group are diagnosed, described and included; dichotomous keys may thereby lead to decisions involving unrecognised species being forced into one or other character choice simply by maximum likelihood rather than because it fits properly. This is clearly not satisfactory, but such mistakes are extremely difficult for a non-specialist to detect. By definition, keys do not contain unknown taxa. Many are out of date, and even for the best-documented invertebrates, such as the British insects and arachnids (for which a guide to identification sources exists: Barnard, 1999), considerable portions of the fauna have not been treated comprehensively.

Barnard noted that identification to species level of some large groups of insects in Britain depended on gathering substantial numbers of isolated research papers, many written in languages other than English, over a period of many decades and to varying standards, and trying to make sense of these. For one of these large groups, parasitoid Hymenoptera, Shaw & Hochberg (2001) noted that 'far too many groups have not received adequate alpha-taxonomic and revisionary attention', so that it is difficult to determine which species occur in Britain. Those authors estimated that knowledge of what may be in the British parasitoid wasp fauna may be about 30–40% incorrect (or 'possibly even more'). This appraisal for the best documented of all national faunas for a group of enormous ecological and applied importance indicates that our knowledge of most other faunas is even more rudimentary. For most tropical faunas, for example, formal identification of most invertebrate groups to species level is, simply, impossible, so that interpreting biodiversity in this way is not a serious practical option – despite its inherent attractions and possible loss of credibility without such definition.

However, clarifying the diversity of invertebrate faunas involves 'identification' at two main practical levels: (1) formal allocation of names as above and (2) consistent allocation to recognisably distinct categories, variously termed 'recognisable taxonomic units' (RTUs), 'operational taxonomic units' (OTUs) or, more broadly, 'morphospecies', each of which approximates a true species, and providing categories that can be appraised in the same way for many aspects of documentation. This second approach is more commonly feasible.

The approach essentially helps to overcome the need for specialist taxonomic knowledge but needs very careful consideration and development. With some background training, many initially non-experienced people can undertake this kind of allocation (Oliver & Beattie, 1993) with sorted categories of 'morphospecies' corresponding closely to real species, as confirmed by examination of vouchers by specialists as a form of quality control. Such 'parataxonomists' or 'biodiversity technicians' are important in facilitating this laborious phase of surveys. However, both obvious kinds of error may occur: (1) 'lumping', the underestimation of RTU numbers by failure to discriminate between closely similar or closely related forms in which morphological differences may be small or difficult to appraise, and (2) 'splitting', the separation of minor variants or morphs (such as colour forms, or different growth stages) of the same species into more than one RTU. Consistent resolution and recognition of RTUs provides a quantifiable basis for assessing diversity and comparing it in different samples, treatments or sites, and its changes over time (references in New, 1998). Ideally, categorisation will be just as rigid as for real species. It may be much more difficult, however, to communicate that information easily and consistently, and responsible deposition of a series of voucher specimens of the morphospecies recognised (ideally accompanied by notes or 'field keys' by which they were separated) should be a routine part of this approach as an archive available to later workers who may need to check the separation accuracy as knowledge accumulates. The aims of this approach are 'accuracy' (placing individuals in the correct taxon/morphospecies) and 'precision' (placing all conspecific individuals in the same entity).

A related problem of immense importance in expensive analyses that may involve examination of many thousands of invertebrate specimens is to optimise the level of taxonomic or sorting discrimination needed. Whereas species (or morphospecies) level is commonly advocated, it may not always really be necessary to appraise diversity to this extent. The term 'taxonomic sufficiency' (Ellis, 1985) implies that the identifications must be at a level sufficient to indicate relevant biological information, but not necessarily fully to species.

The major reasons for pursuing species-level recognition for invertebrate biodiversity analysis, apart from the intrinsic appeal and 'neatness' of doing so, involve three main categories (Resh & McElravy, 1993), as follows:

- Because species are 'basic biological units', the greatest amount of information will come from this level of analysis. Any higher-level amalgamations may decrease the sensitivity of interpretation and miss subtle but important changes.
- Identification above species level may lead to misleading estimates in diversity indices and other measurements that intrinsically rely on accurate, fine-level separations.
- Analyses will be less penetrating – as examples, (1) multivariate analyses usually employ species-level information and (2) comparisons or functional interpretations of communities or assemblages solely in terms of higher taxon (families, orders) may be meaningless because these are widespread; biological variability manifests most widely at the species level.

Identifications solely to family or order level may be an important, financially expedient shortcut in invertebrate biodiversity surveys but are useful only if scientific quality and accuracy are not compromised. Genus level (as is usual, for example as a maximum penetrative level for nematode worms, p. 00) may be a valid surrogate for species in some ecologically diverse groups, in which they may denote consistent, distinctive trophic units or other 'functional groups' (see p. 00). In other cases, it may be little better than nothing because it leads to no meaningful interpretation.

'Taxonomy' of invertebrates to higher levels in such exercises thereby has the twin elements of (1) impracticability of working soundly to specific/morphospecific level without incurring enormous costs and probably also increasing uncertainty and (2) expediency, in a relatively few cases where reliable or consistent higher-level surrogacy can be used without sacrificing sensitivity. Thus, ant genera in Australia are sometimes deemed valid substitutes for species in ecological interpretation (Andersen, 1995; Pik *et al.*, 1999), sometimes simply because many genera are represented by few species in assemblages (Neville & New, 1999). Generic (or even family-level) separation of soil nematodes is usual, with some important implications on relationships in soil processes and agroecosystems based on this level of interpretation (Yeates & Bongers, 1999). In this case, the authors noted that 'the specialised input required to compile species lists means that most work is done at the less determinate level; for this reason, practical identification is typically limited to genera with some distinctive species being separated.'

Interpreting ecological variety

Formal systematics, as above, is a clue to ecological variety. The various groups of invertebrates range from those for which broad generalisations on ecology may be possible – thus, within the insects (and despite innumerable nuances at finer levels) most larval Lepidoptera are herbivores and most Neuroptera are active predators – to others in which such ordinal-level recognition masks considerable variety. Trophic groups of soil nematodes, for example, are listed in Table 2.4. However, Yeates & Bongers (1999) commented that 'many nematodes are allocated to these groups on the basis of inadequate evidence and, as many species have various food sources, ideally the feeding activity of each species in each particular soil: crop: management: climate combination should be quantified.' Variety of feeding habits is frequent within families and genera of nematodes. Genera (or other higher taxa) become useful as ecological surrogates for species only when they can be determined to be functionally equivalent units.

Table 2.4. *Ecological diversity of nematode worms in soil: the major feeding groups delimited by Yeates* et al. *(1993)*

1 Plant feeders
2 Fungal (hyphal) feeders
3 Bacterial feeders
4 Substrate ingestion (especially ingestion of substrates on which bacteria are growing)
5 Predation on Protozoa and soil invertebrates such as rotifers, enchytraeid worms and other nematodes
6 Unicellular eukaryote feeders: bacteria and algae
7 Dispersal stages of animal parasites, of both invertebrates and vertebrates (some have a bacterial- or fungal-feeding cycle in soil)
8 Omnivores (combinations of 1–6, above)

Many aspects of invertebrate distribution and ecology on farmlands remain intractable, largely because the patterns of heterogeneity are complex and difficult to elucidate by relatively simple or rapid surveys. As Thomas *et al.* (2001) emphasised, experimental work generally is feasible only at the scale of field or within-field plot, but the resulting data may need to clarify processes occurring over larger scales – plots for fields, fields for farms, and farms for regions, as examples. Extents of aggregation and spatial patchiness are important, particularly for sampling to obtain estimates of population size and density, and their spatial and temporal variations. 'Contour plots' from sampling grids provide one avenue to clarifying these problems (see p. 234). Nevertheless, in seeking to use invertebrates as measures of change in environments, or as measures of ecological integrity, some broad knowledge of ecological variety within the major groups is needed. Most practical interpretations, particularly those involving agroecosystems, have devolved on one or more of four major groups (Nematoda, Mollusca, Arthropoda, Annelida; see p. 44), commonly on only small segregates of any one of these. These phyla are also ubiquitous and generally abundant in natural ecosystems, and each is taxonomically and ecologically diverse. Other invertebrate phyla (see Table 2.1) have been used to only minor and more irregular extents in quantified surveys for agricultural and other biodiversity. For each of the four major groups, adequate analysis of samples can provide clues to ecological functions within the host ecosystem, and changes in the group representation reflect changes imposed by human activity.

The amount of information available, particularly on Arthropoda, is encyclopaedic. For functional interpretation within samples, important needs are to be able to recognise taxa involved in each major ecological role and to appraise the relative diversity and importance of these – the latter is often taken as a reflection of abundance, so that assessment of biodiversity in such terms necessitates quantitative analysis of samples rather than simply relying on taxon presence or absence information.

Sampling invertebrates

Interpreting biodiversity using invertebrates, as noted above, depends strongly on samples that are taken in ways that are standardised sufficiently to furnish information on the assemblages/communities/environments under investigation. The parameters for many analyses include (1) which invertebrates, usually of particular higher taxa or focal groups, are present; (2) their relative abundance; and (3) their overall taxon richness. In many cases, data from any sample series will be compared with other samples, so that the sampling methods must be standardised sufficiently for valid comparisons to be made. Because many invertebrates are short-lived (or, at least, have particular life stages that are present for only short periods each generation or year), any need for broader inventory or diversity estimation may need to be a long-term exercise, at the least involving continuous or interval sampling over a year or more. In addition, the very large numbers of individuals obtained commonly necessitates sub-sampling rather than total analysis of the catches.

Planning a sampling programme for terrestrial invertebrates is a complex exercise, and the aims and scope of any survey should be formulated very clearly (New, 1998). Almost always, surveys target particular invertebrate groups rather than their totality; sampling replication may be needed for pertinent analysis; and each taxonomic group of invertebrates may be especially amenable to capture by different sampling methods and need individual approaches to sample treatment and preservation. No single sampling method may be entirely suitable for a given purpose or for the entire set of invertebrates to be studied, so that a 'sampling set' of complementary methods may be needed. Compendia of sampling methods, such as those by Edwards (1991: soil biota), Southwood & Henderson (2000) and New (1998), should be consulted for background information on the considerable range of techniques available or suitable for any given study. Duelli *et al.* (1999) emphasised further the need for standardisation of sampling methods and sampling effort in any study in which comparative data are needed.

The sampling regime may also need to reflect approaches to so-called 'rapid biodiversity assessment' (RBA), which, overlapping with the other themes noted above, recognises that selectivity is an important part of interpreting invertebrate biodiversity simply because of the logistic prohibitions imposed by seeking to examine total biodiversity. RBA may necessitate selection of particular groups, taxonomic approximations of various sorts, restricting numbers of samples or sampling methods, each as a way to reduce costs and effort without losing sight of the level of information needed from the survey.

In practice, these three topics overlap in important ways, but all must be considered in appraising the needs and strictures of any biodiversity survey involving invertebrates. In direct contrast to surveys of many vertebrate animals or vascular plants, material usually cannot be identified in the field and released or (with a few exceptions, such as some butterflies and dragonflies, particularly in northern temperate regions) identified by viewing through binoculars. Analysis of invertebrate faunas, other than for such relatively well-known groups of larger, conspicuous, diurnal insects, and the distinctive context of monitoring pests whose identity is not in doubt, depends on collecting, killing and preserving, and examination of bulk samples from various habitat units, each of which commands different methods and approaches. Much of the material accumulated in invertebrate surveys is never utilised, simply because of the impracticalities of processing it in other than superficial terms and lack of knowledge over the values of the information to be obtained, other than the simple impressiveness of very large numbers.

Invertebrate biodiversity may be viewed as a hierarchy: in a few cases, family-level recognition may be sufficient (Hutcheson (1990) was able to characterise different forest associations from family-level interpretation of New Zealand beetles), but progressively more information is available from generic or species/morphospecies analyses. The balance is to decide which of these is the minimum level needed to provide adequate response to the questions being asked. Preliminary investigations may well be advisable before embarking on a full-scale survey.

Lack of ability to interpret invertebrate faunas provides a serious problem in conservation, despite the importance of these animals as ecosystem engineers and maintenance agents being acknowledged widely. Lack of specific level names and imprecision in studies is equated by many managers as equivalent to lack of interest or lack of importance. Basically, invertebrates are often considered not only too hard but also too trivial to have value in conservation, despite the range of fundamental and applied values they manifest (New, 1995). Yet, conservation or other ecosystem management based solely on the diversity and features of vertebrates and higher plants may incorporate only a small proportion of the information that even limited appraisals of selected invertebrates can provide. The widespread perception that invertebrates are secured and conserved effectively under the umbrellas of these more conspicuous groups is attractive but superficial (see p. 282). In terms of conservation of each of the three main axes of ecosystems – structure, function and composition (Noss, 1990) – invertebrates predominate both in diversity and abundance and in the suites of roles they fulfil.

Invertebrates in agroecosystems: the major players

The broad focus on invertebrates in agroecosystems has been on members of four phyla, with by far the most attention paid to two of these with massive perceived applied relevance. These are Annelida (as mediators of soil quality) and Arthropoda (including the bulk of pest species and those valued as pollinators and natural enemies), together viewed as beneficial biodiversity. Nematoda and Mollusca have also been emphasised, predominantly because of their more intermittent roles as pests or, more rarely, beneficial organisms. Collectively, these four phyla dominate much of what we know about the ecological maintenance of both agroecosystems and natural ecosystems. This section enlarges on their broad roles summarised in Table 2.1 to provide further grounding for later chapters.

Annelida

Most of the 3000–4000 species of earthworms (Oligochaeta) inhabit soils, with others constituting elements of the benthic fauna in marine or freshwater environments. True earthworms are distributed among about 18 families, one of which (Lumbricidae) is viewed widely as 'typical earthworms'.

The five general ecological categories into which earthworms are conventionally divided (after Bouché, 1972, 1977) reflect their primary habitats, burrowing capabilities and food preferences, as follows:

- Epigés. Surface active earthworms that generally do not burrow but live in litter on the ground surface. Some taxa found under bark of rotting wood are also allocated to this category.

- Anéciques. These large, usually deep-burrowing forms sometimes have semi-permanent burrows. They tend to draw litter down their burrows to lower soil strata.
- Endogés. These worms produce mainly horizontal galleries, predominantly in the organic horizons near the soil surface.
- Coprophagic species. As their name implies, these worms live in manure or compost.
- Arboricolous species. Relatively few earthworms occur in suspended soils in wet forests, mainly in the tropics.

The groups of most concern in agriculture and in environmental maintenance are the first three of this list. They are embraced also within the earlier, somewhat different functional classification of earthworms (specifically, of Lumbricidae) by Gates (1961), who recognised three groups on dietary preferences:

- Geophagous – those that pass considerable amounts of soil through their gut.
- Limiphagous (or limicolous) – those that inhabit mud or saturated soils.
- Litter feeders – those that feed on various kinds of organic matter, such as leaf litter, compost and manure.

Earthworms are major contributors to the breakdown of organic matter and increased mixing of this with mineral soil. Epigés are the most significant group in this process, whereby soil humification is also enhanced by reducing size of particles, increasing organic matter movement, and soil aeration through their burrowing activities. They occur in most soils, other than in extremely arid regions.

Most literature on earthworm biology deals with Lumbricidae, whose castings can return mineral content to the soil in forms more accessible to plants than when it was ingested. For example (Edwards & Lofty, 1977), soil fertility is increased in this way by increasing the availability or amount of calcium, potassium, magnesium and molybdenum. Castings are also higher in microorganism densities than soil.

Although not commonly associated with true earthworms, the small pale-coloured Enchytraeidae are also abundant, reaching their greatest representation in acid soils with high organic matter content (Dash, 1990). They are also important in decomposer systems. Their major roles are three-fold (Burges, 1967):

Table 2.5. *Variables integrating the effects of soil fauna at different hierarchical levels in agroecosystems (Crossley* et al., *1989)*

Variable	Level of integration	Effects of fauna
Primary production, yield	Agroecosystem	Direct/indirect; increase/decrease
Energy flow	Ecosystem or process level	Increased respiratory losses
Decomposition	Process	Accelerated carbon cycling; organic matter formation
Nutrient dynamics	Process	Altered rates of mineralisation, immobilisation
Soil microflora	Community	Decrease/stimulation: altered community structure
Soil structure	Abiotic/biotic interface	Aggregate formation, water and nutrient retention/loss Destruction of microstructure

- Ingesting decomposing plant material with high microbial content, digesting part of this and partially breaking it down, so providing food for other decomposer organisms.
- Providing faecal material in the soil, so providing sites for microbial activity.
- Encouraging higher levels of activity of microflora.

Collectively, earthworms participate in most of the important faunal influences on soil structure and fertility (Table 2.5).

Mollusca

Terrestrial Gastropoda, snails and slugs, are found almost everywhere but depend on adequate moisture and (for snails) a source of lime for shell construction. Many are found in soil and on vegetation and sometimes are regarded as pests in agricultural and horticultural crops. Despite the group's ubiquity, many species are very restricted in habitat, some requiring particular kinds of soil, being associated with particular kinds of vegetation, or depending on substantial leaf litter. Most vegetation-associated species appear to prefer angiosperm vegetation, but a few are more abundant on conifers. Landscape topography, particularly slope and other features influencing drainage patterns, may be an important determinant of local distributions, as also are many microclimatic factors, such as temperature and degree of exposure. 'Refuges' from such extremes may be particularly important for gastropod wellbeing.

Most species are generalised herbivores, some feeding particularly on fungi. Relatively few species of gastropods are predatory and feed on a variety of other invertebrates. A few predatory species have been employed as biological control agents against other, pest snails, and a few have fostered conservation concerns as serious threats to native snails, particularly in the Pacific region and albeit mainly in non-agricultural contexts.

Local diversity of Mollusca is commonly relatively low in relation to that of the other groups noted here, but regional faunas can be much larger. Stanisic (1999) suggested that this relatively lower species richness can be a considerable advantage in biodiversity assessment, in rendering snails (in particular) taxonomically more accessible than most other, hyperdiverse invertebrate groups. He also noted the important adjunct that snail shells are persistent and are often sufficiently distinctive at the species level to enable recognition from these alone, without need for whole specimens.

Nematoda

Nematodes are the most abundant organisms in soils and also are ubiquitous in litter and on vegetation. Despite their vast abundance, though, coupled with very considerable ecological diversity (Table 2.4), very little is known about the biology of most free-living taxa, with the bulk of information derived from those that are economically important as pests of plants or parasites of domestic animals.

Nematodes feed on a wide variety of foods, and many of the main feeding categories can be recognised by the structure of their digestive system and mouthparts. Predatory nematodes, for example, often possess denticles, or 'teeth', which help dismember the small animals on which they feed. The main trophic groups distinguished sometimes differ a little from those listed in Table 2.4. Freckman & Baldwin (1990) listed the following categories:

- Phytophages:
  - Endoparasites, feeding within roots or (rarely) plant shoots.
  - Ectoparasites, feeding on surfaces of roots.
- Microbivores, feeding on bacteria and microflora.
- Fungivores, feeding on mycelium of fungi.

- Omnivores, feeding on several categories of foodstuffs, such as fungi, bacteria, algae, rotifers and protistans.
- Predators, feeding on other invertebrates, including other nematodes.

Nematodes feed on living protoplasm, so their activities affect many ecological processes. Their abundance is often correlated positively with primary production, because phytophagous roundworms depend directly on roots. They thus respond directly to some forms of agricultural management. Calibrations by Yeates (1979), for example, showed that phytophagous nematodes can constitute up to 98% of the nematode assemblages in agricultural fields, compared with only 20–30% of those in uncultivated soils. Nematodes also play important roles in decomposition, as fungivores or microbivores, and can be the most abundant soil animals in these roles. Both categories may feed on harmful (pathogenic) and beneficial organisms and can have major influences on nutrient recycling. Some are effective vectors of bacteria, fungi or plant pathogenic viruses. However, their overall contribution to total soil respiration has been considered small (Sohlenius, 1980), probably because most nematodes appear to be rather inactive and to move around little.

The general roles of nematodes in ecosystems are clearly diverse but difficult to quantify generally. In soil assemblages, for example, most studies have been on presumed or actual plant-pathogenic taxa, and attempts at broader recognition of the fauna are still relatively scarce.

Arthropoda

'Since arthropods comprise most of the organismal variety in practically all habitats, they are good candidates for quantitative biodiversity evaluation' (Duelli *et al.*, 1999). However, summarising this diversity briefly is almost impossible. The groups shown in Table 2.1 differ substantially in their impacts and apparent importance, and those of major relevance to ecosystem studies can be considered by either major habitat or by feeding habits or ecological roles. Many groups defy simplistic compartmentalisation in this way. Thus, spiders are all predators, but their variety within this basic practice enables definition of 'spider communities' based on the relative representation of different functional groups (Marc *et al.*, 1999), with their abundance in many litter- and vegetation-based habitats giving them ecologically strategic importance in many food webs.

The four major groups of Arthropoda are all important, but some (especially Arachnida, Araneae and Insecta) predominate in most terrestrial ecosystems. Myriapoda are also numerous, with millipedes (Diplopoda) being the most diverse class, but the various crustacean groups (Isopoda, Amphipoda, Decapoda) found on land are much less prominent than their aquatic relatives. Thus, Isopoda (slaters, woodlice) are the only group of Crustacea to have become well adapted to terrestrial life and are completely independent of water bodies. They can occur in very large numbers in accumulations of decaying vegetation or in 'cryptic habitats', such as under bark or stones, or in rotting logs, and their main feeding roles are as decomposers through omnivory and general scavenging. Paoletti & Hassall (1999) described them as 'key system regulators of the ecosystem functions of decomposition and nutrient recycling', and they can respond to land-use changes and chemical pesticide uses.

Mites (Acarina) and springtails (Collembola) occur in vast numbers in most agricultural and other soils and participate in a variety of functional groups. Edwards (2000b) cited Brussaard *et al.*'s (1990) study on these groups in Dutch agroecosystems, in which the following functional groups of these microarthropods were recognised:

- Omnivorous Collembola, feeding on fungi, algae and other organic matter.
- Cryptostigmatid mites, feeding on fungi and organic matter.
- Omnivorous non-cryptostigmatid mites, feeding on fungi, organic matter and possibly nematodes.
- Bacterivorous mites, feeding on bacteria only.
- Nematophagous mites, specialised predators on nematodes.
- Predatory Collembola, feeding on a variety of prey taxa.
- Predatory mites, feeding on a variety of prey, including predatory Collembola.

These divisions led to construction of the food webs shown in Fig. 2.5 (Edwards, 2000b, after De Ruiter *et al.*, 1994), in which the complexities of distinguishing the roles of any particular group of soil invertebrates are readily apparent, notwithstanding the implications of their collective importance. Many different groups of arthropods participate in such food webs, and their presumed key positions in food chains, in addition to their influences on microbial activity, are often difficult to quantify.

Although groups such as mites are documented most completely from soil and surface litter, where decomposition processes have been a major focus for investigation, more recent studies on arboreal fauna have revealed them to be very diverse and, indeed, sufficiently abundant to be termed 'aerial plankton' in forest canopies (Walter & Proctor, 1999;

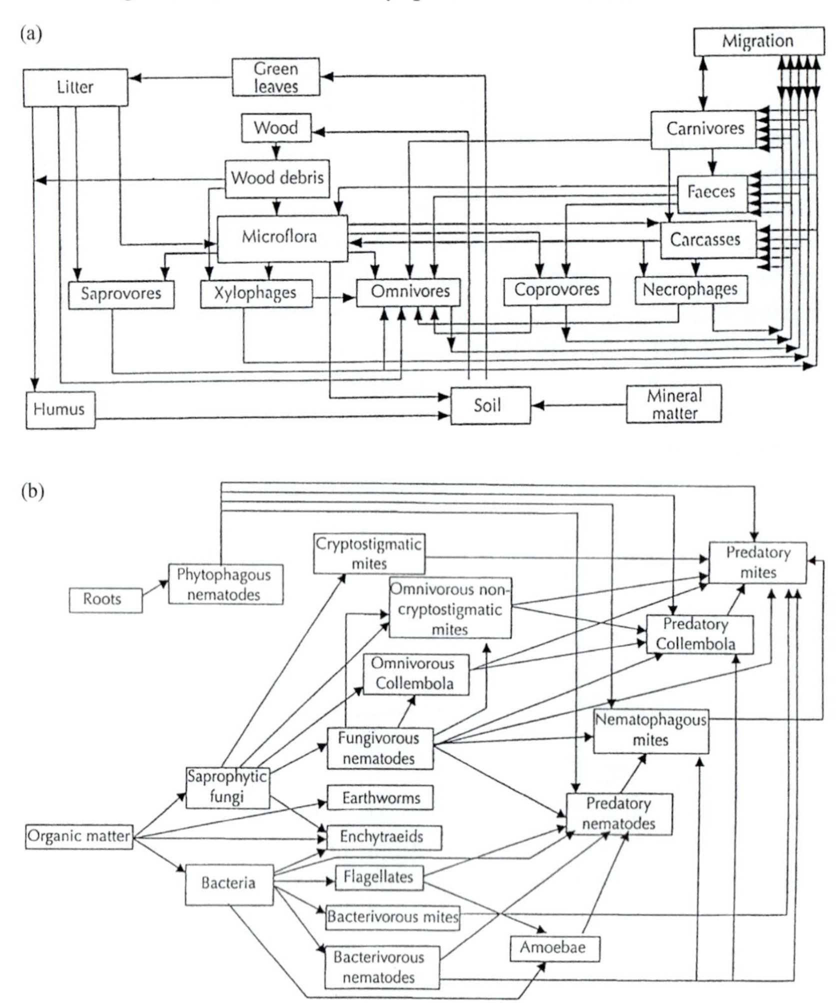

Figure 2.5. Food webs in soils (Edwards, 2000a,b): (a) general patterns of energy and nutrient paths through soil populations, with major trophic groups involved; (b) example of a soil food web in the Netherlands (after De Ruiter *et al.*, 1994) to indicate some of the numerous roles of invertebrates.

Behan-Pelletier & Walter, 2000). Most arboreal mites are Oribatida, many of them feeding on fungi. However, as Behan-Pelletier & Walter (2000) noted, the term 'mycophagous' does not distinguish between undesirable effects such as spreading of pathogens through dissemination of live spores and more beneficial effects such as decomposition, so that the 'true' ecological influences of such taxa are largely unknown. Gut-content analyses of oribatid mites from soil and from trees suggest that their food spectra are basically similar. As with many other groups of small invertebrates, such as Pauropoda (most of which feed on decaying vegetation or fungi), many such organisms are generally ignored, except by specialists, and largely disregarded when appraising ecosystem functions.

Of the larger myriapod groups, Diplopoda generally also feed on decaying plant material, whereas most centipedes (Chilopoda) are believed to be carnivorous. Other than for the trophically varied mites, most other groups of chelicerate arthropods are predators or, at least, believed to be predators.

Terrestrial insects also have many impacts and roles, with their importance perhaps being manifest by the impacts of some phytophagous species as economically damaging crop pests. In Chapter 1, the high proportion of primary production lost to insects and related consumers in agroecosystems was noted, and their roles in natural ecosystems are probably little less dominant. Likewise, their importance at higher trophic levels is demonstrated by their diversity – with, for example, parasitoid Hymenoptera amongst the most speciose of all insect groups, and the predatory habit having arisen on many independent occasions in at least 12 entirely or predominantly terrestrial orders (New, 1991).

Indeed, it is difficult to express (even superficially) any terrestrial food webs in agroecosystems or elsewhere in which insects are not key players, and many workers have emphasised the ecological importance of virtually every insect order. Thus, termites (Isoptera) and ants (Hymenoptera, Formicidae) are viewed as keystone groups (see p. 31) in desert ecosystems, in which the soils may be too dry to support large earthworm populations (Whitford, 2000), and are important determinants of soil quality and productivity. They thus fulfil the roles of other invertebrates in some environments too extreme for those other groups to flourish. Above the ground, very few plants are free from insect attack, often in some intimate and specific association with strong suggestion that (at least in some cases) the interaction between insect and plant may be involved in regulating production. Pollinators, likewise, can be viewed as a specialised category of herbivore exploiting mutualistic associations with plants and are a prime functional guild for manipulation in crop-husbandry practices. Particular insect herbivores feed in a variety of different ways; many chew foliage or other plant parts (Lepidopterous larvae, Coleoptera, Orthoptera and Phasmatodea, for example), while others suck sap by way of specialised piercing mouthparts (Hemiptera) or rasp surfaces of plants to gain access to cell contents (Thysanoptera). The latter groups exemplify those that can act as vectors of plant diseases by their feeding activities.

Large insect groups, such as Coleoptera, include massive trophic variety. In addition, most insects are holometabolous, with the consequence that their larvae and adults manifest very different lifestyles and occupy different feeding roles. Thus, larval Lepidoptera (caterpillars) are typically chewing herbivores, whereas adults of many moths and butterflies feed on nectar and can effect pollination of many angiosperms. As a group, the insects are involved in a greater variety of trophic interactions, and therefore in nodes for determining patterns of energy flow, than any other group of terrestrial animals. This variety extends to decomposition processes, with well-known groups such as dung beetles (Scarabaeidae) facilitating breakdown of mammalian dung as valuable agents to counter pasture staling in some parts of the world, particularly Australia.

With their wealth ofbiological and morphological categories, combining to produce communities or assemblages comprising up to thousands of individual species and with different representations and relative abundances in each habitat or ecosystem on land, invertebrates still intrude on human consciousness to only limited extents. The importance of conserving them, and the need to do so as the major players in ecosystem maintenance and sustainability, is still one that many people find difficult to accept. Invertebrates (as 'creepy-crawlies'or 'bugs') are viewed widely as 'bad' and as targets for extermination. Relatively few are viewed as 'good' with defined positive values as pollinators or natural enemies in agriculture or simply because people like them (butterflies, dragonflies). Other values that impinge on the importance of invertebrate conservation, particularly in relation to agricultural practices, are discussed in the next chapter.

3 · *Agriculture: effects on invertebrate diversity and conservation*

Loss of ecosystem functions previously provided by invertebrates may need to be compensated for in agroecosystems by costly human intervention, such as increased applications of agricultural chemicals. Threats to invertebrates from agriculture are outlined and discussed in this chapter, with particular emphasis on the influences of habitat change and simplification and the effects of exotic species. Particular invertebrates may be endangered by agriculture and thus become targets for species-focused conservation programmes; others may be used as monitors of the wider effects of such changes and are sensitive functional indicators of wider community changes at levels otherwise difficult to determine.

Introduction: a central paradox

There is no reasonable doubt that agriculture has had dramatic effects on invertebrates and has been associated with decline and local extinction of many species. Although the causes of these losses are diverse, much recent attention has devolved on issues related to crop protection, predominantly the control of pests. Crop protection in agriculture is involved largely with measures to kill insects and, more rarely, other invertebrates as major pests. These organisms commonly are related closely to many other species in natural environments, and some of those may be of specific conservation concern in addition to being constituents of wider biodiversity without adverse effects on human wellbeing. Many insects disperse actively and, indeed, such dispersal can be important in crop and pest management so that optimal management – although focused largely on cropping or other production areas – commonly may need to incorporate wider considerations. The paradox to be addressed in much of this book is how to harmonise measures to suppress pest invertebrates in particular areas or contexts whilst simultaneously safeguarding the wellbeing of close relatives elsewhere, including adjacent areas such as field margins. There is also practical need to foster increases in pollinators, predators and parasitoids in the very environments in which undesirable pests, as targets for suppression, are present. Thus, for example, pest species of grasshoppers (locusts) or moths comprise very small proportions of taxonomically related (even congeneric) and biologically similar species within such large groups, and some of these may occur frequently in crop environments in small, non-damaging numbers. Essentially, practical and ethical problems can arise over the simultaneous needs to reduce pests, increase other populations of insects, and safeguard wider spectra of species in the same environments.

Agriculture depends on fostering productive environments; invertebrates are important components of those environments, and ecological losses resulting from their loss may need to be countered by costly human intervention. However, concern over such wider losses is generally low amongst agriculturists, so that agriculturists and environmentalists are often viewed as opposing categories in their priorities for landscape use. This unfortunate polarisation in attitude devolves on the different positions on practical conservation. Simplistically, both groups are committed to sustainability. On the one hand, the priority is to conserve simplicity and uniformity through (at the extreme of low diversity) plant monocultures; on the other hand, the environmentalist priority is to conserve ecological complexity (equated, broadly, with high diversity) with need for minimum human inputs.

Threats to invertebrates from agriculture

Threats to invertebrates in terrestrial ecosystems fall into relatively few broad groups (Wells *et al.*, 1983; New, 1984, 1995; Samways, 1994), but the ramifications of agriculture extend also to those in freshwater ecosystems and even to the marine environment. Recent concerns for parts of the Great Barrier Reef, Queensland, Australia, for example, have arisen through increased chemical washout and soil erosion entering the sea from rivers due to intensification of nearby coastal sugarcane growing. Whereas some threats are clearly local, and their ramifications do not extend beyond the immediate area of impact, other practices have much wider-ranging influences.

The major categories of threats to invertebrates are:

- effects of habitat loss or change;
- effects of pollution and pesticides;
- effects of exotic species of animals and plants;
- effects of overexploitation and overcollecting.

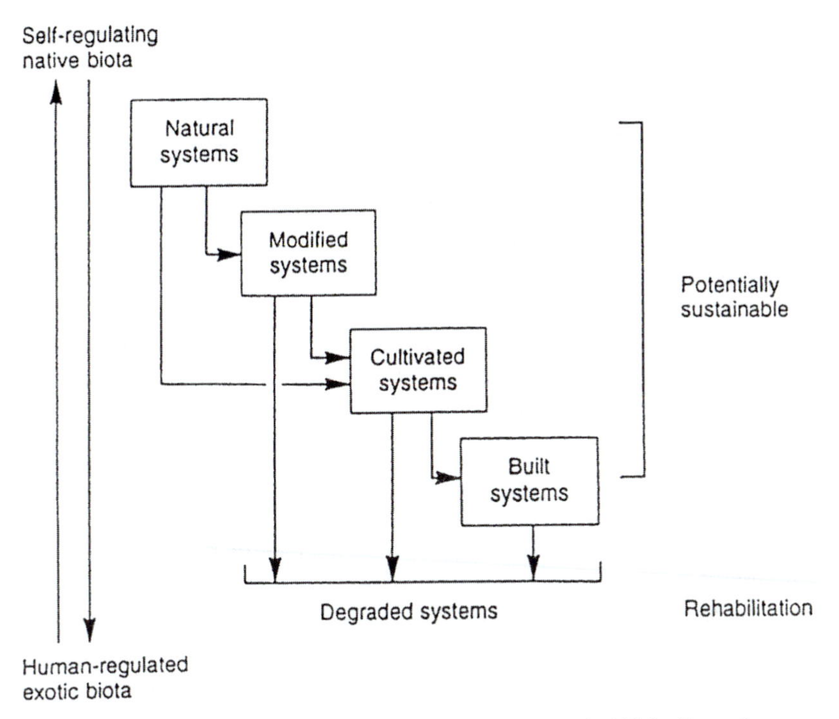

Figure 3.1. Habitat changes wrought by people (IUCN *et al.*, 1991). General scheme linking the extent of habitat change with potential for sustainability and need for subsequent management to maintain the systems.

Each of these broad categories is general, and the first three are of particular importance in considering wider effects of agricultural establishment and maintenance of agroecosystems. They are introduced briefly here and exemplified further in later sections.

Habitat effects

On a broad system of habitat changes wrought by people (IUCN *et al.*, 1991) (Fig. 3.1), agroecosystems are viewed widely as 'degraded', with the level of degradation from natural ecosystems reflected in the extent of change and intensification that has occurred. At the extreme, all aboveground traces of the parent system are removed and the underground system modified substantially. Whether such changes occur in grassland or forest, the result is an obvious loss of the resources formerly present and reduction of the environment's carrying capacity for the earlier resident taxa. At least locally, loss or displacement of taxa is almost inevitable. Such habitat 'destruction' has occurred over wide areas simply by the act of clearing native vegetation to free land for cultivation. The effects are thereby proximal; habitat loss is the paramount threat to many invertebrates (and others), and the speed and intensity of such changes often leaves little opportunity for species to adapt to more gradual changes.

Over much of the tropics, for example, the clearing of forests to produce croplands has been severe but – despite widespread documentation of such habitat change – few groups of invertebrates have been studied along gradients from natural to heavily disturbed environments in those regions to determine the extent of changes within assemblages. The loss of termite species accompanying forest clearing, through sequences of primary forests to disturbed or secondary forest or agroforestry systems (see p. 299) and thence to field cropping areas in the tropics, can be substantial. Thus, in Sumatra, termite richness fell from 34 species in primary forests to only one in a cassava garden (Jones *et al.*, 2003), with decline in soil-feeding taxa being greater than that in wood-feeding termites along this gradient of increased intensity of change. This probably reflected the use of heavy machinery compacting the soil and increasing its bulk density, a measure known to adversely affect soil macrofauna elsewhere (Radford *et al.*, 2001: Queensland, Australia). Similarly, in Cameroon (west Africa), forest clearance and conversion to farm fallow led to reductions to around half the numbers of butterfly and termite species (Watt *et al.*, 1997). In the same survey, complete clearance of vegetation before replanting to establish forest plantations had negative effects on additional groups studied, namely leaf-litter ants and arboreal beetles.

In another instructive example of effects of agricultural disturbance on diversity, Perfecto & Snelling (1995) compared the diversity of ants with changes from traditional, low-input coffee production to intensive, high-input (monoculture) coffee production in Costa Rica. The traditional regime (Fig. 3.2a) is vegetationally diverse and resembles in structure some of the 'forest-garden' (kebun) systems of perennial cropping practised in Indonesia and elsewhere in the tropics. Intensification, as elsewhere, leads to elimination of plant species other than the crop (Fig. 3.2b). Perfecto & Snelling examined ground-foraging ants (using baits) and those active on coffee bushes (sampled by sweeping). The species pool comprised 37 species (ground) and 28 species (on coffee), with nine found only on the ground and 14 found only on the coffee bushes; 14 species were found in both habitats. The transition from conditions of Fig. 3.2a to those in Fig. 3.2b was associated with significant decrease in ground-foraging ant diversity, but no significant change was found in the coffee-frequenting ants. In addition, the lowered species richness in

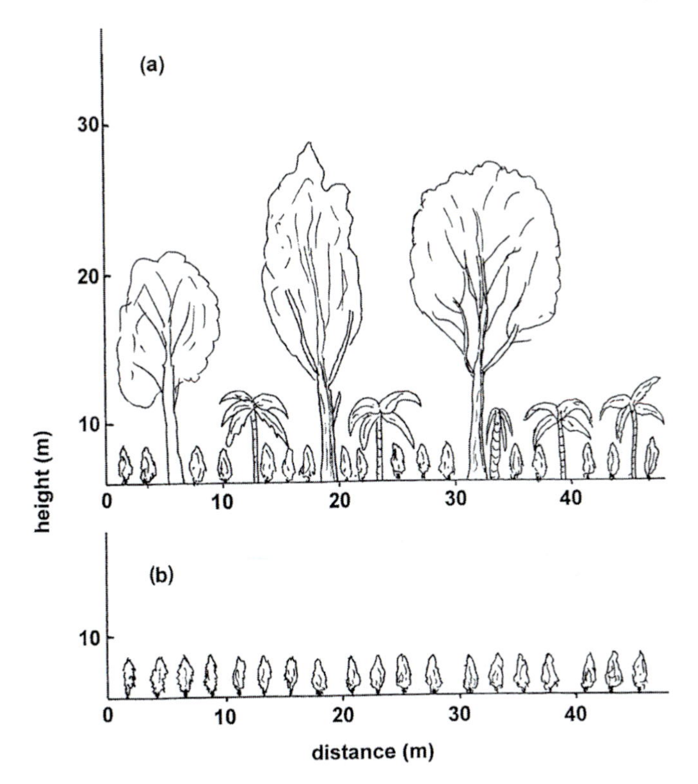

Figure 3.2. Ants in coffee plantations in Costa Rica (Perfecto & Snelling, 1995). Two types of coffee-management systems: (a) traditional agroforestry with diverse shade retained and (b) unshaded coffee monoculture.

coffee monoculture was accompanied by greater uniformity of assemblages, with a few dominant or codominant species, whereas assemblages in the more structurally diverse cultivation were more varied. However, as Perfecto & Snelling (1995) emphasised, detecting such differences may depend on use of particular sampling methods, so that particular habitats or species guilds may be influenced by such changes but might not be appraised unless targeted specifically. The wider implication, reflected in many published accounts, is that the ecologically more specialised taxa are the more likely to be displaced.

Habitat loss for invertebrates is often not as obvious or dramatic as in the above cases, in which natural habitat was removed to a large extent, because relatively small changes in habitat quality may have high impacts on ecologically specialised taxa with limited resource requirements. Many invertebrates are also highly localised, so that changes to quite small areas (of, sometimes, a hectare or less) may be sufficient to extirpate populations or taxa. Such changes as modifying the extent of exposure to desiccation or insolation, changing the amounts of litter on the ground surface, removal of dead fallen timber, and in the distribution of food plants and other species may be important vectors of habitat changes sufficient to endanger numerous species. Habitat changes thereby include 'macrochanges' and 'microchanges' that were previously buffered within the system. Likewise, changes to or within a particular habitat may influence the hospitality or carrying capacity of other habitats not modified directly or intentionally.

Many invertebrates differ from most vertebrates in that they do not disperse much over their lifetime, and populations can thrive in very small habitats. Thus, even some putatively vagile species such as many butterflies live as resident colonies in very small areas (Thomas, 1994). This means that even very small natural habitat patches in largely degraded landscapes can have considerable value for invertebrate conservation. However, much agricultural intensification, as noted earlier, has involved trends such as increasing cropping areas by making larger cultivable areas through removal of field boundaries such as hedgerows and other native vegetation remnants simply to facilitate access for large machinery. In such ways, agriculture has led to landscapes becoming much less heterogeneous than previously in many parts of the world, inducing a contrast with more natural landscapes referred to as 'polarised' by Keesing & Wratten (1997) (Fig. 3.3). This trend has important adverse effects for invertebrates, not only by direct loss of suitable habitat (important though this is) but also by rendering any remaining suitable habitat patches increasingly isolated and subject to external or stochastic influences.

'Habitat fragmentation' is frequently used as a descriptor for this process; it implies the loss of continuity of previously widespread habitat as it is alienated progressively by human changes and the remaining patches are spread more widely over the landscape. Fragmentation of habitat is presumed to be accompanied widely by fragmentation of populations, so leading to likelihood of genetic modifications by reproductive isolation. One general scheme by which fragmentation arises is summarised in Fig. 3.4. An intact habitat may be bisected initially by an access road or similar human-made conduit, so rendering the area more accessible for settlement or exploitation. This increased chance of disturbance can lead to loss of further habitat for cropping (or mining, urbanisation, etc.) so that it is reduced progressively to isolated patches: fragments. In extreme cases, not unusual in modern extensive agriculture, even those fragments are

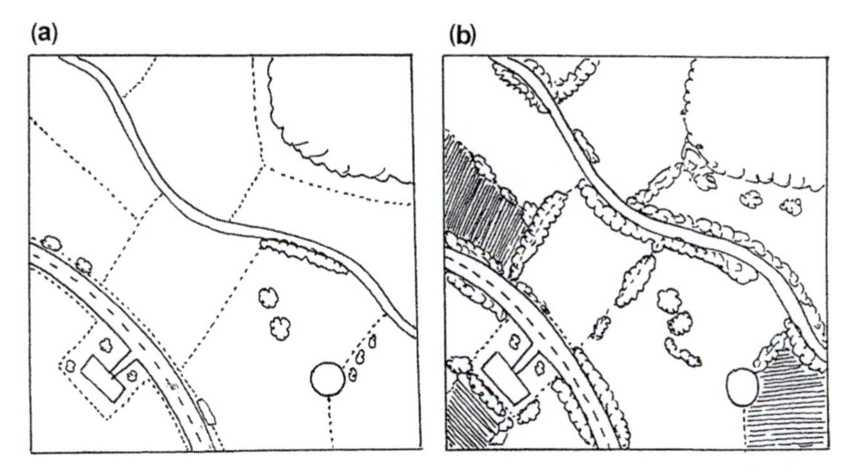

Figure 3.3. The idea of polarised landscapes (after Keesing & Wratten, 1997). These authors drew the contrast between agricultural ecosystems in New Zealand and the UK: (a) low-conservation-value landscape (New Zealand) in which little natural vegetation is retained, connectivity is low and land use is relatively uniform over large areas; (b) much higher conservation-value landscape (UK) with diverse land uses and extensive presence of corridors and nodes of natural vegetation. (cf. Figure 10.1; see text).

lost or minimised in extent so that the landscape is described as in a state of attrition, bearing little resemblance to its former natural condition.

Consequences of fragmentation are very diverse. Even construction of a road or machinery access track can have substantial implications for invertebrates, either by (1) providing an alienated barrier that they cannot cross, so isolating populations on either side, or (2) by constituting a corridor along which they can move to increase their distribution or increase interchange between segregates, or through which invasive species may enter the ecosystem. The dynamics of habitat fragments (often paralleled in terminology with 'remnants', although remnants are not necessarily consequences of fragmentation) have received considerable attention. Some aspects are discussed further later in this book (see pp. 252 and 263). Most studies on invertebrates have involved surveys (inventory) of parental habitat and fragments, with differences purported to show effects of fragmentation, even though previous surveys to determine the extent of heterogeneity in the area investigated (which is often substantial) are relatively rare. Studies on ground-dwelling beetles and eucalypt forest fragmentation in south-eastern Australia have illustrated well the difficulty of making generalisations about such effects. Particular

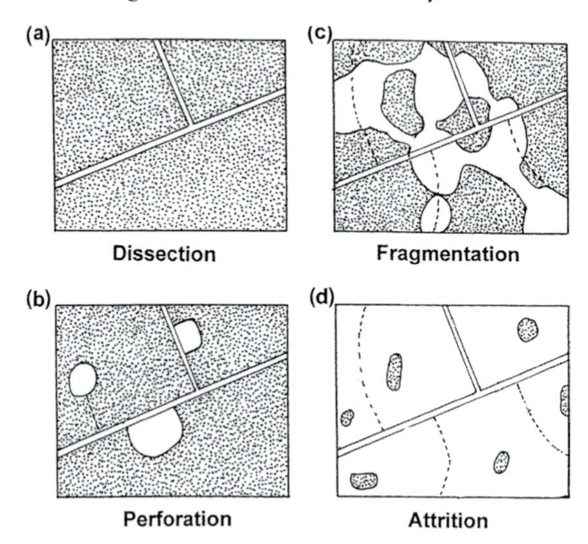

Figure 3.4. Example of the process of habitat loss. (a) A natural habitat is dissected by a road or other access way, which divides it into two or more compartments. (b) Such access may lead to local settlement, with local clearing for farming or residence to produce a perforated landscape. (c) Continuation of this process, with increased area of change, leads to only fragments of the natural habitat remaining, increasingly isolated from each other. (d) These fragments are then diminished progressively to constitute only a small proportion of their original extent and probably increasingly subject to edge effects and decline by attrition (after Hunter, 1996).

beetle species variously decreased in abundance, increased in abundance or changed little in response to fragmentation (Davies & Margules, 1998, 2000). Responses must therefore be evaluated at the species level.

The conventional (perhaps simplistic) interpretation of fragmentation is that because (1) fragments have only limited area, often small, (2) the number of species, and the sizes of their populations, supported is necessarily smaller than for larger areas, so that (3) internal changes may occur to diminish genetic variety, (4) stochastic influences may have greater relative effects, and (5) external effects (edge effects; see p. 245) whose intensity related to fragment size through length of the habitat boundary are increased on the resident species, together with increased chances of community invasion by external species. In short, the biota of small fragments are regarded as more vulnerable than those in larger areas or to have increased chances of becoming so. Once fragments become too small or too isolated to sustain internal or near-internal communities, many extirpations are regarded as largely inevitable. The fragments are, in essence, 'islands' in the altered anthropogenic landscape dominated by agriculture and other changes.

Habitat loss or change is, as implied here, a potentially universal threat to specialised invertebrates, and relatively few have demonstrated capability to thrive in highly altered environments. Those that have done so are a prime focus in agriculture. However, habitat change is often linked intricately with the other threat categories listed above.

Pollution and pesticides

'Pollution' encompasses all the by-products of human activity in anthropogenic systems, but it is assessed most frequently as chemical presence in environments other than those intended or in excessive amounts – or, more broadly, the totality of non-target effects of those chemicals. Agricultural and industrial chemicals are prime contributors to pollution, from local to global scales, with pesticides and fertilisers among the major categories involved. Pollution can thus occur as a by-product of deliberate applications of chemicals, as in agriculture, or by the accidental spillage or puposeful discard of wastes or other non-inert chemical materials. Pesticides are often considered an individual category of pollutants. They differ from most other chemicals applied deliberately in that they are designed to kill, so that their roles in conservation are highly relevant and, sometimes, controversial. We see later (Chapter 4) that much of the impetus leading to more effective harmonisation of agriculture and conservation has been driven by concerns over uncritical pesticide use. Many other industrial chemicals have been intended for more benign uses, such as in manufacturing processes, but pose problems for safe disposal in the environment.

'Ecotoxicology' is a broad suite of topics of major concern for conservationists, with agricultural chemicals a major contributor to those concerns. Invertebrates are particularly susceptible, and the considerable attention they have attracted as indicators or monitors of pollution effects and the consequences of using pesticides has been accompanied by attempts to develop predictive models to estimate the probability and extent to which non-target biota may be affected.

Exotic species

Agriculture is founded largely on exotic species, and many of the pest species of highest concern are, likewise, exotic. In this sense, 'exotic' means that the species are alien to (and did not evolve in) the communities and areas they now occupy. Most commonly, they are 'foreign' and as non-native entities in ecosystems are analogous to forms of 'biological pollution' from the viewpoint of many conservationists.

Many exotic species are confined to limited areas where they were introduced deliberately, and many such species are thus non-invasive and may pose little risk to other environments. Others are aggressive invaders of natural ecosystems and can have damaging impacts on native biota through providing consumer pressure (as predators on native species or herbivores on native plants, for example) or competitive rivalry for limited resources in the receiving environment. Thus, many of the most aggressive environmental weeds are exotic species, as are many of the most damaging invertebrate agricultural pests. Many of these species could not establish in new areas without the facilitating environments constituted by agricultural change (New, 1994). They can occasionally become major constituents of local biotas. Aphids (Homoptera) in Australia, for example, comprise about 150 species, of which some 120 are exotic species, mostly restricted to agricultural crops or other, ornamental or horticultural plants.

In many cases, exotic species were introduced deliberately to their new areas – plants as crops or ornamentals and for sand-dune stabilisation and other purposes; animals for stock or companion animals, as biological control agents or other beneficial purposes, or simply because people like them. Many such species have adapted in unexpected ways to their new environments, with the consequences that (1) exotic species are viewed widely as second only to habitat loss as a threat to native organisms and ecosystems; (2) quarantine measures are in place in many countries to help prevent or reduce ingress of further exotic species; and (3) concerns over proposals to deliberately introduce exotic species, such as non-native biological control agents (see p. 162), have led to development of formal protocols to appraise risk to the receiving environments. The concerns over such classical biological control agents are discussed in Chapter 5, but concerns extend also to introduction of pollinating agents for crops.

Introduced pollinators

The concerns over conservation implications of exotic or otherwise introduced pollinators are exemplified here by noting recent concerns in Australia; they involve possibilities of the introduced insects becoming invasive and displacing native invertebrates. 'Pollination services' are an important component of so-called migratory beekeeping in Australia, by which hives are shifted seasonally to a number of different sites to track seasonal nectar flows and supplies. The practice commonly incorporates enhancement of pollination of field and orchard crops by seasonal placement of up to 100 honeybee (*Apis mellifera*) hives near or within the crop, and the economic benefits have been evaluated as high as, or higher than, the more familiar honey crop from the industry. Many details and concerns of this practice were summarised by Paton (1996) in a wider appraisal of the controversial effects of honeybees in the Australian environment. More than 500 000 hives are maintained in Australia, and bees forage at distances up to at least 2 km from the colony. They visit flowers of at least 200 native plant genera (Paton, 1996), and many possible adverse effects have been implied from this dominant exotic species. Several studies have indicated that honeybees can be the most frequent visitors to native flowers and consume a significant proportion of the floral resources produced; they have thus been suggested to displace native pollinating insects and to diminish the food supplies for nectarivorous birds. In addition, they can affect patterns of seed production and pollination by numerous plants. As Paton (1996) noted, there is still insufficient information about interactions between honeybees and Australian biota, but additional concerns arise from the frequent proximity of managed apiary to conservation areas and the industry demands for access to protected areas for nectar as part of the seasonal needs for migratory beekeeping at times when adequate nectar is not available on private or agricultural lands. Honeybees also have large feral populations in Australia, and these provide further conservation concerns, such as usurpation of nesting hollows otherwise needed by parrots and marsupials; there is some supposition that use of managed hives near conservation areas may spawn further feral populations.

The demand for increased agricultural pollination services has fostered demands for introductions of other native bees, with fears that these might also have adverse effects in the wider environment through becoming invasive. The European bumblebee *Bombus terrestris* has been dispersed widely for increased pollination, especially of greenhouse crops (particularly tomatos) but also of some field crops (Hingston *et al.*, 2002). This species is a proven invasive in New Zealand and in Japan. Its discovery in Hobart, Tasmania, in 1992 has been followed by invasion of much of the state, including national parks within the World Heritage Area of the south-west. Competitive displacement of native bees has been reported (Hingston & McQuillan, 1999), as occurred also in Israel

(Dafni & Shmida, 1996). Interactions of *Bombus* and native fauna in Israel are summarised in Fig. 3.5.

Despite the proven invasibility and potential effects of such social Hymenoptera, industry groups continue to lobby for their introduction. Thus, strong pressures occur for introduction of *B. terrestris* to the Australian mainland for tomato pollination (Goodwin & Steiner, 1997).

At a somewhat different geographical scale, within-country translocations or movements of species can render native species 'exotic' in having artificially extended ranges and establishing them in new communities with which they have had no previous association.

Overexploitation

Rational harvesting, or the sustainable use of natural resources by not overexploiting them, is an important aspect of the conservation of species that are used or collected by people. Its relevance to agriculture, in which maximum harvesting is commonly a prime aim, is often small except for two general scenarios that may arise from time to time. First, local drought or crop failure may lead to increased exploitation of wild food species that are normally taken in only much smaller amounts, simply because they become more important as a food supply. Hockey & Bosman (1986) noted this trend for food invertebrates on rocky shores in Transkei, for example. Second, where collectable species (such as commercially desirable butterflies or snails) have been reduced in range by habitat loss to occupy only small remnants (as noted above), unrestricted collecting could increase their vulnerability considerably.

Threats and agricultural practices

Conventional agriculture, according to the views of many ecologists, implicitly threatens much wildlife. Agriculture involves alienation of natural habitat though its loss and modification over large areas. As noted earlier, diverse vegetation systems are replaced by low-diversity systems, commonly of monocultures and commonly involving exotic species and additional uniformity through standardisation of varieties or stocks and imposed genetic uniformity. Simplification is thereby accompanied by predominance of non-natural species and stocks, commonly with the ensuing likelihood or reality of further exotic species entering the system as pests or (more deliberately introduced) as natural enemies of these. Maintenance of desirable exotic species commonly involves uses of chemicals, with potential for undesirable side effects on non-target species, and

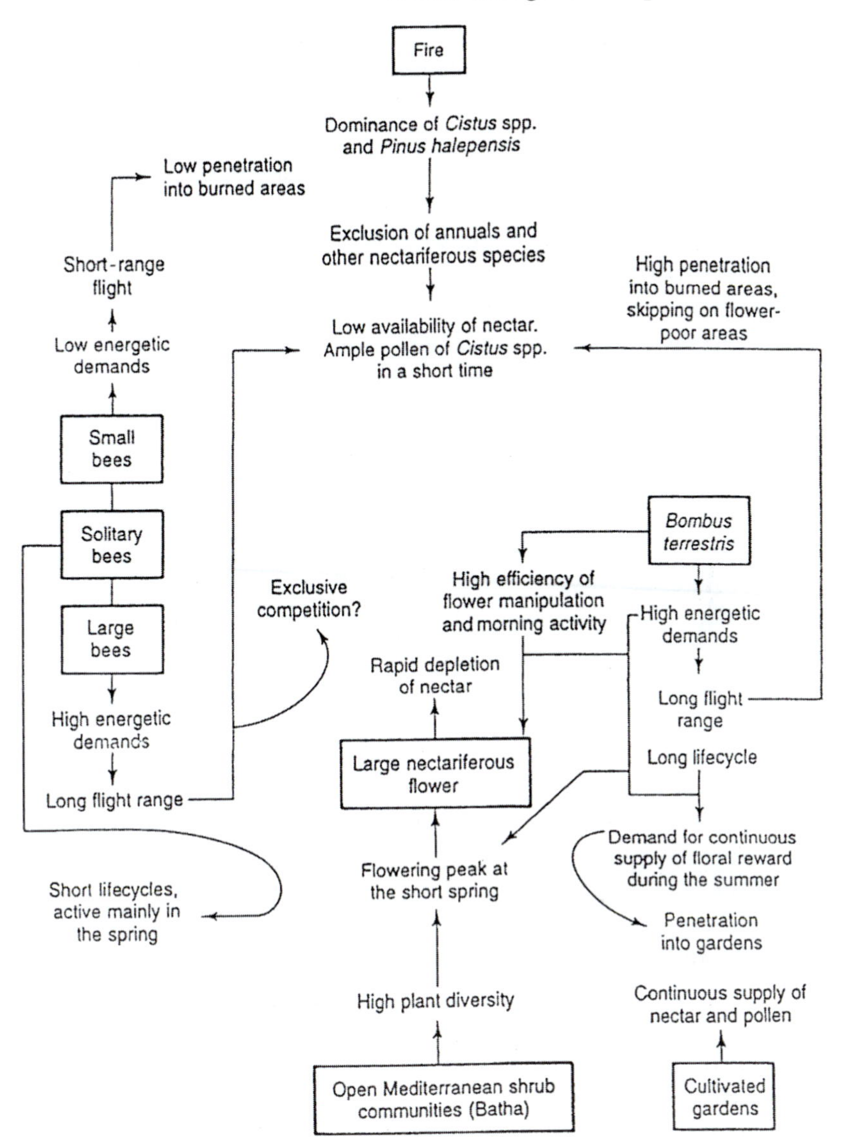

Figure 3.5. Example of ecological interactions of an introduced pollinator. Some possible interactions of the bumblebee *Bombus terrestris* with native fauna in the Mediterranean ecosystem at Mount Carmel, Israel (Dafni & Shmida, 1996).

of wider pollution of local environments. Intensive agriculture manifests all these factors to produce highly unnatural habitats and species associations novel to many parts of the world in which they are now major landscape components.

The categories of threat listed above emphasise what might be considered as direct threats to invertebrates. Many agricultural practices lead to changes in local environments that may have ramifications well beyond the areas subject to direct impact. However, resulting changes in hydrology, the quality of nearby waterbodies, and the relatively intangible effects such as influences of changed floristic balance on constitution of natural phytophagous invertebrate assemblages are not treated in detail. They should, however, be considered in wider aspects of conservation of assemblages. Later in this chapter, the sensitivity of many invertebrates to even small changes in their environments is noted, so that virtually any disturbance or alteration may cause some changes to their welfare. It is clearly impracticable to avoid all of these in an industry as widespread and diverse as agriculture, but many more tangible effects (as threats arise) can be appraised in much more detail, and measures to ameliorate these are also likely to incorporate wider benefits. Many aspects of agricultural practice and intensification thus constitute real or putative direct threats to non-agricultural invertebrates or change the structure of invertebrate assemblages. Whereas some aspects of agricultural manipulation are indeed motivated by practical conservation needs for those invertebrates seen as directly beneficial to production systems, others – particularly some involved with aspects of crop protection – are not so driven. These have historically caused concerns for their unanticipated side effects, both on agricultural land and well beyond the areas of direct application, both geographically and in the variety of biota they may influence. Detecting, monitoring and ameliorating these effects is perhaps nowhere as important as for invertebrates, despite their incidence and influences being so poorly documented and unremarked. The future of much invertebrate conservation devolves squarely on continuing to change practices within the agricultural estate and increasingly realising that this is necessary to sustain species and coevolved assemblages.

Invertebrates: targets and tools in conservation

The twin functional strands of invertebrates in wider conservation agendas differ considerably in emphasis and, to some extent, are complementary. On the one hand, and as the aspect most familiar to many people, invertebrates parallel all other groups of animals and plants in that particular species can become the foci (targets) of individual programmes to conserve and protect them. This act results from perceptions of threat of extinction brought about by declines in distributional range, in the numbers of populations and in size of populations, or combinations of these as consequences of human activity. The species thus become 'threatened' with the above-listed factors variously implied or specified in this process of increasing vulnerability. For European butterflies, for example, van Swaay & Warren (1999) assessed 'agricultural development' as the primary threat (constituting a threat to 63 of the 69 most threatened butterfly species, and followed by 'habitat isolation and fragmentation' (62 species) and urbanisation ('built development', 58 species)). Most (probably all) of these butterflies are not pests or otherwise intrusive on agricultural wellbeing, but the realisation and implication that they can be affected by agricultural practices helps to open up the true dimensions of the intensification of agriculture and its influences on invertebrates. Butterflies are simply the most conspicuous and best-documented group of these animals. Not least, the recovery plans designed for conservation, based on need for threat amelioration for those species, have helped to (1) clarify the variety of threats posed by agriculture and their relative importance and (2) indicate some possible pathways toward 'less threatening agricultural practices'.

Not surprisingly, in view of the general inconspicuousness of many invertebrates and the difficulties of detecting and evaluating changes in their numbers and distribution, most terrestrial invertebrates reported for such individual needs belong to the more conspicuous groups; the global compendium of such needs, the World Conservation Union's *Red List of Threatened Animal Species* (IUCN, 2002) lists 1854 such invertebrates, as summarised in Table 3.1. The greatest numbers are for 'popular groups', particularly Lepidoptera and Mollusca, with members of many large invertebrate groups not appearing at all. Such coma are representative of and indicative of individual species' needs, but they are by no means a complete appraisal.

The second aspect is the wider conservation of invertebrates in communities and their use as tools to estimate biodiversity or as indicators of environment condition and change. Changes in species incidence, relative abundance and assemblage composition can often be useful early-warning systems of wider undesirable or unplanned change and monitors of planned changes. Because many invertebrates respond to environmental changes rapidly and in subtle ways, and the great variety of invertebrates

Table 3.1. *Numbers of species of various invertebrate groups listed as of conservation concern on the IUCN* Red List of Threatened Animal Species *2002 (as accessed June 2003)*

Phylum/class	Number of taxa listed
Mollusca	1222
Arthropoda	614
Insecta	595
Arachnida	18
Chilopoda	1
Annelida	18
Nematoda	0

facilitates interpreting a variety of different responses, some have uses analogous to the miner's canary. Imposed changes to an ecosystem can be very diverse and often not very obvious to human observers, whereas more sensitive responses may occur among invertebrates.

Selection of optimal invertebrate groups as indicators for particular contexts of disruption is often possible, and this approach gives invertebrates considerable importance in developing indices for evaluating and monitoring changes of many kinds to biodiversity and the environment. Instead of simply concentrating on individual 'rare' species, the emphasis then moves more to the wellbeing of communities or assemblages (the context in which any species must persist and evolve) and to helping to prevent common species becoming rare as well as conserving the notable taxa present.

Both these contexts have considerable linkages with agriculture and in evaluating its impacts both locally and on the wider environment. They represent somewhat different emphases on invertebrate importance, and their scope and roles are noted further below.

Species focusing: targets

This approach has clear limitations for practical conservation, despite its wide appeal as concentrating on entities to which people can relate readily and in paralleling the most frequent approach to conservation of popular vertebrates as individual targets for management and protection. Indeed, the most popular invertebrate targets, as noted above, are of the more 'charismatic' groups that people like rather than fear or disregard.

The limitations derive from the very limited expertise available, limited financial/logistic support and capability, and the problems of applying these to the most deserving cases in the sure knowledge that in doing so, other species will surely be ignored or bypassed for such support. In short, if all of even the relatively small proportion of needy invertebrates so far signalled for attention were funded sufficiently to undertake basic recovery and management actions, there would still be a massive shortfall in practical capability.

Long lists of such 'threatened species' have both advantages and disadvantages. They can indicate dramatically the need for conservation measures: long lists are politically impressive and may confer formal obligation for investigation and necessary action if they transit from advisory to formal legislative status as lists of protected species. Some of the shortcomings of species focusing in this way for invertebrates (New, 1999a, 1999b, 2000b) include the following: (1) the species listed for protection are often difficult to evaluate objectively for relative or actual need because of lack of knowledge; (2) they are merely the 'tip of the iceberg' of needy species, as the few that have attracted notice; (3) in consequence, lists are indicative rather than fully representative; and (4) formal listing of a species may not necessarily be accompanied by conservation action. Indeed, in some cases, the converse can occur. Most knowledge of the lifecycles and biology of the most popular invertebrate flagship group, the butterflies, is in the hands of non-professionals (including many hobbyist collectors and naturalists), whose interest may be thwarted by restrictions such as prohibition of collecting without evidence that this is a threat or risks of prosecution unless complex permit application processes are pursued. Alienation of such people can seriously diminish the flow of information so badly needed to document such insects and their status (Sands & New, 2002). Schemes for 'ranking' species on schedules of protected taxa and for subdividing long lists into more manageable categories have been discussed extensively, with the former approach relying on assessment of severity of threat of extinction using quantitative criteria wherever possible. For invertebrates, knowledge of population structure, size and dynamics is almost invariably highly inadequate (or non-existent), and criteria such as those developed by the IUCN (1994) are impracticable without recourse to considerable and perhaps misleading approximations. Holt (1987) suggested several reasons for subdividing long lists of protected species into more easily managed categories, as follows:

- To convey an impression of which species are most threatened.
- To draft and eventually implement legislation on threatened species.
- To set priorities for funding and action to preserve species.
- To break up long lists into groups that can be perceived easily by people using them.

Emphasis on severity of threat of extinction, and losses that have occurred already, for listing merit and priority treatment generally means that those species listed are demonstrably endangered so that conservation has a flavour of crisis management, with high emphasis on needs for threat abatement and recovery. Intensive species management taken in isolation may have serious ramifications in relation to multiple species wellbeing within the community, and Carroll *et al.* (1996) have emphasised the need to seek more 'inclusive benefits' from species-management programmes.

Wider focus

Indicators

Emphasising the conservation of the communities in which invertebrates participate is clearly a major avenue to such inclusive benefits, in that success in conserving diversity through assuring 'typicalness' or 'representativeness' (Usher, 1986) conserves the context for sustainability of its members. In the case of keystone species (see p. 31), simply conserving these (or other strongly functional individual species) may be important, and any such species listed for individual attention merit high priority for action as tools whereby greater ecological protection can be gained. More often, such species are unknown or their roles have not been detected, and wider investigation and monitoring are necessary to find or predict changes at the community level. Because of the impracticalities of surveying all invertebrates present, particular taxa (be they species, genera, families or even orders) are invariably the prime focus of surveys with the implication (or suggestion) that changes in their abundance, diversity and species composition may mirror wider aspects of community change or help forecast changes as early warning indicators.

Selection of good focal groups is therefore an important topic of such wider surveys, with the general principles that data must be (1) obtainable easily and in a standard manner, (2) intelligible and amenable to logical analysis and (3) placed into a relevant context. Aspects of alpha-diversity (species richness) and beta-diversity (the extent to which species

Table 3.2. *Ranking of criteria for selecting and using optimal indicator taxa in different contexts; seven criteria are ranked from least important (1) to most important (7) for the two kinds of study (after Pearson, 1994)*

	Ranking for	
Criterion	Monitoring	Inventory
Economic potential	1	1
Occur over broad geographical range	2	4
Patterns of response reflected in other taxa	3	5
Biology and natural history well known	4	3
Easily observed and manipulated	5	6
Well-known and stable taxonomy	6	7
Specialisation to habitat	7	2

are replaced by others along environmental gradients, as a reflection of compositional change) are both important, so investigations should provide information on either or both of these features. They may be 'baseline' documentation or, variously, measure or imply changes that have occurred or (through repeated sampling) are planned to occur or are continuing. The information can therefore be used to (1) help to define the state of a community, as in natural ecosystems and (2) help to define the changes that have occurred, are occurring, or will occur from modification, ideally in relation to some unmodified reference standard. Invertebrates used in this way are 'bioindicators', tools whereby condition and change can be monitored, quantified and appraised. The concept of bioindicators is 'a trivial simplification of what probably happens in nature' (Paoletti, 1999) and is an applied use of biodiversity to evaluate landscape or ecosystem condition and change, including the numerous changes due to human activities, as well as appraising the success of management for conservation.

Selection of bioindicator groups depends on (1) the purpose or level of responsiveness or sensitivity to particular change(s) and (2) the ease and expediency of sampling and interpreting the samples consistently. The properties they manifest may be ranked somewhat differently in different contexts (Table 3.2). Indicators may be selected also on pragmatic reasons, such as being of known functional importance (for example, pollinators) or of individual interest, but some such reasons may not be sufficiently objective in leading to optimal group selection for monitoring environmental changes effectively.

Table 3.3. *Broad functional categories of indicators (Spellerberg, 1993)*

Category	Role
Sentinels	Sensitive organisms used/introduced as early-warning systems
Detectors	Taxa that show a measurable response to a given environmental change or stress. Changes may be in abundance or incidence, behaviour, demography or other feature
Exploiters	Taxa whose presence indicates pollution or disturbance, as taxa that have the capacity to become abundant under such regimes; they are normally kept at low abundance by species eliminated by the disturbance imposed
Accumulators	Taxa that take up and accumulate chemicals in measurable quantities
Bioassay	Laboratory use of organisms in studies of pollution

Kremen *et al.* (1993), McGeoch (1998) and others have emphasised the great values of arthropods in evaluating terrestrial ecosystem changes as bioindicators. However, the term 'bioindicators' is used very broadly and contains a number of different functional categories within the wide ambit of environmental monitoring (Spellerberg, 1993) (Table 3.3). Much work on indicators has not given firm definition of the rationale and parameters to be 'indicated', and McGeoch (1998) suggested redressing this imprecision by defining three broad groups within terrestrial insects, as follows:

- Environmental indicators. Taxa that respond predominantly in observable, quantifiable ways to environmental disturbances or other environmental changes.
- Ecological indicators. Taxa that demonstrate the effects of specific environmental changes such as habitat fragmentation or climate change, instead of being gauges of broader environmental changes, as above.
- Biodiversity indicators. Taxa that are focal groups recognised as surrogates for broader biological diversity, and whose evaluation is a 'short cut' for interpreting wider biological changes. Hammond (1994) distinguished four groups within this broad category:
  - Reference groups, valid for extrapolation to other groups for which less information is available.
  - Key groups, used to provide focus for studies to document wider species richness.

Table 3.4. *Desirable features of invertebrate groups to be used as indicators (Kremen* et al.*, 1993; Noss, 1990; New, 1993)*

1 Be amenable to simple (cheap), easily replicable sampling techniques, if possible those that can be used reliably by non-specialists, and preservation.
2 Be taxonomically tractable, if possible nameable at least to the level needed for analysis. Availability of recent taxonomic keys/regional handbooks/guides is a considerable advantage.
3 Be sufficiently well understood biologically that functional interpretation is possible and that responses to various forms of environmental condition or disturbance are reasonably well understood. Normal seasonal variation should not be confused with impact trends.
4 Be reasonably abundant and diverse within the systems to be studied, and be widespread within those systems.
5 Show direct trends or changes in response to the particular (defined) changes being investigated.

- Focal groups, used as a reference standard and selected also for abilities as 'predictor groups'.
- Target groups, i.e. simply those that are made a focus of attention, without assuming any wider surrogacy values.

The term 'indicators' is often used without qualifications such as these, so it can be difficult to interpret meaningfully. Nevertheless, numerous studies have been made on invertebrates as bioindicators, with the usual approach involving field assessment of selected taxa, with comparisons between different environments, sometimes (and usefully) with a reference standard of 'good' conditions and associated species representation against which to evaluate differences. As with other studies on invertebrate distributions and diversity, scale of assessment is important. Thus, use of bioindicators may variously (1) be related to a particular context or site of change or (2) be a more absolute measure of condition over a wider geographical range. As with inventory surveys (Chapter 2), repeated sampling may be needed, and sampling effort should be standardised over treatments.

Selection of optimal indicator groups of invertebrates has been discussed extensively, and the features listed in Table 3.4 are generally considered desirable. These factors can interact in various ways, and clear understanding is needed for their practical use. Thus, many invertebrate groups remain abundant in disturbed environments such as agroecosystems, but the spectrum of species found is far different from that found in natural environments. Without, for example, recognising the loss of ecologically specialised species and compensation by increased numbers of generalist taxa, often including exotic species, complacency may be induced because of lack of any apparent effect on the numbers of species present.

The extent of taxonomic penetration (Chapter 2) needed may be an important determinant of which taxonomic groups can be used, simply because reliable taxonomic resolution may be critical in ecologically varied suites of taxa designated as functional groups (see below). Further, it may effectively limit and dictate which groups it is possible to employ and eliminate others as serious candidiates as bioindicators if species-level responses are sought. Three very broad categories of invertebrate groups can be distinguished on their taxonomic tractability (New, 1999a):

- Well-known groups. Most species described and recognisable unambiguously in a local fauna, often with field guides and well-illustrated handbooks or keys to facilitate identification by competent non-specialists.
- Catch-up groups. These are the groups with a strong species-level framework and understanding in place but with further work needed to complete this to a level where they can be used reliably – perhaps to revise outdated keys or to sort out particular complex segregates. They are groups that, with relatively little taxonomic attention, could be converted to well-known and thereby have their use in measurements of biodiversity markedly increased, so increasing the portfolio of groups available for this. They could be considered a priority for allocation of limited taxonomic expertise.
- Black-hole groups. These are the groups of invertebrates that are the most poorly known taxonomically and for which only low proportions of species have been diagnosed or described. They cannot be utilised at species level or near analysis, and such groups need massive taxonomic attention to transfer them to functional equivalents of the other categories. At present, they cannot be employed for detailed analyses involving species-level data, except (possibly) by a handful of specialists. Keys, handbooks and the like are simply unavailable or misleading in their simplicity. Notwithstanding this, some such groups may furnish useful information at the generic or family level, but using these as ill-defined and uncritically appraised surrogates for species needs to be considered carefully and undertaken very cautiously.

Perhaps not surprisingly, biologists specialising in particular groups of invertebrates have commonly advocated the use and importance of 'their' group for use as bioindicators or in other applied contexts. At some level, everything indicates something, but simply that organisms are present in high numbers or high diversity does not necessarily mean that they are good bioindicators in any particular context. With greater ecological understanding, it is certain that the spectrum of useful groups will be increased, but relatively few groups can be appraised comprehensively at present. Thus, some groups of insects that satisfy several of the criteria listed in Table 3.4 have little proven value as indicators. Hoverflies (Diptera, Syrphidae) in Europe have a variety of larval feeding habits, are common and conspicuous in many kinds of terrestrial habitats and are taxonomically very well known. Their high dispersal levels, allowing for rapid recolonisation of disturbed areas, offsets their practical uses as indicators, notwithstanding their sensitivity to agricultural chemicals, simply because they may 'disappear' for only short periods (Sommaggio, 1999). In contrast, carabid beetles in northern temperate-region agroecosystems are more useful, with many of the species being associated intimately with those areas. However, again, caution is needed as 'many of the studies in which carabids were used as indicators of the environmental impact of farming practices or systems have failed to produce conclusive results' (Hance, 2002; Holland *et al.*, 2002).

Single species can be very relevant and sensitive indicators of particular processes and be effective harbingers of threat, of course, but broader groups have the advantage that compositional features related to ecological traits can also be evaluated and, perhaps, mirror a wider variety of changes.

The usual distribution pattern of abundance in invertebrate assemblages is that few species occur in large numbers, more species are found in lesser numbers, and a great number of species are 'rare' and found in only very small numbers (Fig. 3.6). This pattern necessitates care in ensuring that sampling effort is sufficient to distinguish the species that are 'responding' rather than simply rare species whose presence or absence in samples may be simple chance. Appraisal of a portfolio of different groups (perhaps particularly if these differ ecologically, such as a herbivore group being complemented by one at a higher trophic level) may increase sensitivity of analyses but, as Kremen *et al.* (1993) emphasised, arthropod indicators could be selected to represent (1) a diversity of higher taxa, (2) a diversity of higher taxa and of functional groups, or (3) a diversity of functional groups within the same higher taxon. Collectively,

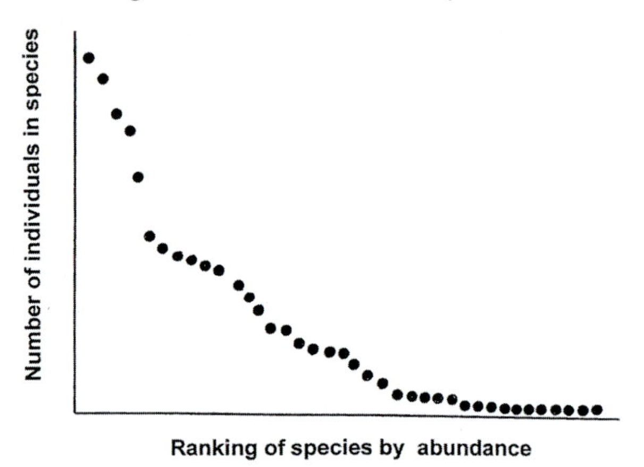

Figure 3.6. Distribution of abundance of individuals among species in natural communities. In this example, typical of many sampling results for invertebrates, few species are 'common' (having numerous individuals) and many more species are 'rare' (having individuals or singletons). This basic pattern is usually retained with increased sampling effort: more individuals of the common species result, and more rare species are added.

these should respond to human impacts well before those impacts ramify through ecosystems to produce wide, even irreversible changes.

For example, declines in some butterflies and day-flying moths, taxa that can be evaluated easily by simple observations based on transect counts (Pollard & Yates, 1993) without need for more complex methodology, can indicate changes in habitat quality well before those changes are evident in changed proportions of plants (Erhardt & Thomas, 1991). Likewise, particular arthropods and others may be especially sensitive to low levels of certain chemicals or pesticides; even methods such as collecting dead honeybees at commercial hives may provide valuable information on a variety of pesticides to which they are particularly susceptible (Johansen, 1977). As well as response differences between different groups influencing the invertebrate groups to be used, the test environment may also be very heterogeneous. Greenslade's (1997) comment on patterns of distribution of Collembola and other groups in Australian native grassland (namely 'intrinsic variation in these ecological systems is too great for a linear response to disturbance to be shown by the invertebrates studied here, which makes it unlikely that a single, reliable indicator can be found') is also salutary in considering the likelihood of reaching conclusions that are not justified, based on patchiness and trends from limited sampling.

Taxonomic and functional-group bases for indicators

Within certain large taxonomic groups, ecological variety is common and is often founded in trophic variety (see Table 2.4, for soil nematodes). Different trophic categories, or other 'functional-groups', may respond in different ways to a disturbance or change in their environment. Instead of changes in the overall diversity of the major group, a change may manifest in some proportional change or relativity between the various ecological groups, whereby some groups decline or disappear and others increase their abundance and influence. In some groups, this variety may be interpreted at levels higher than species, but the functional-group approach for some invertebrate groups is merely speculative. However, its promise can be exemplified by discussion of two invertebrate groups in which existence of valid functional groups seems unambiguous, and outlining the kinds of application that have occurred and the descriptive indices that may result. Taxonomic sorting of samples here is linked intricately with ecological sorting.

Soil nematodes

Samples of nematodes in soil cores up to 20 cm deep have revealed that their numbers can reach around ten million per square metre, with substantial differences in numbers and diversity (number of genera) between management practices on different soil types (Yeates & Bongers, 1999). A variety of trophic groups (see Table 2.4) are usually represented. Because of the effects of plants, soil types and biogeographical patterns on distribution of the various species, functional groups have been considered to be a better indicator of soil condition than particular nematode taxa per se.

In agroecosystems, such functional groups have comprised cyst nematodes (Heterorhabditidae), virus vectors (Longidoridae, Trichodoridae) and root-lesion nematodes. Yeates & Bongers (1999) suggested that permanent grassland may be a useful baseline for appraising the composition and diversity of soil nematodes from a particular soil type, and also that the relative abundance of fungus-feeding and bacteria-feeding nematodes is sensitive to management changes and may indicate satisfactorily the underlying changes in wider composition of the nematode fauna. Grassland at least three years old (i.e. not cultivated over that period) provided fauna typical for that soil type and region. In grasslands, there appears to be a clear relationship between nematode diversity, soil texture, climate, plant species, pesticide use and management practices (Yeates & Bongers,

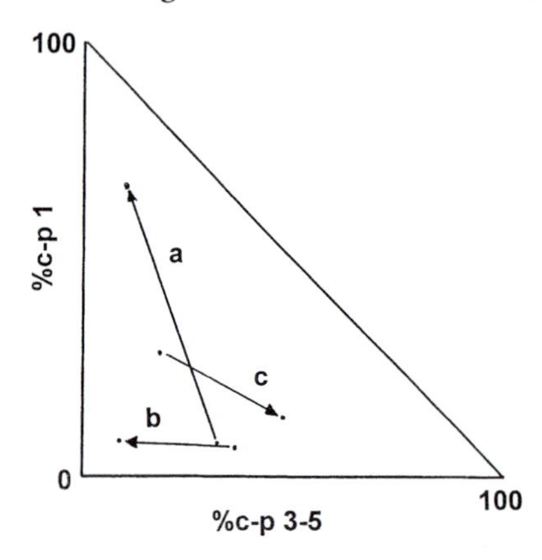

Figure 3.7. Nematode 'c-p triangle' (see Yeates & Bongers (1999) and included references for examples). This example shows effects of (a) eutrophication with initial situation followed by that two weeks later after adding powdered cow dung; (b) artificial acidification of coniferous forest soil; and (c) recovery 33 and 44 weeks after organic manuring. See text for explanations of terminology.

1999) potentially a very useful suite of correlated features for setting basic standards of documentation and demonstrating change or condition of soils.

With the realisation that nematodes are likely to be good indicators of the all-important soil system, Bongers (1990) devised a 'nematode maturity index' (MI) based on non-plant-feeding nematodes. For these, MI is the weighted mean of the coloniser-persister (c-p) values for samples. These values indicate the perceived place of taxa on an *r*–K continuum (see p. 11), and parallel indices for those nematodes feeding on higher plants (plant parasite index, PPI) have similar properties. Yeates (1994) (see also Yeates & Bongers, 1999) remarked that all nematodes should be included in a collective maturity index (Σ MI), and currently accepted standard values for c-p for nematode factors are graded from 1 to 5.

The general form of calculation of the MI is

$$\mathrm{MI} = \sum_{i=1}^{n} c - \mathrm{p}i \cdot \mathrm{p}i$$

where c-p is the c-p value, p is the proportion of individuals in the *i*th taxon, and *n* is the number of taxa in the sample. Changes in the nematode fauna can be expressed graphically, most simply by a c-p triangle (Fig. 3.7). In this diagramatic form, changes of condition are reflected: succession by moves to the right, stress by moves to the lower left, and enrichment by moves towards the upper left. Yeates & Bongers (1999) noted that data derived from a series of successional or treatment sites may help to reflect changes over time (succession) or recovery following disturbances in the nematode fauna. Several specific applications are possible, as exemplified in Fig. 3.7.

Ants

Ground-dwelling ants are often diverse and ecologically varied. They have attracted attention as indicators, particularly in Australia, where more than 100 genera are found (Shattuck, 1999). Many ecological categories of ants can be defined at the genus level, so helping to overcome the impediment caused by numerous undescribed species in interpreting the fauna meaningfully. A combination of ecological variety, reliable generic identification and reasonable segregation of genera into morphospecies (see p. 34) has prompted categorisation of Australian ants (in particular) into functional groups (following Greenslade, 1978) of value in helping to appraise ecological conditions. The basic scheme (as modified progressively by Andersen, 1990, 1995, 1997) is shown in Fig. 3.8. The approach is based on broad ecological roles, particularly on the competitive interactions and habitat requirements/relationships of the various genera. Andersen noted that functional groups have been identified that vary predictably with disturbance, vegetation, soil and climate, collectively providing a broad suite of potential response measures through which community composition can be analysed, spanning scales ranging from local to major geographical regions.

Ants, categorised by functional groups in this way, show clear successional patterns of considerable values in monitoring restoration, for example during rehabilitation of mine sites (Andersen, 1993), where their patterns are mirrored by other invertebrate groups. Wider aspects of land use are also reflected in changes in ant functional-group balance, with forestry practices (Vanderwoude *et al.*, 1997) and grazing (Read & Andersen, 2000) both studied in some detail in Australia. An important consideration, probably of much more general relevance, is that functional groups recognised at one spatial scale may be inappropriate at others (Andersen, 1997).

Many kinds of human land use can reduce ant diversity and/or change the composition of ant communities. In Western Australia, Majer & Beeston (1996) found that road construction had the greatest long-term

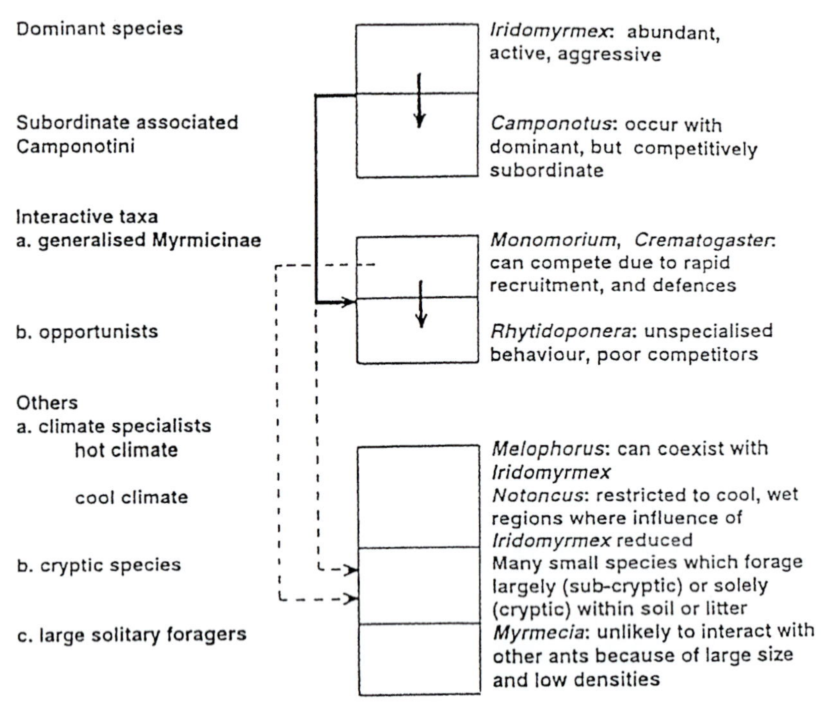

Figure 3.8. Functional groups of ants, as devised initially for Australia and refined progressively by Andersen (1990, 1997) (see text). The diversity of the various functional groups, and the balance between them, differs according to levels of disturbance and condition of the ecosystem. Strength of interaction between the groups is indicated by the thickness of linking lines; lines ending in a box refer to all component fauna of that box category.

effect on ant species richness, followed by agricultural clearing, mining, urbanisation and rangeland grazing. Combining changes in ant richness with the extent of land use showed that agricultural clearing (followed by rangeland grazing) was responsible for the greatest loss of ant 'biodiversity integrity'. This approach linked biodiversity with landscape condition (quality).

Values of the taxon approach to indicators

Most invertebrate groups suggested to be useful as bioindicators in the broad contexts of agricultural change are those found on the ground or in soil, with those found on vegetation used less critically or frequently. Association with any particular vegetation forms or taxa subject to modification (1) renders loss of that vegetation an 'automatic threat',

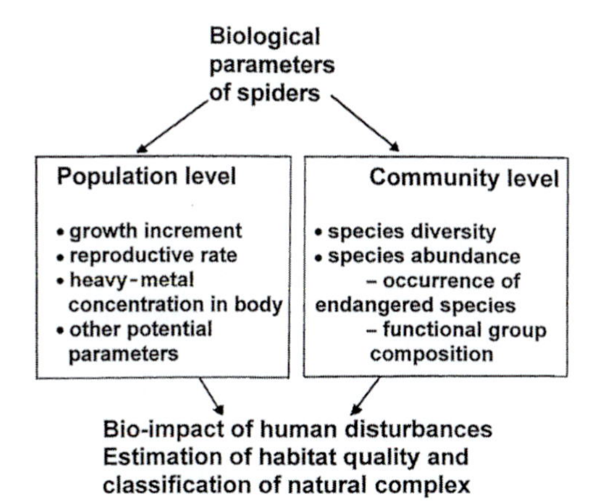

Figure 3.9. Spiders as indicators (Marc *et al.*, 1999): some of the parameters that facilitate use of spiders at population and community levels.

with the vegetation itself being more generally obvious than the associated invertebrates so that (2) the plants may be monitored for change far more easily than accompanying invertebrates. Changes in the herbivore guild can indeed indicate environmental changes (such as from the side effects of pesticides, see p. 99), and their overall considerable diversity renders them very important in wider biodiversity evaluation. Attention has also been paid to predators on vegetation. These may have strong linkages with particular vegetation types but may not be as specifically related to particular plant taxa as are many herbivores. They may, therefore, reflect a wider spectrum of habitat features.

For example, spiders have been considered a useful bioindicator group (Marc *et al.*, 1999) for several reasons:

- Spiders are present at high densities from the ground to the highest vegetation layers.
- Spiders have specific ecological features and relationships with their natural habitat.
- Spider community variation and changes can be detected even within small areas within a particular biotope.
- Spiders hold strategic positions in food webs, as predators and prey.

Fig. 3.9 (from Marc *et al.*, 1999) indicates the possible uses of spiders, as polyphagous predators, as bioindicators with facility to respond to changes at the population level and the multispecies (community, assemblage) levels. However, because many spiders have strong association with the architecture of the vegetation on which they live, simply changing major vegetation form may influence the spider community directly, with features such as structure and cover being far more important influences than vegetational species composition. According to Wise (1993), habitat selection by spiders can be influenced in complex ways by vegetation, the variables also including prey availability and composition, microclimates and the possible exchanges of spiders with adjacent areas. These are, equally, considerations for other predatory arthropod groups, and others – together emphasising that the quality of the environment is often of paramount importance to invertebrates, whose existence there may mirror changes or differences not easily discernible by people.

Many phytophagous insects may be grouped in similar ways. Thus, Hemiptera: Heteroptera (the true bugs) often show clear preferences for particular structure of vegetation, within which many of the species are not strictly host-specific (Fauvel, 1999). As a broad biodiversity indicator of the above-ground fauna, Duelli *et al.* (1999) included Heteroptera with Symphyta (sawflies) and aculeate Hymenoptera as groups that appeared to be useful in Europe. Those authors emphasised that a major focus or tendency to concentrate on surface-dwelling arthropods for appraisals in agricultural areas reflected three main assumptions:

- That they are easily collected in standard ways using pitfall traps, a simple trapping method that can be employed cheaply and used by non-expert practitioners.
- That catches in most habitat types contain sufficiently high numbers of species and individuals to enable standard statistical analyses and comparisons. Within agricultural habitats, pitfall traps rarely take species that are protected or that have been signalled to be of conservation significance, simply because most of these are confined to more natural habitats.
- That many of the species are polyphagous predators, so that the broad taxonomic groups involved may, in general terms, be considered beneficial, and their evaluation has applied interest to the land owners.

The problems of using pitfall traps uncritically are numerous (see papers cited in Yen & New, 1997), largely reflecting the inadequacies of interpreting relative abundance data based on a simplified measure of activity alone. With standard sampling effort, binary data (presence/absence of taxa) may be reasonably valid, but abundance data may be less reliable. Ground-dwelling predators are sometimes poor correlates of other diversity (Duelli & Obrist, 1998) and, following McGeoch's (1998) admonition to recommend an indicator group only if there is a significant, strong, positive correlation between its diversity and the diversity of other taxa, extrapolation from catches to the wider environment may be possible only with strong reservations.

On this basis, many studies that purport to demonstrate the values of particular groups as indicators fall short of doing so. A number of important studies have demonstrated very clearly that the numbers and diversity of group *x* differ, often greatly, between sites with different features and extrapolate to claim indicator values. Many fewer studies have evaluated spatial heterogeneity within those systems and shown causative relationships between environmental change and a changed pattern/abundance/richness of those groups, be they taxonomically or trophically based. The use of bioindicators in terrestrial ecosystems lags well behind that in freshwater environments, and strong efforts to achieve comparable protocols have occurred over the past decade or so. Many invertebrate groups appear to be promising, rather than having unambiguous proven predictive value, in particular contexts. For example, Baldi & Kisbenedek (1997) used Orthoptera to evaluate the structure of native grassland remnants in the Hungarian steppes, confirming North American studies that grasshoppers can be good indicators of disturbance in grassland systems. In North America, about two-thirds of the variation in grasshopper species composition on rangelands could be related to vegetation variables controlled by human activity (Fielding & Brusven, 1993).

In Hungary, the distribution of Orthoptera across a series of 27 steppe patches was not random, and these insects seemed to be good indicators of naturalness and disturbance (Baldi & Kisbenedek, 1997).Their density was greatest on disturbed sites, and species richness was greatest on natural sites. This kind of pattern (involving a form of density compensation by relative generalist species after the loss of specialists) may be much more general.

A somewhat different application, as bioaccumulators, is feasible for European staphylinid beetles (Bohac, 1999). Staphylinids are often abundant in European agroecosystems where (with Carabidae) they are amongst the most important groups of ground-dwelling invertebrates in terms of activity (fostering sampling by pitfall traps) and abundance, and are significant generalist predators. Some kinds of response to agricultural management are summarised in Table 3.5, but inadequate taxonomic knowledge of staphylinids lessens their value as indicators in many other parts of the world, outside the northern temperate zones. Bohac (1999) (after Bohac & Pospisil, 1988) noted their use as indicating accumulation of heavy metals, with three main groups discernable: macroconcentrators (k2), microconcentrators (1k2) and deconcentrators (k1), where k is the

Table 3.5. *Influences of some management actions on staphylinid beetles in agricultural landscapes (after Bohac, 1999)*

Management	Influence on agroecosystem	Influence on staphylinids
Landscape structure	Mosaic replaced by larger units Fragmentation of habitat	Decrease in distinctiveness of communities Increase in generalists
Manure	Increase soil organic matter Increase humidity Increase insect larvae	Increase predators and saprophages
Chemical fertilisers	Decrease soil humidity	Decrease moisture-sensitive species
Pipe drainage	Decrease water-table depth Decrease organic content Decrease moisture	Decrease high-moisture species Increase generalists
Crop change	Change in microclimate Change soil humidity	Increase high-moisture species
Pesticides	Direct effect	Short-term eradication of sensitive species
	Indirect effects: removal of prey, absence of cover plants, microclimate changes	Decrease in predators and high-moisture species

ratio of the metal concentration in the beetle's body to that in the soil. Macroconcentrators are typified by some predatory species, but fungal feeders can also concentrate elements, such as mercury, which are accumulated by some kinds of fungi used as food.

Collectively, terrestrial invertebrates exhibit responses to most environmental variables and changes, and many can be used to help appraise impacts and effects of human intervention. At one extreme, their representation in different management regimes can be affected to the level of extirpation and massive loss of taxa. Under less extreme influences, some invertebrate groups may be powerful tools for appraising and monitoring changes, thus helping to guide management in agroecosystems towards greater chances of sustainability. Assessing such effects necessitates both characterising assemblages and using changes in their composition (taxonomic, functional, or both) as indices of change related to impacts of particular factors. Many other examples of their use are noted in later chapters.

4 · *Agricultural disturbance: diversity and effects on invertebrates*

Discussion of the characteristics of agroecosystems and their conventional management for crop protection helps us to focus more closely on the threats to invetebrates introduced in the previous chapter. Understanding the needs and principles of pest management is a critical theme in invertebrate conservation in agroecosystems, and the rationales of pest definition and evaluation underpin crop-protection measures. This chapter also introduces a major concern in crop protection and its incompatability with conservation: the use of pesticides that may have substantial undesirable side effects, and how these may be appraised and overcome through development of wider integrated pest-management programmes for both economic and environmental benefits.

Introduction

Agroecosystems are, essentially, 'disturbed' ecosystems. The disturbances are equated to varying extents with threats to the natural biota that they subsume through initial establishment and continuing management and diversification. Disturbance, very broadly, may alter the diversity within an ecosystem directly (by killing individuals) or indirectly (by changing resource quality and accessibiity). If disturbance is severe, then biodiversity tends to be reduced, because only the relatively low proportion of taxa that are insensitive to the disturbance can persist. 'Disturbance' can be defined as the cause of a perturbation – that is, of an effect or change in the state of the ecosystem relative to an undisturbed reference ecosystem (Rykiel, 1985).

In Chapter 1, some of the differences between agroecosystems and more natural ecosystems were discussed, but it is important also to recognise the similarities and parallels between them. Thus, many natural ecosystems are by no means constant and are subject to continual disturbance and change. Wildfire, for example, is an integral part of the ecology of many natural Australian ecosystems and, without it, many native Australian plants could not survive. Both directional change (succession) and more irregular non-directional changes are frequent in most ecosystems. Both in agroecosystems and natural ecosystems, disturbance fosters productivity and sustainability. As Wood & Lenné (1999) noted, agriculture continually mimics the roles of fires in nature for clearing vegetation and sometimes has strong parallels with natural fire regimes. Many of the invertebrates (and other taxa) in natural ecosystems may be expected to show sufficient adaptive capabilities to resist disturbances of many kinds, with their fate often depending on the scale of disturbance.

Wood & Lenné (1999) also pointed out that that monospecific plant stands paralleling monocultures, often taken by conservationists to be purely a human artificial construct and ecologically undesirable, are not uncommon in natural ecosystems. They are sometimes maintained by disturbances such as short fire intervals. Such 'natural monocultures' are not necessarily restricted to early successional plants (such as bracken, *Pteridium aquilinum*), and they can include also forest trees, particularly in parts of the northern temperate zones. Some of these systems appear to be ecologically stable and so confuse the putative (and widely debated) paradigm that low-diversity systems are inherently unstable in relation to higher-diversity, more stable ecosystems (Wood & Lenné, 1999, and included references). The differences between agroecosystems and natural ecosystems listed in Table 1.2 (see p. 7) can therefore sometimes be less polarised than implied there, reflecting the varied origins of agroecosystems and that the habitats they replace and contain are indeed varied in the amounts of 'biodiversity' they can support, together with the extent to which 'natural biodiversity' is disturbed or displaced.

Although disturbance is universal in ecosystems, many agroecosystems are characterised by well-defined, regular and extensive disturbances (Edwards *et al.*, 1999), in some cases sufficiently regular that organisms can adapt to them reasonably well. Edwards *et al.* (1999) cited traditional hay meadows in Europe, some of which have had regular management for centuries, so that a diverse flora is present together with numerous associated invertebrates and others. Disturbance regimes can create temporary habitats (sensu Southwood, 1962) and may favour the evolution of characteristics of *r*-strategists (see p. 10) associated with 'weediness' and ability to capitalise on such regimes, as do many pest arthropods and plant weeds. Indeed, some botanists believe that many weeds have evolved directly from wild species that frequent habitats disturbed by people

Table 4.1. *Some comparative aspects of processes in agroecosystems and natural ecosystems (after Edwards* et al., *1999)*

	Agroecosystems	Natural ecosystems
Site characteristics	Selected sites; disturbance frequent and predictable; high levels of resources	Total range of sites, mostly with long history; disturbance less frequent, unpredictable; resources may be limiting
Biota	Planned and unplanned; recent modifications	Defined by habitat and biogeography
Community interactions	Much simpler	Complex; multiple interactions
Evolutionary processes	Recent history important; rapid evolution from strong selection pressure; increasing external control	Often slower
Ecosystem processes	Strongly influenced by external management; nutrient cycles open	Nutrient cycle mostly closed, sustainable system
Landscape structure and dynamics	Planned, minimum variation; often simplified, externally managed	High structural diversity; hierarchical structure due to autogenic processes

(Edwards *et al.*, 1999). As with other pests, however, perception of 'a weed' may more or less fortuitously incorporate ecological features into more subjective conditions such as 'being in the wrong place'.

Nevertheless, with these correlations, we can view agroecosystems and natural ecosystems as perhaps having more in common than implied earlier. Table 4.1 (cf. Table 1.2) exemplifies this further. With this emphasis, the sometimes substantial numbers of invertebrates and other species found in some agroecosystems become more understandable. However, this does not diminish the severity of impacts that many agricultural processes inflict on natural biota. Rather, it helps to indicate that many invertebrates are sufficiently versatile to benefit from even limited resources and hospitality, despite the disturbances imposed by agriculture. However, without those disturbances, it is presumed widely that even more invertebrate species would thrive.

This chapter considers further the spectrum of threats posed to invertebrates by modern agriculture, focusing on the general themes introduced in Chapter 3 and, in particular, on the consequences of intensification and pest management for crop protection. Despite the reassurance noted above, cultivation leads to major changes in soil biota, and reduction in plant diversity (commonly to an extensive monoculture crop) is usually associated with corresponding reduction in diversity of associated invertebrate richness and functional diversity, other than for a few species whose abundance (density) can increase greatly so that they may constitute pests through directly or indirectly causing loss of crop yield or quality in other ways that lessen economic returns. Intensification (Chapter 2) has been identified as a major contributor to loss of global biodiversity, but Edwards *et al.* (1999) noted the confusion that has arisen in considering its impacts in relation to market forces and trends, the socioeconomic context of agricultural benefits. Both loss of natural habitat (affecting the spectrum of species in an area) and the various levels of intensification (reflecting the extent to which persistence may be possible) affect the wellbeing of organisms on the treatment site and in the surrounding environments. The implicit need for conservation is to harmonise continuing and accelerated agricultural productivity with less harmful effects at all levels. Most commonly, suites of characteristic invertebrates are ignored in agricultural practices, other than the few that command individual focus and attention as pests or other noteworthy species. This reflects more than simple neglect; it endorses the widespread lack of knowledge and appreciation of invertebrate function and importance and of the difficulties of defining and promoting these features.

Intrinsically and ethically, any loss of biodiversity is undesirable, but some losses may need to be accepted as a reality in managed agricultural systems. However, writing on soil organisms, Pankhurst & Lynch (1994) noted: 'Whether it is necessary or feasible to emulate in an agricultural system the biodiversity of soil organisms that may be present in a natural ecosystem is a question that really has no answer.' They noted that there is no case in which the full diversity of a soil fauna has ever been documented, that no optimum level has been defined (or, even, any method developed to determine what such an optimum may be) and, even if it could be defined, how it might compare with a similar optimum for a natural ecosystem. Similar statements could be made for virtually any group of invertebrates in any agricultural system or subsystem. Problems of agricultural management are thereby themselves difficult to formulate other than in strictly profit-oriented terms, but reflect patterns of (1) intensity, (2) frequency and (3) spatial arrangement of disturbances as the major causes of threat to native biota.

Invertebrates as pests in agriculture: an ecological framework

Agricultural intensification and the need for crop protection against pests are linked in many ways. The great majority of pests involved are invertebrates, a high proportion of them insects, whose effects are related to their numbers and the favourable conditions that agricultural crops provide for their wellbeing. Measures to control pests also involve intensive management, with impacts that can extend well beyond the target species or arena. The term 'pest' is universally familiar, but it is more difficult to define objectively in scientific terms because, to a large extent, it is, as Jones & Kitching (1981) observed, 'a human construct rather than an ecological reality'.

Very generally, a pest is a species whose interests overlap those of people or that conflicts in some way with human needs or interests. It may cause heavy economic loss through eating or despoiling crops or by carrying or causing disease; or it may simply be a nuisance. 'Pests', therefore, does not encompass any single ecological entity, but transcends the *r*-K continuum of ecological strategies. However, and again as Jones & Kitching (1981) emphasised, pests (at least those that cause major economic or other losses to humanity) are amongst the most intensively studied of all invertebrates. Some have been subjects of long-term biological studies, and realisation that optimal management of these species depends on application of sound ecological knowledge ensures that this bias will continue, not least because comparable levels of funding and interest are difficult to foster for most 'innocuous' invertebrate species. Many of the so-called 'natural enemies' of pests, including their predators and parasitoids, are also well studied (Chapter 5). Otherwise, perhaps the best-studied invertebrates are some of the highly threatened species of conservation significance and have thereby attracted 'conscience funding'. With some exceptions, then, our knowledge of the autecology of invertebrate species in terrestrial environments has accumulated from studies of two rather different focal categories – abundant pest taxa and scarce and vulnerable taxa – with the aims of devising better ways to respectively suppress and conserve these through planned management. For both categories, knowledge of the species' 'life system' (Clark *et al.*, 1967), namely the species and its effective environment, the latter including all other species that interact with the species, is important, with increasing emphasis on quantitative interpretation of population dynamics and changes in numbers and distribution to help predict future trends.

Whereas many pests provide massive amounts of data, simply because they are abundant, widespread and obvious, and so are relatively easy to study, many conservation targets do not. The latter are usually, by definition, scarce and limited in distribution. The core strategies of pest assessment, involving destructive or removal sampling, cannot be used for these organisms, because any additional mortality imposed may be harmful to the species. The scientific bases for management may thereby differ considerably in the extent to which they can be interpreted reliably. Figure 3.6 (see p. 76) exemplifies the common scenario confronting managers and highlights aspects of the central paradox introducing Chapter 3. The major need is often to reduce abundance of the most abundant species present (the pests, typified by species to the left of the figure) whilst simultaneously conserving rare species (those to the right of the figure), with one possible component of this involving increasing the numbers of some of the intermediate abundance species as natural enemies of the pests (Chapter 5).

Despite the difficulties in generalising about pest ecology, some background information is given here as a foundation to understanding how the needs and strategies to manage agricultural pests have developed and diversified in response to increased ecological understanding and awareness of the wider concerns that those strategies may engender for invertebrate conservation.

Many agricultural pests are indeed thought of as '*r*-strategists'. They colonise a field crop, increase rapidly in numbers through their high fecundity and rapid developmental rates, and may continue to do so for as long as the crop persists. That crop is commonly grown as a monoculture, essentially providing a large, concentrated and so highly visible 'target' for colonising invertebrates, and a superabundance of food – commonly of a species not suitable for consumption by many other naturally occurring invertebrates in its vicinity. The quality of the food may also be sustained at unnaturally high levels through addition of chemical fertilisers to promote rapid growth and high levels of productivity. It may also be uniformly high-quality, through selection of particular plant cultivars or genetically modifed varieties. Collectively, such crops constitute a near-ideal milieu for a specific consumer to thrive, especially as they are commonly also environments in which predators and other natural enemies are naturally scarce or absent. Pests may capitalise very effectively on such environments.

Many pests therefore have the form of population growth (a rapid rate of increase, *r*) schematised in line A of Fig. 4.1; in contrast, many other

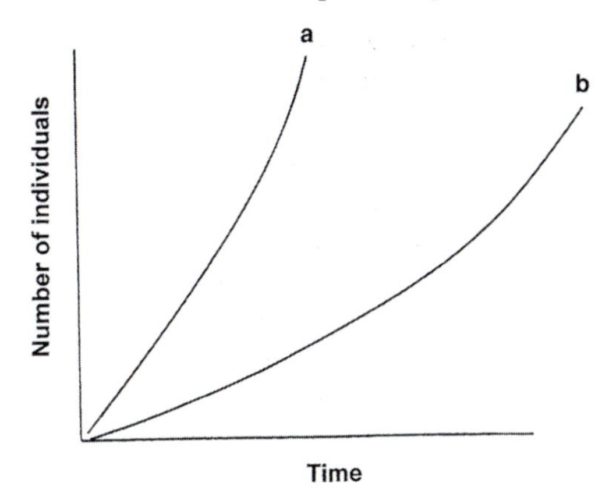

Figure 4.1. Rates of population increase: (a) typical *r*-strategist, with population increasing rapidly over time; (b) more typical K-strategist, with low rate of population increase.

species (and some pests of long-lived crops such as plantation trees) may have a much lower rate of increase (as reflected in line B) as strategies further from the *r* extreme, and can persist for longer in their environments. One characteristic of *r* species is that their rapid expansion may render their environment unsuitable within relatively short periods, after which they must disperse and colonise anew rather than remain permanent residents within the area. If species adopt the strategy of persisting in their environment, they are likely to participate in increasingly diverse and complex communities in which energy may be diverted from feeding and growth or population increase to competitive interactions with other resident species. Lower fecundity or slower population growth is a frequent correlate of persistence in more permanent habitats, as is a lower level of migration.

However, population growth in nature is rarely unrestricted, and long-term population studies of individual species tend to manifest continual changes in numbers around a long-term general equilibrium position (Fig. 4.2); downturns from higher peaks reflect mortality imposed (for example by natural enemies, see Chapter 5) or the limits of carrying capacity of the local environment (such as limited food supply), both of which may have reduced influences in artificially established low-diversity cropping systems, as noted above. Changing seasonal conditions are also important. In most places, pest or other species population growth does not continue uninterrupted throughout the year. Periods of dormancy may occur during cold winter periods, for

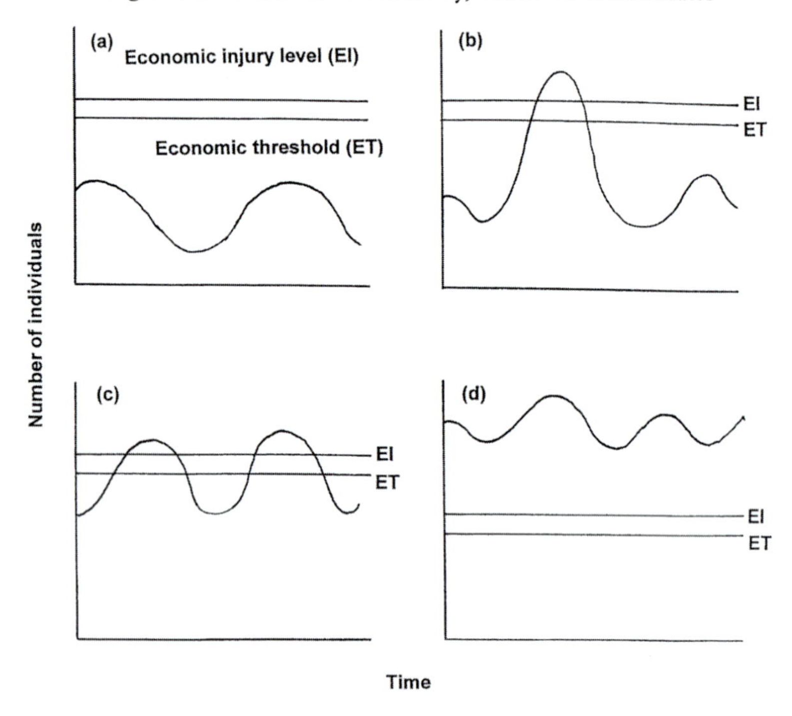

Figure 4.2. Population fluctuations and pest status: over time, each population varies continuously in abundance, ranging about a long-term general equilibrium position, and numbers may exceed an economic threshold level, beyond which economic injury occurs: (a) permanent non-pest species, in which even the highest numerical peaks remain below an economic injury level; (b) occasional pest, in which some of the peaks result in damaging levels of abundance, but this not occurring every peak (or season); (c) regular pest, in which every peak is of damaging abundance, this perhaps occurring every season; (d) severe, permanent pest, in which even the lowest normal levels of abundance are above the economic threshold level.

example, and food may not be available over such times. For evaluating pest importance, both the extent and rate of population increases and the frequency with which these occur are dictates of management needs.

Many pests, mainly those of field and orchard crops, are highly predictable in incidence. They occur every growing season, perhaps every generation, so that their numerical peaks are a phenological expectation in crop production and management. Other pests are much more sporadic. They occur in low numbers for much of the time but, in response to factors such as unusual weather, can increase suddenly to attain damaging levels. Their incidence is thus often unpredictable, at least without very close monitoring. For crop protection, the incidence and abundance of key pest invertebrates is related to levels that cause economic injury (see Pedigo *et al.* 1986, for review), so that monitoring of their populations indicates likely economic threshold levels of abundance at which measures should be taken to prevent economic damage such as yield losses. These parameters, and the above pest categories, are indicated simplistically in Fig. 4.2. Also shown (Fig. 4.2d) is the most serious category of pests, the so-called 'permanent pests', in which numbers are always (even at the lowest troughs of the generalised sinusoidal pattern of fluctuations) above economic injury levels. The usual or most frequent strategies used to control pests are to inflict mortality in some way (at the points where populations rise to the economic threshold levels). The extreme of Fig. 4.2d may be very difficult to remedy as the general equilibrium position must be re-established well below the economic threshold level. Such a move may be very difficult to achieve, and very expensive, and other options (such as changing crop species or changing the market so that some damage can be tolerated – for example, by diverting fruit crops from prime individual market quality to juicing or other processing grades) may be preferable. For completeness, Fig. 4.2a shows a species that is never damaging, because even its highest numerical peaks are well below the economic threshold, and that can be tolerated or supported within the system.

Economic injury is not always related to high numbers of pests. A single codling moth (*Cydia pomonella*) caterpillar in an apple can cause total loss of that fruit, for example. Conversely, thousands of cabbage aphids (*Brevicoryne brassicae*) infesting the outer, looser leaves of a cabbage may cause little or no loss, as those leaves are trimmed in normal processing before the product is sent to market.

Pests may be regarded as permanent (resident in or near the crop environment and so 'on hand' to infest it), mobile (many *r*-strategists move around easily on a local scale and colonise local crops as these become available) or migratory (moving over larger areas, and intergrading in scope with mobile pests). Many members of the latter two of these groups are polyphagous and multivoltine and can range over a variety of different crops in a region (Hirose, 1998). The epithet 'migratory' is often restricted to more extreme dispersers, such as locusts, which may be entirely absent from the crop area for most of the time and which disperse over large geographical scales rather than just in the more local environment. However, these groupings are useful in indicating possibly different management needs, and the extent of pest population movement is a key element of their ecology.

Determining the economic injury level is usually context-specific, but it may be guided by the wider consideration of the relationships between damage and commercial returns. Rabb's (1972) list of features essential in determining economic injury level is still very pertinent; he distinguished six main factors, as follows:

- The amount of physical damage to the crop, and its relationship to pest density. It may, therefore, be possible to use 'simple' indices such as the number of caterpillars per plant to provide an estimate of economic damage and a threshold number at which control measures are needed in order to prevent subsequent loss.
- The monetary value and production costs of the crop at various levels of physical damage.
- The monetary loss at various levels of physical damage. This may be total or proportional.
- The amount of physical damage that can be prevented by control measures. This recognises that some kinds of damage are rapid and irreversible (invasion of individual fruits), whereas others (such as early yield loss) may be compensated for over a longer growing season.
- The monetary value of the portion of the crop that can be saved or rescued by control measures.
- The monetary cost of the control measures.

The simple economic sum is that costs of pest control (more broadly, of crop protection) should not exceed the benefits to be gained, otherwise economic loss ensues. Even equalling the benefit is merely 'breaking even' and usually not worthwhile – except in cases where public relations benefits are important and other non-economic benefits (including wider conservation) may accrue.

Determining the losses from pest attack, and concomitant needs for crop protection, involves considerations of patterns of pest numbers, times of incidence, frequency and duration of occurrence, and the costs of any protective measures in relation to losses caused by those pests. Losses may sometimes equate simply to yield reduction, but they can also reflect traits such as enforced changes in market niche; in many cases, markets may be closed to blemished commodities (fruits in highly competitive supermarkets) or the treatment may affect marketability, including potential for export (for example, presence of pesticide chemical residues). Likewise, premiums for high-quality produce may be offered, such as for organic produce, whose production methods may impose severe limits on the protective measures that can be used and for which preventative strategies may be the only realistic alternatives. At least some of these potential losses, highly relevant in costing crop protection, as noted above, are moderately predictable, notwithstanding short-term vagaries of market pricing. For longer-term crops, predicting market returns several years or even decades in the future is much more difficult. However, calculations of economic injury levels (and economic thresholds) can be a difficult exercise. For polyphagous pests (such as *Helicoverpa* moths in Australia), damage to some crop species may be severe while that to others may be trivial. Adequate protection of the sensitive crops may command a broader approach to incorporate treatment of others in order to prevent cross-over effects. Treatment aims to prevent losses, with costs also reflecting the frequency and intensity of treatments needed. Treatments may also extend beyond cropping areas, with resultant vulnerability of other species.

Sporadic pests are sometimes seen to undergo outbreaks; because outbreaks may result in some of the most serious losses to agriculture – albeit not commonly – considerable effort has been made to understand the causes of these phenomena. A working definition of an outbreak, after Berryman (1987), is 'an explosive increase in the abundance of a particular species that occurs over a relatively short period of time'; since outbreaks in pests are often attributable to some form of 'stress' altering their normal population dynamics, agricultural simplification has been regarded as one of the facilitators of their occurrence. Berryman (1987) emphasised the ecological variety of outbreaks and the possible correlates of their incidence (Table 4.2), but specific causes are attributed only rarely to any particular pest outbreak. 'Stress', for example, can simply reflect growing plants in monocultures rather than in diverse assemblages, or artificial stimulation by addition of chemical fertilisers, and it is well-known that loss of vigour in plants can sometimes render them more susceptible to insect attack (see Speight, 1997). Pest outbreaks can thus be related to any of a variety of physical and chemical factors influencing 'stress'. They may be short-lived, lasting for a season or less before an equally spectacular decline or more gradual decrease in numbers, or they may persist for much longer.

Pests are thus ecologically diverse and can also originate in various ways. Several workers (Cherrett *et al.*, 1971; Clark *et al.*, 1967) have designated broad categories of pests based on their mode of origin, and these are helpful in understanding their ecological diversity and interactions. Such schemes can be summarised as follows:

Table 4.2. *Hypotheses implicated in the causes of pest outbreaks (Berryman, 1987)*

Hypotheses
Outbreaks are caused by dramatic changes in the physical environment.
Outbreaks are caused by changes in the intrinsic properties (for example, genetic or physiological) of individuals in the population.
Outbreaks result from interactions between plants and herbivores, or between predators and prey.
Herbivore outbreaks (as the major category of concern in crop-based agriculture) are related to changes in host plants, often influenced by environmental stress.
Outbreaks are particularly common amongst *r*-strategists, as reflecting life-history characteristics.
Outbreaks occur when pest species 'escape' from regulation by their natural enemies.
Outbreaks occur when pest populations overcome the defence systems of their hosts.

- Naturally occurring pests. Native species previously innocuous and largely unnoticed by people but that have changed in habitat or characteristics, such as switching hosts or increasing in abundance, so that previously non-existent or negligible interactions with human interests become significant.
- Exotic species that have entered new environments. In Australia, for example, many of the most serious pests of agriculture have arrived, or been introduced, during the 200 years or so of European settlement and are not part of the native fauna. They depend on other introduced resources, including most of the agricultural crops, as a facilitating environment in which they can establish and thrive (New, 2002).
- Native or exotic species responding to anthropogenic changes. These are species that have not themselves changed in habit to any major extent but that have attained pest status through changes in human activities. These include changes to natural environments such as agricultural simplification providing large targets of previously scattered resources for herbivores, introduced crops attractive to native consumers, and changes in social habits (such as establishment of waterfront holiday properties and resorts increasing exposure to biting flies associated with brackish or freshwater habitats).

Each of these broad categories represents a change of ecological balance or human perspective, or both; and each may demand remediation or other management.

The above has emphasised arable systems and crop protection as the major agricultural arena of concern. Pest problems in agricultural systems extend also to the livestock sectors, through ecological changes associated with pastoralisation and by effects on livestock. As with crop systems, pastoralisation can have severe ramifications for conservation (Chapter 9).

One well-studied case, from Australia, has been the diverse causes and consequences of *Eucalyptus* die-back in parts of inland New South Wales, reflecting a considerable change in the balance of natural resources caused by pastoralism. This case is important in demonstrating how an initial disturbance (clearing of eucalypts to provide land for pasture establishment) and subsequent pastoral enrichment have together led to a major imbalance with drastic outcomes for the landscape and many native invertebrate biota. Die-back is a complex phenomenon. Its causes are diverse and, as Heatwole & Lowman (1986) emphasised, a combination of causes is likely to be involved in any particular case. The major victims are large eucalypts, left isolated in paddocks and intended for use as shade trees under which stock can shelter during hot or wet weather. Such trees are exposed, far more than those within forests, to climatic extremes (such as storm damage) and can also become a focus for intensive defoliation by native herbivorous insects. The major obvious change in the New England area of New South Wales was decrease of formerly widespread large eucalypts to such isolated trees and increase in pasture area. Heatwole & Lowman (1986) quoted the comparative figures for a local area of (in 1860) 60% forest and 40% open woodland, compared with (in the 1980s) <1% forest, 20% woodland and 79% cleared pasture. In addition to clearing, addition of the chemical fertiliser superphosphate to increase amount of feed (and thereby stocking rates) occurred. This form of intensification was associated with additional stress to the surviving trees, with increased likelihood of death and increased susceptibility to grazing insects and/or to fungal pathogens. A very simplistic partial explanation of the process involves Christmas beetles (native species of *Anoplognathus*, Scarabaeidae). The young adult beetles feed on eucalypt foliage, a step necessary for their reproductive systems to mature. They then move to pastures to oviposit, and ensuing larvae feed underground on roots of pasture grasses. With the massive increase in high-quality larval habitat, very large numbers of larvae complete development, so that increasing numbers of adults are then concentrated on a highly diminished eucalypt food supply, where severe defoliation may be implicated in die-back. Before introduction of exotic grass species and fertiliser enrichment, the scarab larvae developed in much lower numbers and were probably also kept in balance by natural enemies; at that time, there was plenty of adult food. Defoliation of trees without killing them may stimulate bursts of epicormic flush growth, producing tender young foliage of high nutritive quality, and a variety of other insect herbivores (including other beetles, psyllid bugs, caterpillars, sawflies and stick insects, in particular) may be fostered, all increasing defoliation pressure, extending the stress over longer periods and in a variety of feeding modes. As another aspect of the induced environmental change, removal of most of the eucalypts also helps to reduce local populations of insectivorous birds (through loss of potential nest sites), further exacerbating problems related to phytophagous insects capitalising, however temporarily, on abundant food. Die-back of eucalypts is a very complex suite of processes and is still not understood fully, but it is a clear outcome of agricultural intensification. It threatens large native invertebrate communities linked closely with eucalypts, keystone plants on which substantial radiations of insects and their natural enemies have occurred (New, 1983; Majer *et al.*, 1998). The case is salutary in demonstrating that even when no harm is inflicted on (or intended for) invertebrates directly, they may well suffer from changes in ecological balance. Such possibilities as cascade effects should be considered widely, in addition to simply removing habitat, and avoiding such consequences is a critical focus of optimal management (Chapter 7).

Pest control and pest management

The main ways to ameliorate the impacts of agricultural pests continue to diversify, but several very broad categories of options (following Clark *et al.*, 1967) occur or, at least, may be considered. The four fundamental strategies noted by Clark *et al.* are:

- To evade the consequences of pest attack, for example by diverting fruit to processing, in which cases some damage may be tolerable, in contrast to the blemish-free produce needed for individual marketing.
- To eliminate the characteristics of the target species that render it susceptible to pest attack, for example by developing crop varieties resistant to the pest.
- To suppress the features of a pest species that render it injurious. This is relatively unusual but occurs, for instance, in disease vectors where sources of infectivity may be removed or lessened so that the major 'pest symptom' is removed.
- To reduce the numbers of pests to levels well below the economic injury level (see p. 92).

Traditionally, the last of these strategies has been the mainstay of pest control, with developments involving the continual refinement and diversification of ways to kill large numbers of pests more efficiently and cheaply. In many crop pest contexts, 'control' has been considered synonymous with 'mortality', inflicted by any economically viable means that have been available at the time.

This has been associated most commonly with use of chemical pesticides, to which the treated population may respond in one of three main ways (Clark *et al.*, 1967):

- It may die out, if very high levels of mortality are inflicted, so that the pest (or pest population) is eradicated.
- The numbers fluctuate in phase with the treatments, so that the treatment reduces numbers to innocuous levels, which then increase again to damaging levels, so necessitating further treatment as the population recovers (Fig. 4.3).
- The pest population or species is able to overcome the impacts of treatment, for example by developing resistant characteristics (see p. 105).

The second of these is by far the most frequent outcome and is relevant to two major considerations in pest control, namely (1) the difficulties of eradication and (2) that control may be a continuing exercise, perhaps needed at least every season or pest generation, rather than as a simple one-off exercise. The success of a pest-control strategy may be related directly to both the duration and the reliability of its effects. Less desirable (that is, more costly) options involve treatments at short intervals with limited or transient effects; more desirable options are those with long-term impacts and that do not need frequent repetition or augmentation by other means (Clark *et al.*, 1967).

'Eradication' is a somewhat extreme form of control, difficult to achieve (Dahlsten *et al.*, 1989) but often desired. The aim is to reduce the pest population to zero. It thus contrasts markedly with other strategies in which the aim is usually to suppress the pest, i.e. to reduce its numbers to below damaging (or likely to be damaging) levels of abundance and to continue to maintain it below the economic threshold. For eradication, this threshold is zero, and difficulties of attaining this reflect the considerable effort needed to remove the last remaining members of a population after most have been obliterated. Costs and efforts tend to escalate markedly as the target population tends toward zero, and the increased intensity of treatment then needed may be undesirable from other points of view. As Dahlsten *et al.* (1989) noted, historically very little effort had been made to appraise the side effects of eradication programmes and

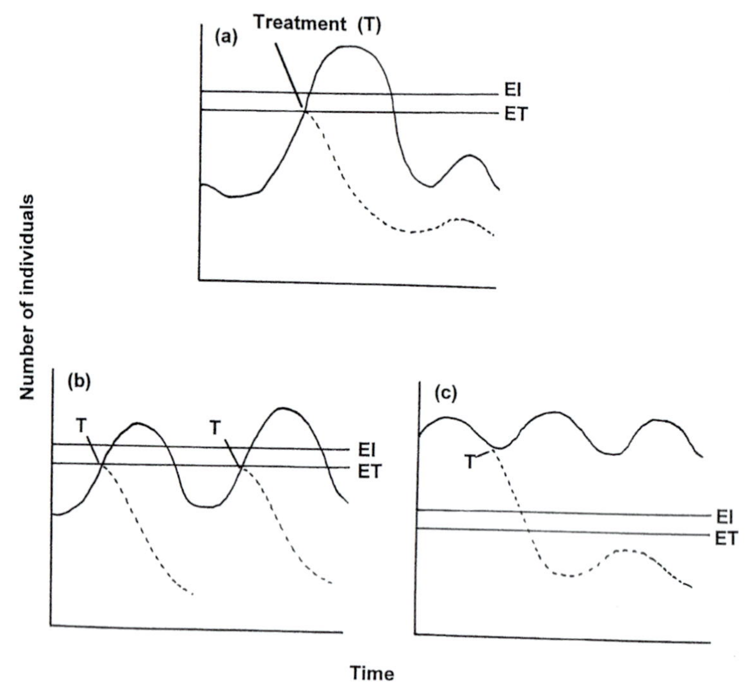

Figure 4.3. Treatment of pests. Treatments (T) are applied to the examples indicated in Figure 4.2 at times when monitoring indicates population increase to the economic threshold level (a, b), with the aim of reducing numbers to avoid economic loss. In the extreme case (c), a drastic treatment may be needed to drive the equilibrium level to a new, lower level, which may be very difficult exercise to accomplish.

their impacts on non-target taxa. Monitoring of such programmes is now commonplace, but the impact of general high reliance on chemical pesticides in many early eradication programmes was largely unmonitored. Most pests are extremely difficult, even impossible, to eradicate permanently, except in small isolated areas such as islands. Many are mobile and can recolonise easily (or be introduced unwittingly), so that any eradication is essentially temporary. Effective quarantine measures may need to accompany such programmes.

The most successful instances of deliberate eradication have involved exotic pest species, for which chances of natural population reinforcement may be relatively low and for which control measures have also been instigated soon after the target species has been detected, so that it might not have reached its potential geographical range or impacts in the treatment area. Pests are much easier to attack in this way if they are still concentrated or in relatively small numbers, for example around a single introduction point. One practical aspect is that little is then likely to be known of the pest's biology in the new area (although considerable background may be available from studies in other parts of its range), so that any method used in attempted eradication may necessarily rely on broad approaches rather than (possibly more sensitive) proven and individually tailored methods.

The major alternative to eradication, suppression, involves reducing pest numbers to an innocuous level, and the great variety of methods available gives high likelihood of success in many such endeavours. Many early attempts to control pests in agriculture relied entirely, or almost entirely, on chemical pesticides (see p. 103), this trait gradually giving way (wherever feasible) to combination of chemical and biological control (Chapter 5) with underlying cultural methods (Chapter 6) as adjuncts to many of these. The concept of simple pest control has thus gradually given way to the more complex integrated pest management (IPM), in which any of a variety of options may be used in concert and tailored to any of a variety of individual contexts. IPM can influence any aspect of the pest, its coexisting and interacting species, and attributes or components of its environment – together the effective environment of the pest and including all features with which it can or may interact.

Integrated pest management

The rise of IPM during the past half-century has resulted from increasing realisation of the shortcomings of relying on chemical pesticides and the problems of resistance (leading to the failure of many pesticides with repeated or intensive use) and of environmental contamination and side effects. Together, these constitute a formidable practical and ethical rationale for seeking to reduce and refine pesticide use in conjunction with development of effective alternatives. IPM thereby has a strong conservation ethic component, which has become enhanced progressively; it was endorsed by UNCED (1992) as part of section 14 ('promoting sustainable agriculture and rural development') as 'the best option for the future, as it guarantees yields, reduces costs, is environmentally friendly and contributes to the sustainability of agriculture' (UNCED, 1992, paragraph 14.74).

A number of different perspectives of IPM have ensued, but these generally agree in:

Table 4.3. *Desirable attributes of an integrated pest-management (IPM) system (after Dent, 1995)*

Provide effective control of the pest species or pest complex
Be economically viable
Be simple and flexible
Utilise compatible control measures
Be sustainable
Have minimal harmful effects on the environment, producer and consumer

- the need to reduce chemical applications and use of fossil fuels in pest control;
- the need to refine use of pesticides, not only to reduce them but also to target their uses more accurately to avoid non-target effects and environmental contamination;
- the need to provide working alternatives to pesticides wherever possible, particularly by manipulations of biological and other environmental components both to prevent and to ameliorate pest damage;
- the need for all such measures to be 'safe' and based on sound biological knowledge and understanding of the target species and the environment in which treatment is needed;
- the need for all available measures to be coordinated for maximum, most effective and most sustainable effect.

The major needs for an effective IPM system are noted in Table 4.3.

Integration can also occur above the single target species level, as recognised by Prokopy (1994), Kogan (1988, 1998), and others. Thus, Kogan's level 1 encompasses single species IPM; level 2 integrates management for multiple species (or of different taxonomic groups of pests) within the same cropping or other production system; and level 3 integrates management for these species within the total cropping system or agroecosystem. The last level is very complex, and most IPM is limited to level 1, sometimes extending toward level 2. Level 3 is very rare in practice but remains as an ideal standard for achievement.

The main phases of crop protection and intensification have historically run closely parallel. At subsistence levels of agriculture, no organised or coordinated crop protection measures are needed. With expansion of the crop and marketing enterprises, increased needs for protection arose, with impetus to safeguard and increase profits through increased yields. A strong correlation was evident between the amounts of protection afforded (in practice, largely the amounts of pesticide used) and the benefits obtained. The third of five sequential phases designated by Smith (1969) (for cotton, but of much wider relevance) is the 'crisis phase', reflecting that use of large amounts of pesticides brought problems, for example of resistance. The main counter to decreasing effectiveness of pesticides was to use them in ever-increasing amounts and to diversify their variety, so increasing the production costs of many crops. This trend may continue to the point where it is difficult (even impossible) to market the crop effectively, probably also accompanied by unacceptably high levels of pesticide use. In this 'disaster phase' (Smith, 1969), the control programme collapses. Finally, the phase of 'integrated control' emerged, to overcome the problems of reliance on pesticides. The stages leading up to IPM have sometimes been called 'pest mismanagement' (van Emden, 1983), and the accompanying increase in biological information as a foundation for IPM has been of major benefit in fostering its development (Sotherton, 1989).

Much of the development of IPM has, sometimes tangentially, interacted with aspects of conservation, with considerable recent emphasis on conserving and enhancing taxa from which direct economic benefits are perceived to accrue, namely pollinators of crops and natural enemies of pests, with such measures commonly having wider benefits for other biodiversity. The more pragmatic aims have been to increase productivity and economic benefits through protecting produce and enhancing its qualities and yields, so facilitating access to market niches that otherwise could not be penetrated. IPM is a major tool in agricultural sustainability, and the major challenges are to harmonise pest-management practices with the social responsibility of wider ecosystem and landscape management, whilst basing economic accountability on the more restricted agroecosystem environments.

Pesticides: a universal concern

Uses and concerns

Following the adoption of DDT (dichlorodiphenyltrichloroethane) in the late 1940s, chemical pesticides became the main means for controlling insects and other invertebrates in most pest arenas (Perkins & Patterson, 1997), drawing on the massive benefits derived from DDT during the Second World War by its use against vectors of a variety of human diseases, such as lice, fleas and mosquitoes. DDT rapidly gained the reputation of a 'miracle insecticide' and thus catalysed development of many similar synthetic compounds from the late 1940s onward. Agriculturists viewed such pesticides as a panacea, and some other long-used control methods (such as crop rotation and similar cultural controls) were abandoned in favour of applying ever-increasing quantities of pesticides that were seen to increase and ensure productivity. These were applied with little consideration of any undesirable consequences. However, the first cases of resistance to DDT in houseflies were reported by 1948, i.e. within two years of its use becoming commonplace for their control. Whereas broad-spectrum pesticides were thus the mainstay (and, often, the sole crop-protection strategy) in agricultural pest control, and against some disease vectors targeted for global eradication, similar problems of resistance continued to arise. The initial 'Era of Optimism' (1946–62) was thus followed by an 'Era of Doubt' (1962–76) and the subsequent 'Era of Integrated Pest Management' (1976 onward) (terminology of Metcalf, 1980), with doubts arising through the gradually perceived shortcomings and disadvantages of DDT and other synthetic pesticides.

The major perceived and practical advantages of chemical pesticides in agriculture devolved on the following:

- They were long seen as the only practical, effective and fast-acting means to reduce pest populations approaching or exceeding the economic threshold.
- They are often fast-acting, so that pest reduction is rapid, and high mortality ensues.
- Their collective variety provides for variability in treatments for a wide range of contexts.
- High levels of economic benefit occurred, with many pesticides being made available relatively cheaply, sometimes under various forms of subsidy to encourage their adoption.
- Many pesticides are easy to apply, even by workers with little knowledge or experience, although more recently safety precautions and regulations on use have become more stringent and complex.

In essence, chemical pesticides were seen as the solution to many aspects of crop and other commodity protection in primary industry.

The major problems leading to constraints and concerns over their use fall into two very broad categories, (1) resistance and (2) effects in the wider environment, with concerns over the latter being catalysed by Rachel Carson's (1962) famous *Silent Spring*. Huxley's (1963) eloquent preface to the first British printing of that book summarised the emerging more widespread view that 'it is almost certainly impossible to exterminate an abundant insect pest, but quite easy to exterminate non-abundant non-pests in the process'. From the early 1960s, therefore, issues of concern over non-target and other environmental effects of pesticides led to strong advocacy and effort to seek ways in which their use could be refined, reduced or focused more clearly, not least because of issues of food safety revealed by the abilities of many chemicals to 'bioaccumulate', that is to become increasingly concentrated by passage through food chains in communities. The practical needs to diversify control strategies were driven largely by resistance problems, whereby pesticides simply became ineffective for the purposes for which they were developed and employed.

Resistance

Resistance can be defined as development of a strain that is capable of surviving a dose lethal to most individuals in a normal population; it can develop very quickly, as noted above. The basic cause is the natural variability between individuals within any population (as shown in Fig. 4.4), so that a few (perhaps only very few) individuals have genetic capability to survive a pesticide dose that kills other individuals. They thereby persist and become the parents of the next generation, in which the proportion of such resistant individuals is increased. Pesticide applications then again remove susceptible individuals, so that strong selection occurs for resistant survivors. Repetition of this sequence over several generations eventually provides populations that are wholly (or almost wholly) resistant, and the pesticide is then incapable of reducing pest numbers. Resistance can result from several main mechanisms, which can occur singly or in any combination: (1) physiological resistance involves changes in the pest's physiology so that the pesticide is rendered ineffective; (2) behavioural resistance involves changes in pest-activity patterns, which, for example, reduce its exposure to the pesticide; and (3) biochemical resistance, as the predominant mechanism, is based on the pesticide being degraded or broken down (such as by enzyme activity) before it becomes active on or in the pest.

Cross-resistance is also frequent, reflecting that a pest that has become resistant to one pesticide has simultaneously become resistant to others, even when it has not been exposed directly to them. This reduces the impacts of the major grower response to development of pesticide resistance, namely to change the pesticide used. If, as occurs commonly, cross-resistance reflects the method of pesticide effect (for example, by blocking a particular physiological pathway), then any other pesticide with a similar

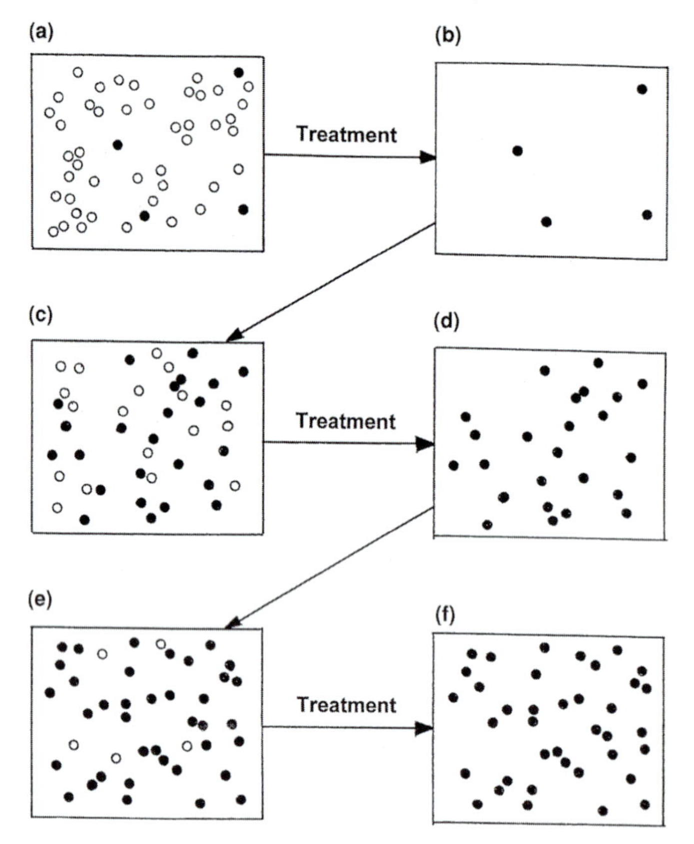

Figure 4.4. Development of resistance to insecticides. The sequence shown is (a) a normal pest population containing few resistant and many susceptible individuals; when sprayed (b), the former survive to breed and the latter are killed, so that (c) the proportion of resistant individuals in the next generation is considerably greater than previously. With further spraying (d), the process is repeated, and (e) an even higher proportion of resistant individuals develops, eventually leading to a totally or almost totally resistant pest population (f).

mode of action may be rendered ineffective as well. In such ways, development of resistance has functionally shortened the useful life of many pesticides in particular contexts and stimulated the development of increasing variety of novel or new-generation pesticides that may be more difficult for target organisms to counter.

The more extreme outcome of resistance is the development of so-called 'superpests', which are not susceptible to control by any pesticide currently available. Diamondback moth (*Plutella xylostella*) is one of the world's most widely distributed agricultural pest insects (as well as being one of the most cosmopolitan species of Lepidoptera) and causes massive losses to cabbages and other brassica crops. It was the first crop pest to develop resistance to DDT (in 1953), but it has also become resistant to every synthetic pesticide used against it, so that there are major constraints on how it may be managed. In some parts of the world, failure of insecticide controls (with *Plutella* also being the first insect to develop resistance to *Bacillus thuringiensis* as an insecticide, see p. 183) has led to economic production of brassica crops simply being impossible (Talekar & Shelton, 1993). Elsewhere, alternatives to pesticides are constantly being sought and trialled in attempts to control it. However, before pesticide use is abandoned for such widely resistant pests, the increased amounts and proliferation of formulations used may lead to increased side effects – and perhaps increased resistance in other members of a pest complex.

Resistance is also associated with diversification of pest impacts, through the processes of replacement and resurgence.

Replacement (or so-called 'secondary pest attack') occurs when suppression of a particular key pest by chemicals leads to its replacement by other pest species, which earlier had little or no significance or impacts and which are not influenced to any major extent by the pesticide(s) used. Although the important initial target pest is controlled effectively, there is no net economic benefit because it is replaced functionally by other species. The other syndrome allied to pesticide use and resistance is resurgence, the situation in which a pest population is initially suppressed strongly by pesticides but eventually recovers to attain numbers even higher than those that caused the initial concerns.

Both these eventualities are attributed most frequently to the effects of pesticides on the natural enemies of the target pest. In replacement, the usual explanation is that the pesticides used to suppress the major pest also kill the natural enemies of other potential pest species in the system (as a non-target effect typical of many that cause concern for conservation), so freeing them from normal constraints on population growth and (perhaps facilitated also by absence of the main pest as a key competitor) to increase to damaging levels far above their usual equilibria. In resurgence, the pesticide kills both the pest and its complex of predators and parasitoids which normally help to regulate its numbers. The residual, resistant population can then increase without opposition from its natural enemies and can reach numbers higher than would normally be the case. In time, the natural enemies may re-establish and resume their regulatory roles, but the increased pest numbers before this may necessitate increased or repeated pesticide uses – which may, once more, kill the natural enemies and prolong the problem.

Each of these possible outcomes may need to be considered carefully in planning to manage the onset of resistance, as a central theme in much modern IPM. Resistance-management strategies are critical components of the overall programme and recognise that they can reduce harmful effects of pesticide by reducing rates of application and prolonging the efficiency of those selected for other aspects of 'safety' in reducing contamination and unwanted side effects (Roush & Tabashnik, 1990).

Insecticides beyond crops

The unwanted mortality of natural enemies of pests is a key concern in developing effective pest management (see p. 193). Most early-used chemical pesticides were broad-spectrum and simply 'killed insects' (or other invertebrates) in the areas to which they were applied. Side effects are a potential, largely inevitable consequence of this practice, with all coexisting species in the crop environment being potentially susceptible. Many studies have investigated pesticide effects on natural enemies (see p. 127), but wider studies of impacts on innocuous taxa are relatively rare. Pesticide effects on Heteroptera in cereal fields in Britain were studied by Moreby *et al.* (1997), with the implication that such bugs are an important component of the diet of insectivorous birds in agricultural lands, so broadening considerations of pesticide use to effects on feeding ecology of farmland birds. Three fungicides each produced very low, insignificant levels of bug mortality, but some aphicides (as the major insecticides used in the system) had much more significant effects. The choice of active ingredient may have important influences on the abundance of Heteroptera.

Similar results have been reported for other groups of bird-food invertebrates, such as carabid beetles and lycosid spiders. Dimethoate has been found to reduce densities of most non-target arthropods in cereals (Vickerman & Sunderland, 1977), acting both through surface deposits as residues and – for sucking bugs – systemic properties. Dimethoate also has substantial impacts on sawfly (Tenthredinidae) larvae (Sotherton, 1990). For some such non-target groups, densities may be greater in field margins than in the crops themselves, so that avoidance of drift by using correct spraying protocols is important for their wellbeing.

Literature on the side effects of pesticides is encyclopaedic, arising from the fact that most pesticide applied against agricultural pests does not actually reach its intended target(s). In one estimate, Pimentel (1995) implied that as much as 99.9% of agricultural pesticides affect the wider environment, so that only a very small proportion reaches the target pests directly. In the early years of expanded pesticide use, many broad-spectrum chemicals were applied very widely (such as by aerial spraying of large forest or urban areas, in addition to cropping areas), with little attention being paid to possible hazards. Similar concerns can arise from spraying of 'biological insecticides', namely insect pathogens (see p. 183) applied as biological control agents but in methods similar to those used for many chemical pesticides.

The main areas of concern for invertebrates in the wider environment are:

- non-target effects of pesticides applied to crops on organisms within the crop communities;
- non-target effects of pesticides applied to crops on biota of adjacent areas such as crop margins, field boundaries and remnant habitat patches within agricultural areas, resulting from drift or transport of pesticides into these areas;
- non-target effects of pesticides applied to widely distributed or migratory pests well beyond cropping areas;
- wider environmental effects due to pesticide residues and bioaccumulation, ramifying through natural communities, sometimes over extensive areas;
- applications of pesticides to non-crop areas.

The first three of these are relatively tangible and can largely be managed by refining methodology and protocols for application – or, at least, possible side effects can be evaluated as part of a risk-assessment procedure. However, regulatory controls and pesticide-registration requirements are most effective in developed countries, so that unwanted side effects may be greatest where such practices are more rudimentary or are of less local concern. The fourth entry in the list reflects that pesticide residues are ubiquitous, with concerns from this playing major roles in the refinement of pest management at all levels. The final point is a particular concern for conservation, as it may result in more direct exposure of a great variety of non-target species to chemicals designed to kill their close relatives.

The greatest and most immediate concerns arise from pesticide applications within or close to crops (points 1 and 2 in the above list). Many of the biota present in annual crops are likely to be those that have moved in following recent disturbance or that are resident within the cropping areas. Despite the additional mortality that can occur to these organisms, few are likely to be species of individual conservation concern and many are relative generalists with considerable dispersal abilities and able to tolerate levels of disturbance and habitat change. However, not all species present are necessarily residents, and the 'tourist' component may also be vulnerable. Point 3 in the list is relatively infrequent but occasionally very important. Some of the general effects reported on soil and aquatic invertebrates resulting from pesticide uses are summarised in Table 4.4.

Invertebrates living near crops, for example on field margins, can be affected by either intentional or accidental pesticide use, and effects may be lethal or sublethal, or both, with sublethal effects often not heeded or evaluated in any detail. Both insecticides and herbicides can be important influences, the latter through removal of food plants (for Lepidoptera, as the most intensively investigated group representing both larval food plants and nectar plants used by adults). A major use of herbicides is to control perennial weeds around crop edges. This is often accompanied by increase of annual grass species, so that a species-impoverished community is established, replacing the former, more diverse plant array. Butterflies in Britain are perhaps the most carefully studied insects that suffer such effects, and Longley & Sotherton (1997) discussed the wide-ranging effects of pesticides on the butterflies of arable farmland; pesticide contamination of field boundaries has been implicated in the decline of many species. Although the impact of pesticides remains poorly understood, the twin scales of effect noted by Longley & Sotherton have much wider relevance. They distinguished 'microscale' effects acting upon single species or habitats and 'macroscale' effects acting upon whole populations over farmland. Microscale effects are very variable and are likely to differ between pesticides, vegetation types, butterfly species, pesticide characteristics, and so on, so that the relative vulnerability of different taxa could be determined only by detailed study – but these difficulties help to emphasise the general principle that particularly important habitats should not be exposed to pesticides more than is absolutely necessary (and, preferably, not at all). Macroscale effects incorporate consideration of population dynamics of the insects, with (in this example) about 64% of the British butterfly species on farmland having closed populations (or defined metapopulation units; see Chapter 8) with minimum breeding areas sometimes as small as 0.5 ha. Pesticides will often have more drastic effects on species that form discrete colonies than on those that range widely over the landscape.

Table 4.4. *Some general effects of insecticides on soil-inhabiting and aquatic invertebrates (after Edwards, 2000a)*

Taxon	Comments
Soil inhabitors	
Nematoda	Not susceptible to most insecticides, but individual effects can occur, for example leading to changes in community composition.
Acarina	Great variation in susceptibility, with active predatory species often more susceptible than sluggish saprophagous species.
Collembola	Susceptible to many insecticides, but some aspects not well documented and difficult to predict. As with insects, strong positive correlations between activity and susceptibility have been reported.
Pauropoda	Seem to be 'extremely sensitive' to many insecticides.
Symphyla	Apparently not very susceptible; repelled by insecticides and can reduce exposure by deep burrowing.
Diplopoda	Intermediate in susceptibility between pauropods and symphylans; increased susceptibility when on soil surface, due to greater exposure.
Chilopoda	Tend to be very susceptible, not least because of high activity as predators; relatively sensitive to many insecticides.
Oligochaeta: Lumbricidae	See text.
Mollusca	Other than methiocarb, insecticides are not toxic to snails and slugs, probably because the mucous coating is protective. Some insecticides can accumulate in mollusc tissues so can then affect predators such as birds.
Insecta	Very difficult to generalise; great variations in susceptibility.
Aquatic taxa	
Mollusca, Oligochaeta	Benthic fauna of these groups appear to have lower sensitivity to insecticides than aquatic arthropods but can bioaccumulate some insecticides.
Crustacea	All seem to be relatively susceptible to most insecticides, with incidences of large-scale kills from spraying or spillages.
Insecta	Most aquatic insect larvae very susceptible to insecticides.

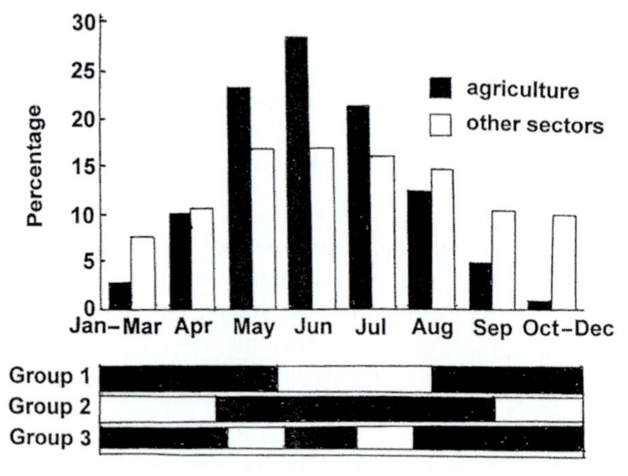

Figure 4.5. Insecticides and susceptibility of butterflies in the Netherlands: the importance of seasonal synchrony (Groenendijk *et al.*, 2002). The seasonal use of pesticides (agriculture, shaded bars; other uses, unshaded bars) in the Netherlands is related to three phenological groups of butterflies (larval presence shaded): group 1, larvae in spring and late summer; group 2, larvae throughout late spring to summer; group 3, more than one generation, with larvae in several seasons.

A more recent Dutch butterfly survey (Groenendijk *et al.*, 2002) emphasised also the importance of seasonality and timing of pesticide use, because most butterflies are likely to be affected mainly during their caterpillar stage on plants. More than 65% of agricultural pesticides used in the Netherlands are used from May to August, when caterpillars of two of the three main phenological groupings are likely to be most abundant (Fig. 4.5). Group 1 species (hibernating as larvae, and univoltine) may often 'miss' a spraying event during summer, unlike the other groups (group 2: larvae present in summer, and more than one generation each year; group 3: hibernating as larvae, and more than one generation each year). A similar principle applies widely to appraising pesticide effects on other taxa (see p. 115).

The full effects of pesticides on non-target species are difficult to conjecture, but many cases emphasise that they can be both drastic and wide-ranging. For invertebrates, our main concern here, Edwards (2000a) commented: 'We know most about the effects of insecticides on insect pests, beneficial insects, and invertebrate predators that live on or are associated with plants and how they affect populations and communities', but otherwise there have been few comprehensive summaries or reviews on particular taxonomic groups – other than for earthworms (Edwards & Bohlen, 1991), to which a number of insecticides are acutely toxic.

However, whereas only about 10 of more than 200 pesticides have been tested for their acute toxicity, all carbamates tested appear to be particularly toxic to Lumbricidae. The relatively large amount of data on earthworms derives not purely from concerns over the worms themselves but chiefly because of the potential risks of secondary poisoning for birds and mammals (Greig-Smith (1992), writing on pesticide effects and registration in Britain). In contrast, toxicity tests on earthworms were not required in the USA (Stavola & Craven, 1992), although these authors also discussed the dangers of carbofuran residues in earthworms to birds feeding on these animals and noted the need for field studies to assess risks.

In some instances, use of chemicals to control migratory or widely dispersed insect pests that invade crops extends well beyond crop or pastoral boundaries. Massive crop and pasture losses occur in eastern Australia as a result of depredations of the Australian plague locust, *Chortoicetes terminifera*, an acridid grasshopper that breeds over wide areas of the inland and undergoes sporadic outbreaks to produce vast numbers. Populations occur over some two million square kilometres of eastern Australia and have been controlled until recently entirely by aerial applications of the organophosphate insecticide fenitrothion in the form of ultra-low-volume (ULV) sprays (from Micronair sprayers mounted on aircraft wings; active ingredient (a.i.) of 267 g a.i/ha or 381 g a.i./ha, at flying heights of 5–10 m and track spacing of 50–100 m: details from Story & Cox, 2001). Two other related locusts, *Austracris guttulosa* (spur-throated locust) and *Locusta migratoria* (migratory locust), generally are less important but are controlled in the same way when necessary, using a higher application rate of 508 g a.i./ha-1. Early intervention in this manner is critical in preventing aggregation of small populations into larger, damaging units and thereby minimising potential for migration and subsequent damage to crops and pastures. Satellite technology is used to monitor build-up of populations in remote areas. Early treatment with insecticides can be, in part, preventive of both locust damage and the use of much higher amounts of pesticide at a later stage. However, in addition to environmental concerns, market constraints have also stimulated the search for alternatives, because of the demand for organic beef production and the need for lack of residues in any produce (Hunter *et al.*, 2001).

Two concerns arise from this practice. First, fenitrothion is a broad-spectrum pesticide used widely for control of insects, including forestry pests (for example, in North America), agricultural pests and disease vectors, as well as direct treatment of grain against stored products pests and

Table 4.5. *Main conclusions relevant to invertebrates from recorded side effects of chemical control of locusts (Everts & Ba, 1997)*

All the current insecticides present a risk to non-target organisms.
The groups of organisms at risk differ for each insecticide.
Long-term effects are rare and, as far as is known, limited to secondary disturbances within insect communities.
Aquatic invertebrates are sensitive to all recommended locust pesticides.
Terrestrial invertebrates are at risk in all cases, sometimes leading to undesirable effects, such as upsurges of secondary pests.

for protection of a variety of other commodities. Second, widespread aerial applications occur over large areas of 'natural' environment and thereby must be applied to a considerable spectrum of non-target insects, including many never associated with crop environments. Fenitrothion has long been the staple (indeed, sole) control used by the Australian Plague Locust Commission, and the side effects have been largely undocumented in Australia. Literature implications, mainly from North American studies, reveal that side effects may last from a few days (e.g. Chiverton, 1984) to more than a year (Freitag & Poulter, 1970). Carruthers *et al.* (1993) examined before and after pitfall trap catches of arthropods from five sites in New South Wales and Queensland, Australia, in which the downwind half of each site was aerially sprayed (at the commercial rate of 381 g a.i./ha) in late winter or spring. Four sites showed an apparent effect of spraying, although considerable variety in sites and the trends in numbers of arthropods/trap/day rendered interpretation uncertain. More data are needed to clarify the side effects of fenitrothion in Australia, which may extend also to vertebrates (Story & Cox, 2001). Laboratory studies, as for many other pesticides, are difficult to extrapolate to field conditions, but most field studies have evaluated only short-term effects, commonly at unstated doses and under a variety of climatic conditions. Table 4.5 notes some of the non-target effects from such programmes involving insecticide spraying against locusts. Tsetse flies (see p. 305) are a parallel example.

Broad-spectrum pesticides, applied aerially, were for a long time the major control method for many agricultural pests. In some cases, the ability to kill many different species has been regarded as an advantage, for example against Orthoptera (Riegert *et al.*, 1997) and other groups in which a number of species may co-occur in infestations. However, the almost inevitable consequence is that non-pest taxa in the same areas may also succumb. Recent modifications to insecticide use for rangeland grasshopper control recognise the importance of conserving non-target organisms, although these have been monitored on only rather rare occasions, despite acknowledgement that large-scale application of insecticides (such as by poison baits) can have severe impacts. Grasshopper chemical control treatments are applied most commonly when the target species are in middle to late instars, a time when maximum effect on some other taxa may also occur (Quinn *et al.*, 1993) – perhaps particularly Carabidae and lycosid spiders in their study, with these groups including likely predators of the grasshoppers. Quinn *et al.* suggested that applications timed for pre-reproductive adults would have minimal impacts on such non-targets, notwithstanding the greater damage that would by then have been caused through extended grasshopper feeding. Despite observation that non-target groups reduced in this way commonly 'rebound' within one or two years, it cannot be assumed that this response is universal. Reducing the 'conservation risk' of insecticides to grasshoppers has received considerable recent attention, and Lockwood (1998) summarised five possible approaches to this. All are of much wider relevance in illustrating basic principles for improved targeting of pesticides:

- Improving target specificity through changes in formulation. For Orthoptera, bran bait formulations narrow the target range, enabling some orders of insects and particular species of grasshoppers to survive.
- Method of application. Land-based application methods can be much more precise than aerial applications. Direct, topical application – for example to locust hopper bands – largely overcome risks of non-target applications.
- Differential toxicity. Particular insecticides may differentially affect even closely related insect species. Lockwood (1998) quoted several instances in which particular chemicals appeared to have low toxicity to some grasshoppers whilst killing others effectively. In a slightly different context, insect growth regulators (see hormone preparations) have few effects on adult grasshoppers, so that timing of application may be critical in avoiding damage.
- Environmental persistence may render sprayed vegetation toxic for considerable periods after application. Formulations that degrade rapidly may thus help to reduce non-target effects.
- Use of some persistent insecticides can be made compatible with conservation through imposition of barriers or reduced area treatments. The basic principle is to create 'toxic barriers' to mobile pests (such as

locust hoppers) or 'toxic strips' for less mobile species, and to leave the major parts of the area as an untreated but isolated refuge (Meinzingen, 1997).

Collectively, these approaches can reduce exposure of non-target taxa by (1) increasing specificity of treatment or (2) reducing exposure to treatment.

The very high diversity of invertebrates renders it very difficult to make broad generalisations on acute toxicity of pesticides to individual species (Edwards, 2000a). However, one widely held generalisation is that aquatic invertebrates are much more susceptible to insecticides than are soil-inhabiting invertebrates, particularly when the pesticide is water-soluble. Lethal doses of an insecticide may be taken up from water passing over respiratory surfaces (such as gills of aquatic insect larvae) and, as Edwards (2000a) noted, it is difficult for aquatic invertebrates to avoid such exposure, and a single 'spill' or run-off can cause widespread mortality in lotic systems. Table 4.4 summarises, very broadly, some trends in insecticide effects on various invertebrate groups. In some taxa, there is clear inference that activity may increase exposure, parallelling the active transport of pesticides in running waters noted above.

Toxicity and screening for non-target effects

Screening of pesticides for non-target effects is a formal prerequisite for their registration and use in any given context, with toxicity and hazard both relevant considerations in this process. Toxicity, an absolute measure, is quantifiable relatively precisely through LD_{50} measurements, and the relevant data are given on product labels. Hazard estimation is much more subjective, as it reflects the local environment and circumstances in which the insecticide is applied. Toxicity testing thus involves assessing direct effects on a variety of different taxa, normally including natural enemies of the principal target pest(s), earthworms and honeybees, in addition to a variety of allergic sensitisation tests, carcinogenicity and teratology tests, and potential for residues and bioaccumulation in the environment. Hazard estimation includes appraising effects of methods of application, weather and other factors that lead toward developing a code of conduct for users. Both themes are involved in determining and applying regulatory standards. Both are thereby highly relevant in affecting the choice of pesticide and its use in ecologically responsible ways. Wider perspective on screening trials than was formerly common is now almost universal; limitations of laboratory studies, for example, are apparent as not providing sufficient information on the fate of the pesticide in the wider natural environment, not detecting the sensitivity of most species under varying environmental conditions, and ignoring interactions with other species. Laboratory studies are much cheaper and often much easier to perform than broader-scale screening trials. However, Everts (1990), writing broadly on the effects of pesticides on wildlife, noted five main reasons why field studies are necessary to assess hazard adequately:

- Validations of predictions of environmental effects made on the basis of laboratory studies.
- Whether observed changes in species abundance or incidence are due to chemical effects.
- To ascertain the harmlessness (or otherwise) of large-scale control operations.
- To study certain species or ecological functions to develop improved test methods.
- To study the mechanisms responsible for ecological effects.

Practical protocols for assessing the side effects and environmental risks from pesticides differ considerably (see below), but their prosecution may depend heavily on pragmatic factors, in addition to the less easily evaluated ecological risks. Thus, adverse effects of pesticides on earthworms are important for three main reasons: (1) reducing the role of earthworms in maintaining soil fertility; (2) reduction in numbers may reduce the quality of habitat to their predators, as a 'flow-on' effect; and (3) that toxic residues in earthworms pose risks of secondary poisoning (see p. 119) to predators such as birds (Greig-Smith, 1992). A framework for assessment of environmental risk to earthworms, incorporating both the main classes of risk, is summarised in Fig. 4.6. The basic food-chain link of earthworm–predator was one of the early classic cases of bioaccumulation and of increasing awareness of pesticide damage to the wider environment, perhaps particularly through broad-scale DDT applications against forest pests leading to residue accumulation in earthworms, and death or reduced breeding performance of a variety of bird species (see Rudd (1964) for much historical background).

A somewhat more detailed sequence for testing pesticide effects is shown in Fig. 4.7. This representative protocol shows clearly the somewhat different approaches that might be needed to appraise effects on populations and risks of secondary poisoning to other organisms. The first (left-hand side of the figure) are the more proximal influences on the

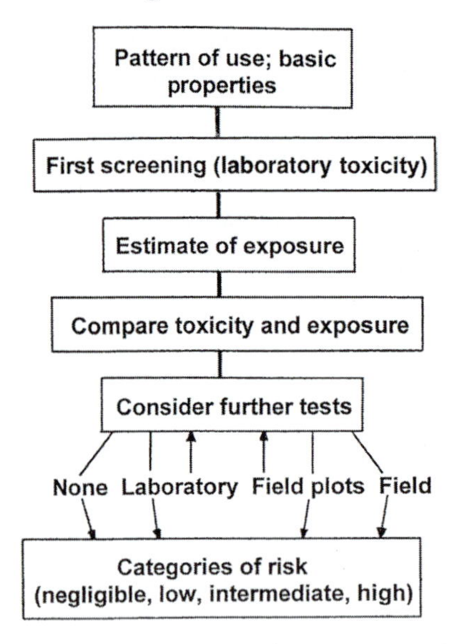

Figure 4.6. Framework for assessment of environmental risks of pesticides (and other chemicals) to earthworms (Greig-Smith, 1992). The stepwise sequence, from top to bottom, appraises the effects of contaminants.

target taxa (here, earthworms) of greater interest to invertebrate conservationists, but the second may be the more politically persuasive option in fostering changes in inadequate current practices, when necessary.

'Accidents' are perhaps inevitable, but the chances of their incidence can be minimised through good practice. One Australian case that attracted considerable attention arose from the drift of the pesticide fenvalerate from aerial spraying of *Helicoverpa* into farm dams with resultant deaths of freshwater crayfish (yabbies) (VFRAC, 1987), which led to recognition of some of the potential hazards associated with contract spraying. These included (1) that pilots may be unfamiliar with local topography and may not always avoid watercourses and dams; (2) that there may be strong incentive to fly in suboptimal conditions, such as windy weather, in order to complete jobs within a limited period (or busy spraying season with other tasks lined up); and (3) that pilots may spray from higher than ideal, because of risks of accidents from very low-level flying over unfamiliar ground, so increasing chances of pesticide drift. Considerable changes in practice to increase safety resulted from that report.

The use of insecticides is therefore increasingly responsible, involving rational scientific decisions largely replacing the more haphazard use

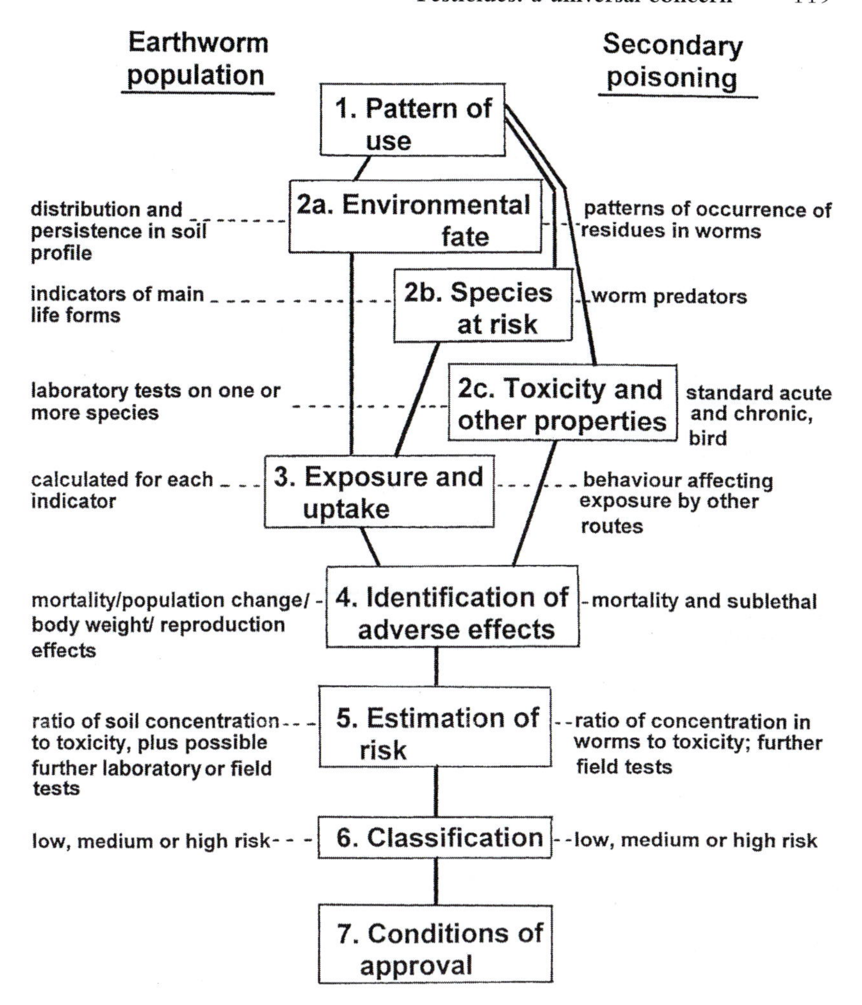

Figure 4.7. Expansion of principles shown in Figure 4.6 to specific context of pesticide effects on earthworms, and risk assessment to non-target taxa (after Greig-Smith, 1992).

of chemicals that predominated in earlier times and aiming to increase economic efficiency and overall safety of use in any particular context. Ecological knowledge of the pest and its environment, including coexisting species, are important components in such refinements.

Risk of harm from a pesticide is a function of susceptibility and exposure, commonly combined as the field parameter hazard (Jepson, 1989), which may be mediated through a series of biological and operational factors (Table 4.6). Exposure and susceptibility are influenced largely by

Table 4.6. *Biological and operational factors that affect the level and duration of pesticide impacts on non-target invertebrates (Jepson, 1989)*

Biological factors	Operational factors
Exposure to pesticide	
At time of spraying	
Proportion of population in sprayed area	Application volume
Degree of protection by canopy or soil refuges	Nozzle and droplet parameters
Droplet capture efficiency	Application frequency
Following spraying	
Residual exposure: distribution pattern and activity cycle	Persistence and behaviour of active ingredient formulation
Dietary exposure: availability of contaminated prey	Environmental influences on bioavailability
Susceptibility	
Genetic, structural and physiological factors mediating uptake, metabolism and toxic effect	Intrinsic toxicity of active ingredient at application rate
Environmental factors mediating toxic effect	
Recovery/reinvasion	
Direct ecological factors	
Mortality, dispersal, phenology	Persistence and breakdown in environment
Reproductive rate/voltinism	
Isolation/proximity to non-crop reservoirs	
Indirect ecological factors	
Degree of oligophagy/polyphagy	Spectrum of activity and toxic effects
Extent of depletion of preferred prey	on alternative prey items in substrate
Sublethal effects	
Repellency	
Behavioural activation	

operational factors and features of the organisms; the recovery phase is mediated by fewer factors and related strongly to the life-history strategy of the organism, including its dispersal proclivity. Jepson (1989) developed a fuller scheme whereby an invertebrate species' relative position in the *r*-K spectrum of ecological strategies (see p. 11) was related to hazard of pesticide applications to appraise broad likelihood of long-term risk. This generalised approach, shown in Table 4.7, requires input from both field exploration and more detailed modelling, and Jepson acknowledged the considerable difficulties in developing the scheme for practical use.

Thus, field experiments demonstrate that some species of spiders are more susceptible to insecticides than others, with foraging mode appearing to affect susceptibility in important ways (Pekar, 1999). As with pollinators (see p. 134) and natural enemies (see p. 89), diurnal or nocturnal activity may affect exposure, because diurnal species may come into direct contact with sprays and nocturnal taxa will contact residues only. Dense webs also appear to give some protection to spiders, whereas active hunting may increase exposure through greater activity.

The choice of the 'best' insecticide is 'one of the most critical decisions in pest management' (Edwards, 2000a). Ideally, an insecticide should minimise possible adverse effects through being target-specific and the lowest effective dose being applied precisely to that target. In particular, many of the more recently developed insecticides tend to have low mammalian toxicities (unlike many of the formerly widespread organochlorines and organophosphates) and are used in formulations that tend to lessen damage to other taxa. Screening against natural enemies and pollinators (such as honeybees) is now widespread (and sometimes obligatory for registering a chemical for use), as noted above. However, highly selective insecticides are expensive and difficult to develop, and their use may be countered by resistance development. The factors of concern in choosing an insecticide are therefore both economic and environmental, with the former reflected in both the variety of applications (species, environments) and the frequency of application. Thus, increased frequency of use can promote rapid development of resistance as well as increase the labour costs of applications, in addition to perhaps increasing hazard to operators through additional exposure to the chemicals. In contrast, more persistent insecticides reduce frequency of application and operator hazard but may increase environmental impacts because of greater (or longer) exposure. Ideally, the persistence should match that of the target pest as closely as possible, and Edwards (2000a) noted that it is easier to increase persistence through changes in formulation than it is to accelerate loss of a more persistent pesticide in the field. Considerable data exist on the minimum effective doses of insecticides, and dosage rates are recommended on product labelling. These recommended doses are often higher than those that will achieve successful field control, because they usually contain an 'insurance factor' (Edwards, 2000a). More

Table 4.7. *A theoretical framework that links relative hazard of different pesticide application practices to non-target invertebrate life-history strategy (assuming commercial-scale applications over more than one season (see Jepson, 1989)*

Relative position of invertebrate on *r*-K spectrum			Probability of long-term effect
r	Intermediate	K	
Dispersive, long potential period of crop colonisation; rapid reproductive rate; probably dietary specialist	Enter crop from other areas; less rapid reproduction; probably less specialised diet	All/part of life cycle within crop; univoltine or bivoltine; polyphagous	
Coincident with time of a broad-spectrum persistent spray application used frequently and on a large scale	Species with one phase of dispersal sensitive within season. Also intermediate between *r* and K	Use of toxic product at sensitive phase of life cycle or frequent use of low-toxicity product. Treatment affects diet. Combined effects of more than one product	High probability
Similar to above, but compound of limited persistence or scale; intensity of use reduced	Intermediate	Exposure to products applied before emergence or colonisation, especially non-persistent compounds. Limited effects on diet	Intermediate probability
Compounds used outside period of colonisation or use of selective products	Intermediate	Selective pesticide compounds or application strategies. Reduced usage via IPM	Low probability

IPM, integrated pest management.

precise applications of such chemicals can often be accompanied by even lower doses, and considerable effort has been made to improve placement, such as by charged sprays (which are attracted to foliage surfaces) or by ultra-low-volume concentrated sprays.

Two other aspects of the roles of insecticides in pest control merit comment in relation to developing the most responsible ways to use them. First, practices such as 'calendar spraying' or other prophylactic applications – that is, the routine use of pesticides on crops irrespective of whether there is any real need for them in terms of monitored pest incidence and abundance levels – have been traditional in many contexts and in many parts of the world. Codling moth control in Australian apple orchards, for example, formerly involved a dozen or more spraying events at different times of the season in relation to the pattern of blossom and fruit development. This practice was thus used as 'insurance', sometimes with minimal attention to whether the sprays increased economic margins through reducing pest effects. Improved monitoring of pest abundance leads to reduced pesticide input, both by reducing the number of prophylactic applications and minimising doses and by using insecticides only when there is a real need to do so.

Second, many human consumers demand fruits and vegetables of prime, unblemished quality, and there is considerable pressure (and market competition) for such produce. As noted earlier (see p. 93), achieving these levels of perfection is expensive and environmentally costly, particularly when combined with calls for clean, uncontaminated produce. This also increases pressures (in many parts of the world) for development of pesticides with low mammalian toxicity and for which active ingredient doses are reduced to even lower levels. Economic pressures can become very complex in relation to increased specialisation.

The tangible benefits of insecticide use to the grower naturally reflect the priorities of increased crop yield and greater economic returns, perhaps through access to prime markets, and these often do not take into account the indirect costs associated with environmental impacts. Indeed, most of these cannot be evaluated in strict economic terms. Nevertheless, the true costs of pesticide use include both the market price and application costs and the values of any extra damage that results in the wider environment. Conservationists have long argued that manufacturers should bear the financial burden of this extra damage, through such measures as a pesticide tax (Pearce &Tinch, 1998; Robinson *et al.*, 1995), perhaps related to the sensitivity of the area to be treated or to the active ingredient used, so that the higher risk is accompanied by a higher financial imposition.

Environmental impacts are indeed difficult to quantify, and many approaches have been used to evaluate them: Levitan *et al.* (1995) listed 38 different approaches, of varying application and complexity, and implicating different environmental factors and facets of exposure and toxicology. The parameters can include aspects of risk in terms of exposure and transport in various habitats, effects on a range of key taxa and application procedures. They are exemplified by the six representative approaches discussed by van der Werf (1996) (Table 4.8), which differ considerably in the spectrum of topics considered. As van der Werf noted, two of these methods (Higley & Wintersteen, 1992; Hornsby, 1992) do not include the amount of pesticide used; only Reus & Pak (1993) appraised application techniques, and only Vereijken *et al.* (1995) included pesticide volatilisation. A detailed comparative appraisal of assessment methods is not needed here – it is sufficient to emphasise (1) the great variety of approaches, (2) that some of these appear to omit factors of considerable relevance and (3) that each method tends to support a different definition of 'environmental impact', so that there is no overall consensus on the 'best' protocol(s) and comparisons between different studies are commonly difficult or impossible. Levitan *et al.* (1995) suggested the need for a database with comparable data on each pesticide and alternative management options in order to evaluate options systematically; they recognised also that effective tools to advise farmers on choosing the least damaging options for pest management using insecticides are still in their development stages

Four criteria reflect the impact of a pesticide in the environment (van der Werf, 1996):

- The amount of active ingredient applied, and its site of application.
- The partitioning of active ingredient between, and concentration in, air, soil, surface water and ground water.
- The rate of degradation of active ingredient in each of these.
- The toxicity to the species present in each of these.

The term 'degradation', above, refers broadly to the loss of pesticide by microbial and/or chemical action and can be a very complex process, not least because many metabolites of pesticides persist as breakdown products and are transported more easily than the original pesticide. Particularly in dry soils, half-lives of pesticides may be protracted and, in some instances,

Table 4.8. *Six selected methods proposed to assess environmental impacts of pesticide application in agriculture (van der Werf, 1996)*

	Method*					
Parameter	1	2	3	4	5	6
Environmental risk category						
Presence on plant	–	–	×	–	×	–
Presence in soil	–	–	×	×	×	×
Volatilisation	–	–	–	–	–	×
Drift	–	–	–	×	–	–
Run-off	×	×	×	–	–	–
Leaching	×	×	×	×	–	×
Bioconcentration	–	–	–	–	×	–
Plant systemicity	–	–	×	–	–	–
Toxicity to:						
Earthworms	–	–	–	×	–	–
Honeybees	–	–	×	–	–	–
Beneficial arthropods	×	–	×	–	×	–
Birds	×	–	×	–	–	–
Mammals	×	–	×	–	×	–
Humans (acute)	×	–	×	–	×	–
Humans (chronic)	×	×	×	–	–	–
Algae	–	–	–	×	–	–
Crustaceans	–	–	–	×	–	–
Fish	×	×	×	×	×	–
Other parameters						
Role of metabolites	–	–	–	×	–	–
Rate of application	–	–	×	×	×	×
Site of application	–	–	–	×	×	–
Season of application	–	–	–	×	–	–
Soil characteristics	–	×	–	×	–	–
Application technique	–	–	–	×	–	–
Efficacy	–	–	–	–	×	–
Availability of alternatives	–	–	–	–	×	–
Costs	×	–	–	–	×	–

*×, taken into account; –, not considered; 1, environmental economic injury level (Higley & Wintersteen, 1992); 2, screening for adverse water quality (Hornsby, 1992); 3, environmental impact quotient (Kovach *et al.*, 1992); 4, environmental impact points (Reus & Pak, 1993; CLM-IKC, 1994); 5, pesticide index (Penrose *et al.*, 1994); 6, environment exposure to pesticides (Vereijken *et al.*, 1995).

a significant proportion may remain as 'permanent' residue bound to soil colloids. They are often presumed to have then lost their biological activity, but some studies suggest that the products can be released into ground water or taken up by plants, including crops, as a major source of bioaccumulation and possible later toxic effects. Pesticide behaviour in soils, as a representative predominant context for the fate of residues, can depend on numerous factors. Van der Werf (1996) listed:

- degradation by soil microorganisms;
- chemical degradation, such as by hydrolysis;
- sorption and binding by organic and mineral soil components;
- uptake by plant roots;
- volatilisation;
- distributing effects of water-flow processes.

The rates of degradation commonly increase with rise of temperature and soil water content.

Much of the foregoing discussion reflects the need to monitor pesticide effects carefully in order to establish their effects. Monitoring non-target effects can be relatively straightforward, at least to an extent sufficient to detect mortality of non-target species within the confines of crops. Considerably greater problems arise when pesticides are applied over much wider areas, even if these are defined adequately. Thus, applications against tsetse flies (*Glossina* spp.) in Africa helped to emphasise a variety of limitations in monitoring capability, some of which have very wide relevance elsewhere (Grant, 1989). As examples:

- Comparable natural or semi-natural areas to monitor may not be available or definable easily. Unlike agroecosystem studies, where 'controls' may simply be elected as adjacent plots or fields, controls for wider applications may not be available and baseline data against which to appraise changes also may not be available.
- Sampling error has often been too large to detect impacts. From Grant's account, the numbers of insects trapped may simply be too low for statistical analysis.
- Identification of taxa has commonly not been to species level (see p. 36), so analyses lack sensitivity, unless the need for finer-level taxonomic distinction has been appraised carefully.
- Whereas sensitive indicator taxa may be selected for monitoring, these might not occur evenly throughout the large and varied regions to be

appraised. Grant commented that the diversity of ecological zones in which *Glossina* control is undertaken gives high likelihood that species selected for monitors in some places will not be so in others. One possible way to counter this is to use 'sentinel species' as indicators – for example, honeybees, where economic loss from possible mass deaths can be tolerated. A more tangential way to indicate non-target effects involves examining the regurgitated gut contents of an insectivorous bird (the bee-eater *Merops pusillus*) to detect any changes in diet before and after spraying with endosulfan (Douthwaite & Fry, 1982).

- Application irregularities are frequent, and quantitative data on insecticide deposition are often incomplete. Little attention may be paid to differences in hydrological regimes, and groups such as soil invertebrates may be omitted from appraisal in favour of those that are more conspicuous.

The following notes illustrate some of the specific concerns arising from pesticides on non-target invertebrate groups – some of these groups being of major importance in seeking ways in which pesticide uses can be refined.

Pesticides and natural enemies

The abundant studies of the effects of pesticides on natural enemies of pests in cropping systems give many pointers to wider conservation implications of insecticide uses and to focusing this for less undesirable effects. Natural enemies (that is, predominantly arthropod predators and parasitoids; see Chapter 5) can be affected by pesticides at two levels – by direct exposure and by reduction of the critical supplies (prey, hosts) on which they – sometimes specifically – depend to levels at which they are no longer easily accessible. In addition, some predators move away from insecticide-treated fields (Newsom, 1967). Understanding these effects is important, both in pursuing IPM involving combinations of pesticides and biological control (Chapter 5) and in conserving members of these guilds in natural assemblages.

Susceptibility to pesticides (as for other organisms) often varies considerably between different life stages and, as Ruberson *et al.* (1998) emphasised, knowledge of the developmental patterns and biology of natural enemies may be of critical importance in timing pesticide applications to have minimal effects on them. Particular life stages may be more exposed or less exposed to pesticide – for example, pupal stages within cocoons or

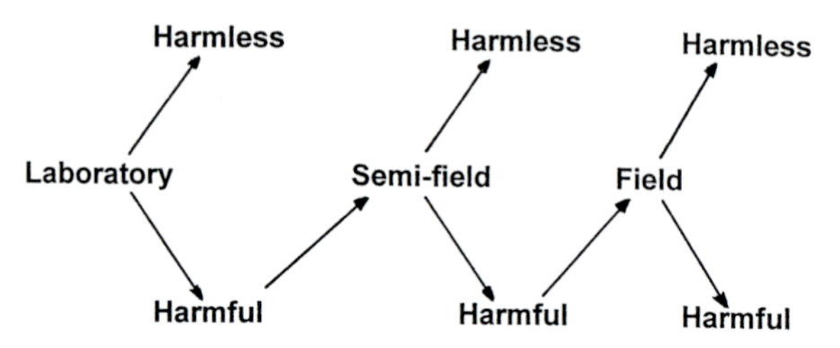

Figure 4.8. Three stages of testing, in sequence, for the side effects of pesticides on beneficial organisms such as natural enemies (Hassan, 1989). Initial laboratory tests, if harmful effects are implied, are followed in turn by semi-field- and field-level testing.

in soil and endoparasitoids within their hosts may be much less susceptible than fully exposed stages susceptible to topical applications. This principle was established firmly by a working group of the International Organisation for Biological Control (IOBC), which developed a testing protocol to help select pesticides that are less toxic to the most important natural enemies of a range of pests in a given crop, in which three scales of assay (Fig. 4.8) are proposed (Hassan, 1989). The evaluation categories used (Table 4.9) comprise five main trial series involving comparison of 'more exposed' life stages (adult parasitoids, free-living predatory larvae) with 'less exposed' stages, as above, in the laboratory, followed by 'semi-field' and 'field' trials. The general principle adopted was that pesticides should be tested initially in the laboratory, and on a variety of relevant beneficial species (four to six species suggested). If the pesticide does not harm (kill) the organisms in the laboratory, then it is unlikely to do so in the field, and no further testing is needed. However, if harm eventuates, then larger-scale tests are necessary. Despite the great diversity of potential natural enemies, the IOBC group agreed the protocols outlined in Table 4.9, and the approach (although derived in the pragmatic context of fostering compatability between facets of IPM) has wider relevance in guiding the wider and wiser use of pesticides (Hassan, 1989). Ruberson *et al.* (1998) also assessed the most important issues in designing assays to evaluate pesticide effects on natural enemies, as shown in Table 4.10. All three of the test levels included in Fig. 4.8 are generally required for practical registration, and the IOBC methodologies have been refined by the Beneficial Arthropods Regulatory Testing Group (BART) (Dohmen, 1998).

The impacts of pesticides on predators and parasitoids relative to target pests involve two broad forms of selectivity:

Table 4.9. *The five pesticide test sequences developed by an International Organisation Biological Control (IOBC) West Palaearctic Regional Section Working Group to appraise side effects of pesticides on beneficial arthropods (after Hassan, 1989), (see also Fig. 4.8.)*

1 *Laboratory – exposed life stages*
  a Exposure to freshly dried pesticide deposits.
  b Recommended concentration of pesticide.
  c Application on glass plate, leaf or sand (soil).
  d Even film of pesticide, standard amount of 1–2 mg fluid/cm^2 on glass or leaf, and 6 mg fluid/cm^2 on sand (soil).
  e Laboratory-reared organisms of uniform age.
  f Adequate exposure period before evaluation.
  g Adequate ventilation.
  h Water-treated controls.
  i Mortality/reduction in beneficial capacity.
  j Four evaluation categories: 1, harmless (<50%); 2, slightly harmful (50–79%); 3, moderately harmful (80–99%); 4, harmful (>99%).

2 *Laboratory – protected or less exposed life stages*
  a Direct spray of organism.
  b Recommended concentration of pesticide.
  c Adequate ventilation.
  d Laboratory-reared organisms of uniform age.
  e Water-treated controls.
  f Mortality/reduction in beneficial capacity.
  g Four evaluation categories, as above.

3 *Semi-field – initial toxicity*
  a Exposure to freshly dried pesticide residues.
  b Recommended concentration of pesticide.
  c Spraying of plants to run-off.
  d Field cages under natural or simulated field conditions.
  e Water-treated controls.
  f Laboratory-reared organisms of uniform age.
  g Use of dense treated foliage to ensure adequate exposure of organism.
  h Food and host/prey near centre of treated foliage.
  i Adequate exposure period before evaluation.
  j Four evaluation categories: 1, harmless (<25%); 2, slightly harmful (25–50%); 3, moderately harmful (51–75%); 4, harmful (>75%).

4 *Semi-field – duration of harmful activity*
  a Exposure to pesticide residues.
  b Recommended concentration of pesticide.
  c Spraying of plant to run-off.
  d Weathering under field or simulated field environment.
  e Experiments up to one month after treatment.
  f Laboratory-reared organisms of uniform age.

(*cont.*)

Table 4.9. *(cont.)*

g Water-treated controls.
h Four evaluation categories: 1, short-lived (<5 days); 2, slightly persistent (5–15 days); 3, moderately persistent (16–30 days); 4, persistent (>30 days).

5 *Field*

a Crops inhabited by beneficials are sprayed directly.
b Laboratory-reared or naturally occurring arthropods.
c Sampling at intervals before and after treatment.
d Recommended dose rates and number of treatments (following good agricultural practice).
e Water-treated controls.
f Dead and/or living individuals counted.
g Number of individuals to exceed a certain limit to allow statistical analysis.
h Four evaluation categories: 1, harmless (<25%); 2, slightly harmful (25–50%); 3, moderately harmful (51–75%); harmful (>75%).

- Physiological selectivity, in which the pesticide is less toxic to the natural enemy than to the target pest at a given application rate or concentration; thus, this is a property of the pesticide itself in relation to the different physiologies of the species affected.
- Ecological selectivity, related to the way in which the pesticide is used and to the region to which it is applied. Systemic pesticides, for example, may affect only foliage-feeders and not directly affect natural enemies and others active on the leaf surfaces. In many cases, though, natural enemies may feed on sap or pollen and thereby become susceptible.

Whereas short-term effects of pesticides on natural enemies have been investigated widely, rather few studies have been made on their longer-term effects. These may arise from (1) separate applications of pesticides with short persistence, so not allowing adequate time for the recovery of exposed populations or (2) the impact of a single application, where recovery is slow or non-existent because of low colonisation or possibilities for invasion from other habitats. One important study, the 'Boxworth project' (Burn, 1989), undertaken in Cambridgeshire, UK, assessed the long-term effects of pesticides on polyphagous predators in cereal crops. The ecological characteristics of predator groups, and the insecticide effects on them, are summarised in Table 4.11. Exposure of the various categories is influenced by their level of dispersal and pattern of seasonal development. Burn (1989) noted that interpretation of the complex data

Table 4.10. *Important issues to consider in designing bioassays to evaluate effects of pesticides on natural enemies (Ruberson* et al., *1998)*

Issue	Specific considerations
Selection of species	Relative importance within particular system Representative of natural enemy guilds within system Known susceptibility to other pesticides
Life stages/sexes	Exposed, active stages that may contact residues Concealed, less exposed stages Gender-specific differences in susceptibility
Routes of pesticide entry	Topical contact, direct Contact with residues on substrate Inhalation Ingestion of poisoned prey or host material Ingestion of poisoned plant material
Life-history parameters to evaluate	Survival, longevity, developmental time Fecundity/ fertility schedules Searching behaviour and rate Consumption/parasitisation rate Dispersal activity and movement Respiratory rate Population growth or reduction
Plot size for screenings	Dispersal capability of natural enemy Proximity of treatments to one another (e.g. possibility of drift)
Pesticide formulations and rates	Anticipated uses Range of recommended rates Possible distribution in environment affecting amount reaching target or target substrate

from this study is an oversimplification but that the data do show the range of trends as a framework for starting to understand such effects on non-target organisms. Four years of pesticide effects were evaluated, in comparison with two years of pre-treatment information, and the findings may indeed merit wider consideration in harmonising pesticide uses with conservation aims.

Thus, group 1 species showed persistent adverse effects, as they were exposed to a wide range of pesticides, including those applied against slugs in autumn, collectively having high direct impacts. Re-establishment is slow and may be possible only when repeated pesticide exposure is removed. Group 2 species overwinter in field boundaries and

Table 4.11. *Four ecological categories of polyphagous predators in British cereal fields, and their relative susceptibility to pesticides; data from the Boxworth project (Burn, 1989) (see text)*

Ecological group*	Dispersal ability	Principal overwintering stage/habitat	Susceptibility to long-term pesticide effects	
			Direct effects	Indirect effects
1	Poor	Adult/mid-field	High	High
2	Moderate	Adult/field boundary	Low	High
3	High	Adult/non-crop	Low	Low
4	Moderate	Larva/mid-field	Low	High

*Representative taxa for each group: (1) *Bembidion obtusum*, *Notiophilus biguttatus*, Collembola; (2) *Agonum dorsale*, *Bembidion lampros*, *Demetrias atricapillus*; (3) Tachyporinae, aphid-specific predators; (4) *Harpalus rufipes*, *Pterostichus melanarius*, *Trechus quadristriatus*.

colonise cropping areas in spring. They were reduced considerably in the Boxworth trials only in seasons 3 and 4, possibly reflecting impacts of a single, unusually late application of pesticide that coincided with spring dispersal. Group 3 species effects did not persist beyond the following season, probably because of ready recolonisation from unaffected areas. Group 4 species are less dispersive and less exposed to direct effects of the pesticides. These trends are discussed in detail by Burn (1988, 1989), and are in response to an 'insurance regime' of pesticide, in which insecticides, fungicides and herbicides were applied at high frequency as insurance against yield loss.

Sublethal effects are much more complex to appraise and, as Elzen (1989) noted, some are difficult to allocate to one or other of the major categories of 'behavioural' or 'physiological' effects shown in Fig. 4.9. Their appraisal and amelioration are important strands of conservation biological control (Chapter 6). Many relevant aspects are listed under 'Life-history parameters' in Table 4.10; long-term consequences of sublethal effects may be mirrored in enhancement or reduction of any of these parameters and others. Most information is available for insecticides, but fungicides are known also to differentially affect host and parasitoid (Sewall & Croft, 1987). In general, relatively long-term sublethal effects (such as those influencing reproductive rate) are difficult to appraise in the field. Not altogether unexpectedly, trials of major groups of pesticides

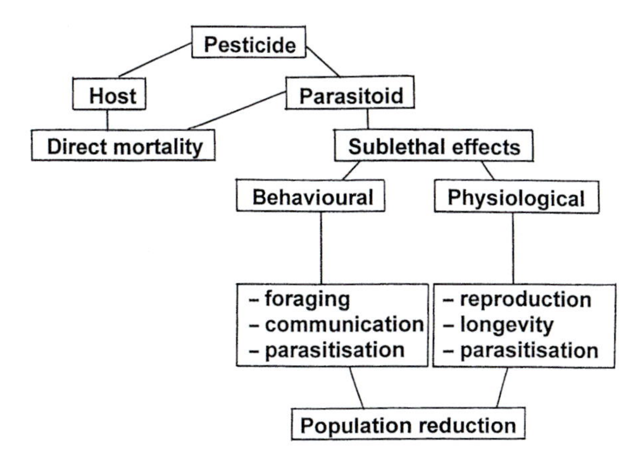

Figure 4.9. Sublethal effects of pesticides: some of the categories that may reduce the effectiveness of parasitoids as biological control agents (Elzen, 1989).

on beneficial arthropods reveal many insecticides to have harmful effects, but significant numbers of fungicides and herbicides also cause substantial harm (Hassan, 1989).

Pesticides may also operate by affecting the food supply of predators (see references in Burn, 1989), with consequences such as increased activity of predators seeking food and increased predation pressure on other species. Thus, reduction of earthworm prey in sugar-beet crops in Britain led to higher levels of aphid suppression by predators (Jepson, 1982, cited in Burn, 1989). These effects were termed 'indirect' by Waage (1989) because they do not involve direct contact between animal and pesticide.

The problems of determining the effects behind the common observation that effectiveness of natural enemies declines after pesticide applications are formidable. Waage (1989) listed five possible explanations for this occurrence but recognised that other factors might also participate:

1. Pesticides kill some natural enemies.
2. Pesticides have a sublethal effect on some natural enemies, so reducing survival or foraging efficiency.
3. Pesticides enhance pest reproduction, by allowing pests to escape control by natural enemies.
4. Pesticides change local pest distribution, causing temporary emigration of natural enemies or reduced foraging efficiency.
5. Pesticides synchronise pest populations, causing the local extinction of natural enemy populations and pest increase, as in point 4.

Pesticides and pollinators

Pollinating invertebrates are amongst the organisms regarded as 'mutualists' in cropping systems. They comprise three main functional groups: (1) species introduced into crops for enhanced pollination services (see p. 62), (2) wild pollinator species for which the local environment may be managed to enhance their beneficial role, and (3) unmanaged wild species (Brown, 1989). Bees are the single most important pollinator group for many crops (with the three major groups of honeybees, bumblebees and solitary bees), with Diptera (perhaps in particular Syrphidae and blowflies), Lepidoptera, thrips and other insects also playing significant – if more sporadic – roles. The relative functional importance of different members of a pollinator guild is often difficult to determine and is often simply extrapolated from the relative abundance of the different taxa found during surveys.

Pollinator importance is recognised by the widespread incorporation of honeybees as a test taxon in pesticide screening trials. Appraising pesticide impacts has been based largely on measures of toxicity, estimated by decrease in forager numbers seen on crops, or by counts of dead bees in the field or at hives. Direct effects are clearly the most important component and are related, perhaps directly, to adequacy of ecological function.

Wild pollinators are often more susceptible to pesticides than are domestic honeybees (Buchmann & Nabham, 1996), and wild pollinators may be eliminated completely from a crop environment and its surrounds or may take several years to recover to normal (pre-treatment) levels (Johansen, 1977). The major pollinators of blueberries in Canada, bumblebees, took up to eight years to recover normal population levels after cessation of fenitrothion (see p. 113) aerial sprays applied widely against spruce budworm (*Choristoneura fumiferana*) in coniferous forests (Kevan, 1977).

Correct timing and selection of pesticide is a critical aspect of pesticide applications to avoid or reduce pollinator losses, and Johansen (1977) recognised four categories of pesticides used commonly on crops in North America:

1. Hazardous at any time on blooming crops.
2. Minimal hazard if applied during late evening on blooming crops.
3. Minimal hazard if applied during late evening, night or early morning on blooming crops.
4. Minimal hazard on blooming crops.

This scheme is similar to a classification of pesticide toxicity to bees summarised by Buchmann & Nabham (1996), with those in categories 2 and 3 above having toxicity levels greatly reduced within three to six hours, so that they are applicable at times when bees are unlikely to be active for that ensuing period. These authors cautioned, also, that many of the pesticides least toxic to honeybees have largely unknown effects on other pollinators, simply because these influences have not been examined.

Parasiticides and dung fauna

One class of pesticides, avermectins (natural fermentation products of the soil microorganism *Streptomyces avermitilis*), are used widely to control parasites of livestock, such as ectoparasites and intestinal worms, as well as for a variety of other pests. Parts of the dosage administered to livestock, whether by injection or other methods such as direct topical application, are excreted to remain largely unchanged in the animal's dung with insecticidal activity largely unaltered. Strong (1992a) noted that more than 90% of the dose given by subcutaneous injection is secreted into the alimentary canal and voided in faeces, with this elimination continuing for several weeks. Residues in cattle dung of two proprietary preparations – abamectin and ivermectin – have been reported to have adverse effects on dung-breeding insects (predominantly Diptera and Coleoptera, as the major groups represented), such as by reducing larval survival, lengthening the period of development, and reducing reproduction of the ensuing adults, collectively implying wider effects on pasture ecology (Strong, 1992b, 1993). These effects reflect both the damage to normal populations of coprophages and their associates and the implications of delayed breakdown of dung resulting in pasture fouling or staling. Strong (1992b) also indicated wider effects such as (1) insects of cattle dung being important dietary components of bats and birds and (2) particular dung-breeding Diptera (such as *Scatophaga stercoraria*) being predatory on other insects. Cowpats may support more than 100 non-target species (Madsen *et al.* 1990), many of which are regarded as beneficial in accelerating the breakdown of dung.

Routine applications of avermectins, for example against nematodes, have the additional benefit of controlling dung-breeding nuisance flies, but effects on other insects may be less desirable and may cause concern for conservation – for example, of rarer dung beetles. As Strong (1992b) put it 'quite simply, we do not know the extent to which the presence of faecal avermectins affects these insects and the roles they play in the environment as a whole'.

For example, a single standard injection of ivermectin to cattle prevented emergence of the South African dung beetle *Euoniticellus intermedius* from dung collected two to seven days after treatment, and fewer beetles emerged from dung collected 1 and 14 days after treatment than from dung from untreated (control) cattle (Kruger & Scholtz, 1997). Emergence of a second beetle species, *Onitis alexis*, was reduced two to seven days after treatment, and the development of both these species was prolonged considerably.

Earlier studies (Wardhaugh & Rodriguez-Menendez, 1988; Houlding, *et al.*, 1991) showed lethal effects of abamectin- or ivermectin-treated dung to a variety of dung beetles when newly emerged. However, no toxic effects were noted for sexually mature beetles. Collectively, these and other studies (most of them laboratory-based) indicated possible adverse effects that might occur to dung beetles in the field. The full ramifications of lethal and sublethal effects of avermectins are yet to be appraised, but Kruger & Scholtz (1997) suggested that any potential risks to local dung faunas might be reduced by treating cattle outside the main activity period of the insects. Relatively little information is available on effects on other groups of invertebrates; not surprisingly, dung nematodes are reduced considerably after ivermectin treatment, but earthworms appear to be affected little. Thus, Madsen *et al.* (1990) found no major effects on earthworms, whereas severe effects occurred on cyclorrhaphan flies, with ivermectin continuing to actively kill non-target organisms in dung for two months or more after deposition. However, the period of persistence can depend on light, temperature and other local factors (Lumaret *et al.*, 1993), so that generalised predictions on the effects of ivermectin are difficult. In some Mediterranean countries, where dung beetles are among the primary agents of dung breakdown, relatively low sensitivity of adult beetles may reduce impacts – but risks may still occur for large burrowers (such as species of *Copris*, *Geotrupes* and *Onitis*), which dig deep burrows and fill them with dung from the surface. The buried dung is then protected from photodegradation and may decompose only slowly (Lumaret *et al.*, 1993).

The above scenarios summarise many of the actual and perceived dangers to non-target invertebrates from using insecticides of various sorts in agricultural systems. Because pesticide use will continue to be a component of many pest-management programmes, refining their use to reduce or remove such side effects is an important theme, together with continuing attempts to reduce the overall amounts used. Sotherton (1989) summarised the threefold approach to reducing non-target exposure to agricultural chemicals (primarily in cereal fields) promoted in Britain, as:

- screening for selectivity against non-target organisms, with development of field-scale screening as a progression from laboratory trials;
- restricting use to only selected compounds, and promoting some degree of pesticide exclusion on certain areas, especially crop margins;
- enhancement of natural enemy activity to avoid unnecessary use of pesticides, and reducing the use of prophylactic applications of broad-spectrum chemicals.

Sotherton noted that 'attempts to date have only begun to provide the necessary data' on pesticide screening. This restriction is related in part to development of partially or wholly unsprayed areas (conservation headlands, see p. 000) comprising on average around 6% of total field area on crop margins. In such areas, guidelines recommend that spring grazing of insecticides is not undertaken and autumn spraying is allowed if drift to margins (such as hedgerows) is avoided. Such practices link with the needs to enhance beneficial arthropods (Chapter 7).

5 · *Biological control and invertebrate conservation*

Pest control by use of 'natural enemies' is the basis for biological control, which may involve the introduction and release of exotic predators and parasites of pests, or enhancement of naturally occurring species that fulfil these roles. Concerns over the former (classical biological control) involve the invasive effects of such introduced species to attack native, non-target biota, and have led to development of convincing protocols to help affirm their specificity (and, hence, safety) in their new environment. Many of the concerns of conservationists over biological control devolve on the difficulties of predicting what introduced species may do after release.

Introduction

Biological control involves the use and manipulation of the so-called 'natural enemies' of pests (that is, most frequently, their predators, parasitoids and pathogens) in the management of those pests. It was developed largely as the major alternative or complementary strategy to insecticide use, and earlier work involved predominantly 'classical biological control' (CBC). This approach dealt with the numerous cases in which the pest (usually an arthropod, and most commonly an insect) is exotic (that is, it has been introduced or arrived naturally but is non-native) to the area in which it causes damage, and that high levels of abundance and invasibility resulted – at least in part – because it is no longer regulated by the complex of associated natural enemies that are associated with it within its natural area of distribution and evolution. CBC seeks to reimpose the earlier balance by introducing selected members of that natural enemy complex to the 'new' area. This process thereby adds further exotic species, but with the expectation that these will lead to a self-sustaining regulatory system for the pest, ideally reducing pesticide use and thereby having considerable economic and environmental benefits in pest management. The major contexts in which CBC has been used extensively are:

- introduction of natural enemies (predators, parasitoids) of arthropod pests, mainly herbivores on crops;
- introduction of natural enemies (herbivores) of plant weeds, some of them associated with cropping areas.

Some major early successes in both these disparate ecological arenas led to CBC being hailed as a widespread panacea for environmental problems associated with pesticide use, with enormous (but difficult to calculate) economic benefits accruing through its 'permanent' nature. However, during the late 1980s, a few ecologists raised doubts over this somewhat complacent perspective, and their views led to a groundswell of wider concerns, with implications that exotic natural enemies have accelerated or caused declines of non-target invertebrates in the areas to which they were introduced. They have thus been implicated as a major threat in conservation of native invertebrates. As Roitberg (2000) put it, 'biological control is now tainted, and will never be the same as before', following 'an emotional and rancourous debate on a number of issues, including threats to endangered species, displacement of native predators, parasitoid drift, and environmental impact'. This statement indeed encapsulates the major concerns of conservationists (Thomas & Willis, 1998). This chapter includes summaries of the rationale of CBC and other forms of biological control, of the variety of biological control agents (including pathogens such as bacteria and viruses), and discussion of the major apparent risks and issues of concern both in introducing exotic agents into new environments and in changing their normal balance in native assemblages. Wider aspects of habitat management encompassed under 'conservation biological control' are discussed in the next chapter.

Classical biological control

The practice of CBC of introduced arthropods and weeds 'has a long and distinguished record' (van Driesche & Hoddle, 2000) and has become a mainstay of numerous agricultural pest-management programmes in many parts of the world.

The need for CBC reflects that, in its native area, any arthropod or weed species is likely to have coevolved with associated species and that a suite of consumer species serve to keep it 'in balance', whereby its numbers are regulated within the interactions of a relatively diverse community or food web. When the species reaches a new environment, geographically distant from its original distribution, it is most likely to be without its associated consumers. In consequence, population growth is no longer opposed by these pre-adapted natural enemies imposing mortality and may build to damaging levels. The strategy of CBC is to reimpose that control by introducing natural enemies from the pest's earlier geographical range, in the expectation that they will remain in association with the pest as their sole or preferred food and, after lowering its numbers to tolerable levels, will prevent its numbers from building up again to damaging abundance.

The basic sequential stages in a CBC programme are as follows:

1. Investigate the need for natural enemies, by sufficiently detailed surveys of the proposed target pest in its area of introduction to determine that exotic natural enemies are indeed absent or markedly under-represented in relation to those in the area of origin, and to find if they have been replaced functionally by local species.
2. Explore the pest's original geographical range (particularly those parts of the range most similar climatically to the area of pest concern) to evaluate the variety and relative impacts of natural enemies present in 'natural association' with the target species, and to gain preliminary impressions on which of these might be candidates for introduction.
3. Select the most promising of these for more detailed appraisal and to evaluate their biology – particularly aspects of feeding specificity and preferences, reproductive performance and potential to build up large populations, and their likely seasonal synchronicity with the target pest. Candidates should also be capable of thriving in the climatic regime of the area(s) to which they are to be introduced.
4. Develop and enhance biological knowledge of the candidate species with a view to facilitating their introduction. The most critical aspect of this is 'screening', i.e. testing each candidate for its feeding specificity and risks of causing environmental damage in the area of introduction. The major purpose of this often complex process is to maximise the likelihood that the consumer will not attack native species or other desirable or commercially significant taxa, such as ornamental plants related closely to target weeds. It is also important that any agents introduced should not carry diseases or other organisms, such as parasites.
5. Release only after screening tests have provided satisfactory evidence of likely safety. Initial release may be directly to the field ('hard release'), but it is often useful to initially undertake a 'soft release' (such as into

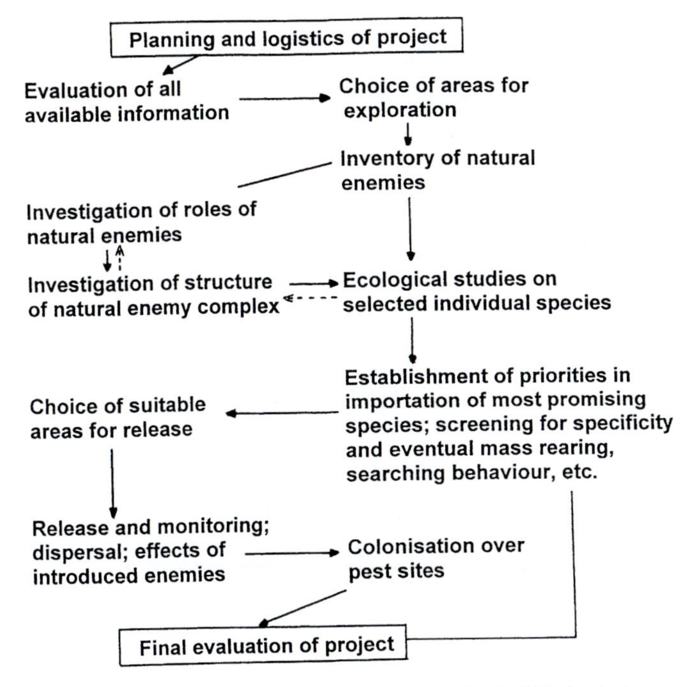

Figure 5.1. General sequence of processes in classical biological control (after Pschorn-Walcher, 1977).

field cages) to confirm that the agent will survive in the new conditions and to help build up numbers in the new environment by preventing overdispersal of released individuals.

6. Monitor the released population and its effects on the target to determine its impact or success as a control agent and to detect any unanticipated interaction(s) with members of the receptor community.

Each of these steps is complex, and the sequence emphasises that careful planning has replaced the previously more haphazard process of introductions, with a view to replacing hit-or-miss outcomes with 'maximum likelihood' of success. The sequence represents a broad temporal process (Fig. 5.1, after Pschorn-Walcher, 1977), which takes at least several years (often a decade or more) to complete. Each step also involves taking important decisions, as emphasised by van Driesche & Hoddle (2000). Even apparently simple initial tasks such as selecting suitable target pests for CBC can be difficult – for example, pests can be highly damaging but also highly localised, so that other control strategies may be more useful; and/or pests may be taxonomically isolated or related very closely to innocuous native species, with the latter situation perhaps indicating or suggesting greater risk of agent cross-over to these relatives.

Each phase in the sequence has been discussed extensively in recent years, but the bulk of conservation concerns has devolved on perceptions of the safety or otherwise of agents introduced into new environments and how they may operate in the receiving environment – for example in establishing new associations with native species. Thus, the above reference to taxonomic relatedness of pest to native species relates particularly to whether the target is likely to be discriminated by the agent. Species of conservation significance, or which are beneficial or otherwise valued, may be of particular concern; in such cases of close relatedness, the 'burden of proof' to be established by screening trials may be very difficult (even impossible) to obtain.

As a corollary to this, and as noted earlier (see p. 37), many agents are themselves very difficult to identify reliably and consistently. Many parasitoid wasps and predatory arthropods are members of sibling groups, morphologically almost identical but commonly differing in significant aspects of their biology, such as host range. Selection of the 'wrong' agent through failure to discern such differences may be disastrous and have severe practical ramifications through failure to anticipate its effects: species *x* may be very different from species *y* in its feeding range and predelictions. Even when there is no such ambiguity, selecting the most suitable candidate agents from a broad range is often difficult; in cases of taxonomic uncertainty, the additional (and sometimes formidable) task of primary taxonomic revision may be necessary.

Using the best available knowledge and research, each species of natural enemy eventually proposed for release must be appraised relative to (1) the benefits it is likely to provide by effective commodity protection and pest suppression and (2) any potential risks it may impose on the receiving environment. The evaluation process also necessitates balancing risks of any side effects with the continuing damage to the local biota or environment that may occur through continuing to use the existing control methods that CBC might otherwise replace. Risk assessment must be undertaken on a species-by-species basis, although some broad 'leads' might be possible (Table 5.1). In general, anticipating the host range or prey spectrum of introduced agents is complex and can easily be misleading when reviewed with the benefits of post-introduction hindsight. Table 5.1 includes only the main natural enemies of arthropod pests, as

Table 5.1. *Relative levels of risk from introduction of different taxa of natural enemies in classical biological control (CBC) programmes (van Driesche & Hoddle, 2000)*

Natural enemy group	Level of risk*	Comments
Terrestrial arthropod scavengers (e.g. dung beetles)	5	Effects depend on receptor community; effects based on competitive rather than trophic relationships
Social predatory invertebrates (Hymenoptera: wasps, ants)	10	Inappropriate to consider as agents; stinging or substantial invasive/competitive effects
Tachinid flies	1–10	Highly variable: some species high risk, others highly specific and 'safe'
Parasitic Hymenoptera	1–3	Host range variable: effects on non-targets need much more investigation
Predatory insects	3	Prey range usually wider than parasitoid host range, but concerns extend also to native species in same guild
Predatory mites	1–2	Prey ranges variable; most species not specialists; effects on non-targets need further investigation

*Scale of 1 (low risk) to 10 (high risk).

the most complex context of predictive CBC, and for which screening protocols are still being developed. In contrast, screening protocols for herbivore agents of weeds have long been better defined. Because of the central importance of screening processes and extent in seeking agent safety, as the core of conservation concern over CBC, this aspect is dealt with in some detail here.

Screening protocols for introduced biological control agents

As with screening pesticides (see p. 116), it is impossible to consider every possible target species and interaction individually, so that logical schemes must be devised to assess realistic risks of transfer to non-target hosts. The central issue of considerable importance to CBC practitioners is that, whereas conservationists tend to regard strict monophagy on the target pest as (to a large extent) a 'guarantee' of its host/food species restrictiveness, some degree of oligophagy may be considered advantageous in the practical context of agent survival in the receiving environment. The agent would thereby be able to establish reservoir populations on other species at times when the pest is scarce and thus be available to attack the pest when needed. Indeed, even broader host ranges have sometimes been advocated as desirable by practitioners, in that the agent can utilise these and 'lie in wait' to exploit the preferred target prey, for example during outbreaks (Murdoch *et al.*, 1995). Polyphagous parasitoids of aphids have been introduced deliberately into Australia, but on the basis that they are entirely (or almost entirely) restricted to a given, definable host range or habitat (in this case, to aphids on cereals), so they are considered not likely to pose any real danger to native taxa or to invade natural habitats. For example, the polyphagous aphelinid parasitoid *Aphelinus varipes* was introduced as a pre-emptive measure (see p. 162) should the preferred host (Russian wheat aphid, *Diuraphis noxia*) reach Australia, and in the expectation that in the mean time it would be sustained adequately on the alternative exotic host, *Rhopalosiphum padi* (Hughes *et al.*, 1994).

The other major dilemma is that screening tests undertaken in relatively restricted laboratory conditions may not provide results that transfer uncritically to the field, as a more complex and less constrained environment. Biologically, many of the features sought in a good biological control agent, and that might elevate its potential for introduction, are identical to those that make a good pest. These include ability to exploit new environments (in which critical resources are often prolific), having high reproductive output and rapid rates of population increase, and being generally 'voracious'.

Views of CBC practitioners on the need for strict testing of candidate agents are divided, particularly for predators and parasitoids of arthropod pests. Major concerns of conservationists include:

- that any introduced agent is intended to be a permanent addition to the local fauna, is likely to exhibit traits that will facilitate its establishment successfully, and may be extremely difficult or impossible to eradicate should unanticipated damaging effects occur;
- that the agent may form new ecological associations, including new trophic interactions that could not have been predicted, even by extensive pre-release host range testing;

- that subsequent range expansion, particularly from agricultural areas into natural habitats, may occur, so that the agent becomes invasive and a possible pest itself;
- that possible genetic plasticity may facilitate or enable unsuspected adaptation to the receiving environment, so that spread or change of host/food range also could not have been anticipated realistically;
- that the agent's behaviour may change, possibly again facilitating expansion or change of host or prey range;
- that the receiving environment may lack competitors or other opposition to the introduced species, even in invaded natural habitats.

The second and fourth of these concerns, in particular, lead to even some comprehensive screening trials being 'best practice' rather than ideal, and with full predictive powers being very limited (Howarth, 2000). The considerable difficulties of improving on this situation are linked with justifications advanced for detection of full host range not being a routine component of a CBC programme in most countries, as 'some argue it is a waste of time and money, and delays carrying forward critically needed introductions' (van Driesche & Hoddle, 2000). These authors noted two issues used as justification for this viewpoint:

- Laboratory testing of host ranges of predators and parasitoids of arthropods is not developed well, so mistakes are likely to occur, with a perceived danger that biological control will be made to appear too dangerous to conduct,
- The general lack of concern over most insects, raising a range of ethical issues in relation to whether attacks on non-target species are necessarily 'bad' (see p. 153).

In short, many screening tests are very difficult to perform in sufficient depth to provide unambiguous results and recommendations, but most are generally adopted only through the general process indicated in Table 5.2. Initially, detailed appraisal of literature and other records is needed, with realistic appraisal of cases of likely mistaken identities (see p. 161). The second stage involves a broader field survey of the agent's native assemblages, and three basic (but rather different) parameters are addressed in laboratory trials before adequate post-release studies to provide for early detection of any changes in host range (van Driesche & Hoddle, 1997).

In contrast to the somewhat laissez-faire attitude noted above, other protocols must be sufficiently defined and convincing to satisfy independent reviewers of the safety of the candidates proposed for introduction and must satisfactorily sort the admitted members of the

Table 5.2. *Steps in a programme to estimate the host range of potential biological control agents of arthropod pests (van Driesche & Hoddle, 1997)*

1. Assemble host records from previous studies, including comprehensive literature survey.
2. Assemble records of non-target species known to coexist safely with the agent at other locations.
3. Pre-release laboratory testing of host range.
   (a) Physiological host range (e.g. those hosts suitable for larval development of parasitoids or sustaining predators; these may include species outside the ecological host range; see below).
   (b) Ecological host range (the range of species suitable for oviposition and attack by adults).
   (c) Taxonomic host range (whether the agent is limited to a given taxon suite of hosts, such as members of a particular insect genus or family, or whether the range is determined by other factors, such as, most simply, availability).
4. Assessment of actual host range in field following release.

natural enemy complex from the admissions pool. Approval for release depends on satisfying national authorities, with regulatory requirements in some countries becoming ever more stringent. Table 5.2 indicates the various ways in which the host or prey range of an agent may be perceived. The problem then arises as to how to test these satisfactorily, and the practical differences are exemplified by cases such as the following, for parasitoids in particular:

- Under laboratory conditions, many parasitoids will oviposit in or on hosts that are outside their normal field host range. Many such trials are based on 'choice tests', in which the agent is presented simultaneously with known host and anticipated non-host species. Oviposition in or on the latter category is recorded commonly, in some instances because they are contaminated by specific chemical searching cues emitted by the true host and that confuse the wasps in an enclosed environment. Predicting field host range from such trials, with unknown internal confounding influences, is unreliable.
- Such ambiguities have led progressively to recommendations for 'no-choice tests', in which the candidate hosts are presented separately to parasitoids and the hosts then maintained and reared to reveal (1) the number or proportion that have been attacked and (2) those that can support parasitoid development to maturity, so helping to differentiate 'physiological host range' from 'ecological host range'.

Barratt *et al.*, (1999) recognised the 'extremely varied nature' of parasitoid–host relationships, together with the vast numbers of taxa involved, and suggested that these factors preclude establishment of a simple pre-emptive set of screening protocols. In the eyes of many conservationists, a biological control agent is 'guilty until proven innocent', an attitude that increases the need for innocence to be proved as effectively as possible, using protocols devised for standard replicable conditions. Criteria for standardisation are enormously varied and may include the cage environment (size, structure, uniformity of content), host–parasitoid ratio and exposure period (to counter variations due to density effects and varying levels of interaction between parasitoids, such as interference when focused on few hosts), food plants (consistent quality, quantity, species or variety, standard replenishment regime), environmental conditions (temperature, humidity, photoperiod), physiological state of parasitoids and test taxa (many biological variables may influence the interaction), and the extent of the data recorded (Barratt *et al.*, 1999). Thus, for the latter point, basic levels of parasitisation are often recorded simply by maintaining hosts until the parasitoids emerge and recording the percentage of hosts that yield adult wasps. Barratt and colleagues pointed out that dissection/examination of other tested hosts, including those that died over the rearing period, can provide much additional and useful information, such as incidence of unsuccessful parasitisation – for example, by detecting eggs or young parasitoid larvae encapsulated by the host's immune system.

'Pseudo-parasitisation' refers to parasitoids attacking hosts without ovipositing, a process that may reduce host viability in many ways. The above list emphasises that parasitoid Hymenoptera can show different levels of host fidelity or specificity as a function of numerous factors, including photoperiod, barometric pressure, nutritional state, presence of conspecifics, egg load and host availability, and the different interactions of these factors (and, probably, others): '. . . the number of potential factors that might affect host fidelity seems endless' (Roitberg (2000), including specific references to the factors listed above). The differential influences of such a variety of factors on different species suggest that a given screening protocol (albeit comprehensive) for one parasitoid species 'may not reveal much about another' (Roitberg, 2000), endorsing the claim by Hawkins & Marino (1997) that there are no 'rules of thumb' to predict whether an introduced parasitoid will colonise native insect communities.

However, other workers – whilst recognising the limitations of doing so – have suggested broad criteria for 'acceptable host quality'

(Sands, 1997). Sands suggested the following possible principles, for example:

- Agents are acceptable if narrowly specific in their host range and demonstrated by testing to be specific to the target pest in their new environment.
- If an agent completes development in or on any non-target taxon, then a decision is needed as to whether this might have any undesirable or detrimental effects.
- The ability of a narrowly specific agent to develop in or on a non-target taxon should not automatically preclude a recommendation for release.
- Development in or on some non-target organisms may be acceptable provided that the host range has been shown to be narrow (such as a small taxonomic group related to the target) and provided that non-targets are not preferred to the target.
- Development in or on some non-target hosts may sometimes be desirable.

The greatest levels of uncertainty arise from evaluating candidate agents known to be capable of attacking non-target species, even though their extended range may be defined and considered desirable by practitioners, as above. Risk assessment then involves balancing the likely benefits against possible undesirable effects in the receiving environment (Simberloff & Stiling, 1996), emphasising the need for broad screening evaluations to realistically evaluate the extent of such effects – together with suggesting means by which these could (or should) be reduced (with the realisation that they can not be eliminated entirely (Bourchier & McCarty, 1995).

The design and inferences of 'choice tests' have major importance in gathering this information, with test design developed most satisfactorily for herbivore agents of weeds (Marohasy, 1998) (see below), influencing interpretation as:

1. False positives: non-hosts are attacked when near to normal hosts, due to contamination, or transferred search stimuli such as host scents.
2. False positives: non-hosts used by agents separated for long periods from their normal hosts.
3. False negatives: less preferred, but acceptable, hosts are ignored in the presence of a more preferred host.

Table 5.3. *Different cases in which both no-choice tests and choice tests are useful in programmes to determine host ranges of arthropod biological control agents for arthropod pests (Sands & van Driesche, 2000)*

	Choice test result (−/+) and interpretation related to result (−/+) in no-choice test	
No-choice test result	Negative result	Positive result
Negative result	Case 1 Test species outside host range	Case 2 Test species outside host range, and positive result likely to be due to 'spill-over effect'*
Positive result, initially	Case 4A Test species in host range, negative result in choice test likely to be due to 'diversion effect'*	Case 3 Test species in host range
Negative result; positive result after several days	Case 4B Test species outside host range, positive result in non-choice test likely to be due to 'desperation effect'*	

*See text.

Applying these principles to agents for arthropod pest control, Sands & van Driesche (2000) referred to these categories as (1) 'spill-over effects', (2) 'desperation effects' and (3) 'diversion effects', respectively. They recognised that neither choice tests nor non-choice tests are necessarily superior to each other, and that there may be substantial advantages in undertaking both test designs on the same agent(s). Four useful case categories were distinguished, as in Table 5.3:

Case 1. Both kinds of test suggest that a given species is not a host, so that the host is excluded from the agent's host range. It may be necessary to confirm the 'health' of the particular agent stock by confirming that it can still exploit the main target host.

Case 2. Both tests confirm that the given species is a suitable host, and it must therefore be included in the confirmed host range.

Case 3. Choice test is positive but no-choice test is negative. In this case, spill-over effects are implicated, and the non-target species is likely to be outside the normal host range.

Case 4. Choice test is negative, but non-choice test is positive. Sands & van Driesche recognised two 'subcases' from this situation. In case 4a, the immediate response to the host in the no-choice test confirms that the test species is a valid host, and the negative choice test result may reflect a diversion effect. In case 4b, the positive response occurs only after some days of deprivation, and it may then be misleading as a deprivation effect. The negative choice test result then confirms that the host is outside the normal range.

Non-hosts included in such tests, which are most easily undertaken mainly before importation to the area of concern, should ideally include species indigenous to the reception area, but this has commonly not occurred (Waage, 1997). Selection of the optimal range of test species is a complex exercise, for which there are few formal guidelines for agents to be used against arthropod pests. Following the example set by more defined protocols developed for weed control agents (see p. 152), phylogenetic ranking has been suggested, with the test species ranging from those closely related to the target to more taxonomically distant taxa. Additional parameters could include examining 'ecological affinities' (Neale *et al.*, 1995) between native fauna and target host by appraising species in the same feeding guild or site. For example, Neale *et al.* investigated specificity of parasitoids of a mining lepidopteron introduced to Australia in part by screening other miners in the community. This approach ignores phylogenetic affinity. Barratt *et al.* (1998) estimated the taxa most immediately at risk from an introduced agent by looking at their co-occurrence with the target in field surveys. These three principles (phylogenetic affinity, ecological affinity, immediacy of risk) may help to guide formation of a priority listing of test species, which should also include beneficial species (Barratt *et al.*, 1999). Need for additional testing may be appraised following an initial test series, but the overall number of species to be included in screening any individual agent will vary on a case-by-case basis.

Hopper *et al.* (1993) noted that even when an agent has been deemed safe and 'worthy' (in the belief that it can perform the anticipated role), unforeseen or uncontrolled variation in the agent population may influence its actual performance after release – so that an additional need for post-release monitoring (as well as detecting possible changes in host range) is to determine how and if it 'works'. Sources of such variation

Table 5.4. *Sources of intraspecific genetic variation in arthropod biological control agents that might influence agent establishment and subsequent performance (Hopper* et al.*, 1993)*

Interstrain variation	The status of an agent from different parts of its geographical range, or reared from different host species or strains, may differ
Common environment	Differences in strains may persist when they are reared in the same regimes/conditions in the laboratory or production facility
Selection	Differences may occur between artificially selected populations, such as mass-reared stocks for release. Differences can occur between 'selected' and 'control' populations
Resemblance between relatives	Differences in variation may occur in quantitative traits within families versus between families
Crosses	Differences may occur among progeny of crosses between different families or strains

are often difficult to clarify, but those noted by Hopper *et al.* are listed in Table 5.4. These factors may need to be considered in assessing reasons for success or failure of an agent in a control programme; but the fundamental point is that wild or 'domesticated' agent populations possess capabilities to change or evolve in response to local environmental conditions (Carriere & Roitberg, 1996). These potential capabilities, paralleling those noted in Table 5.4, include (1) phenotypic variations, (2) genetic components of that variance, (3) consistent covariance between trait value and fitness and (4) limited gene flow between populations associated with different hosts, allowing evolutionary divergence in response to local ecological conditions (Roitberg, 2000).

Screening of herbivores for weed control

As noted above, screening protocols for weed-control agents, most of them herbivorous insects, have been developed into more satisfactory protocols than those for parasitoids and predators of many arthropod pests. Plant-feeding insects may select their hosts at one or other of two stages in their life history, so that effective screening must account for both of these. First, the oviposition habits of the female may dictate the

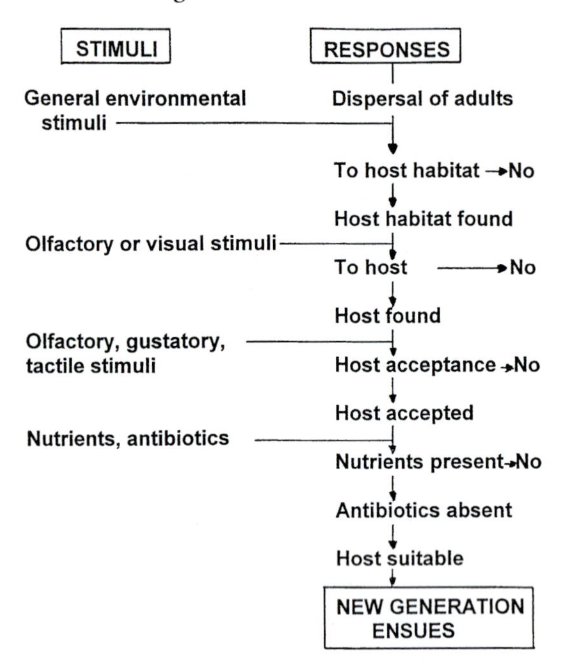

Figure 5.2. Host selection behaviour by phytophagous insects: ecological principles (Kogan, 1982) and the sequence of stimuli and responses that may occur.

food immediately available to hatchling larvae, so that precise deposition of eggs on a particular plant species, and the search cues involved in this selection, are an important component of host specificity. Second, and often linked with less selective oviposition by the parent, the feeding larva may select or reject particular plants that it encounters, so that screening larval selection is also needed. Host selection is a complex suite of processes and may involve a great variety of different stimuli (Fig. 5.2) (Kogan, 1982).

As for arthropod pest agents, host specificity testing protocols should address 'maximum likelihood' host range, and both choice and no-choice trials are used. Sheppard (1999) showed that, despite strong protocols, considerable differences in testing procedures occur – with, for example, no-choice feeding trials being the main standard for North America. Hill (1999) argued that no-choice tests may indeed be more relevant for small, passively dispersed and generally inactive agents attempting to discriminate between large test plants, whereas choice tests could be regarded as more appropriate for relatively large and active agents, capable of sampling a variety of different host plants with relative ease. Much of the debate is based on protocols designed by Wapshere (1974, 1989, 1992) and draws on several basic principles that a potential agent should be tested against:

- Plants related closely to the target weed, usually congeneric species, and genera in the same tribe or family.
- Plants from which the candidate agent has been recorded or reported.
- The host plants of consumer species related closely to the candidate agent.
- Unrelated plants that have morphological or chemical features in common with the target weed.
- Representative crop plants grown in the area in which weed control is contemplated, to help allay fears from farmers about possible agent damage to crops.
- Plants of significant conservation value or relevance in the proposed release area.

If any such trials raise possibilities of harmful effects, additional screening tests involving plants related to the vulnerable taxa may be advisable (Harley & Forno, 1992).

Wapshere (1974) initially suggested the approach of 'centrifugal testing', whereby close relatives of the target weed are tested first, followed by progressively distantly related taxa, and later (Wapshere, 1989) designed a 'reverse-order' testing sequence (Fig. 5.3), with the following sequence:

1. Undertake forced feeding tests using larvae of the prospective agent, if these are the main feeding stage, as in most relevant species of Coleoptera, Lepidoptera and others. Plants on which the larvae can develop are clear candidates for inclusion in the agent's host spectrum.
2. These 'included plants' are tested in enclosed conditions to determine whether they are acceptable by the adult insect for oviposition, under no-choice conditions.
3. If either (a) larval development does not occur or (b) adults will not lay on the plants, then those plant species are considered 'highly unlikely' to be vulnerable in the field.
4. If, on the other hand, trials suggest host suitability, additional trials are undertaken in more natural conditions. These confirm (or otherwise) that the plant is within the natural range of the agent or that the laboratory acceptance was simply due to some more artificial effect such as 'desperation' (see p. 00).

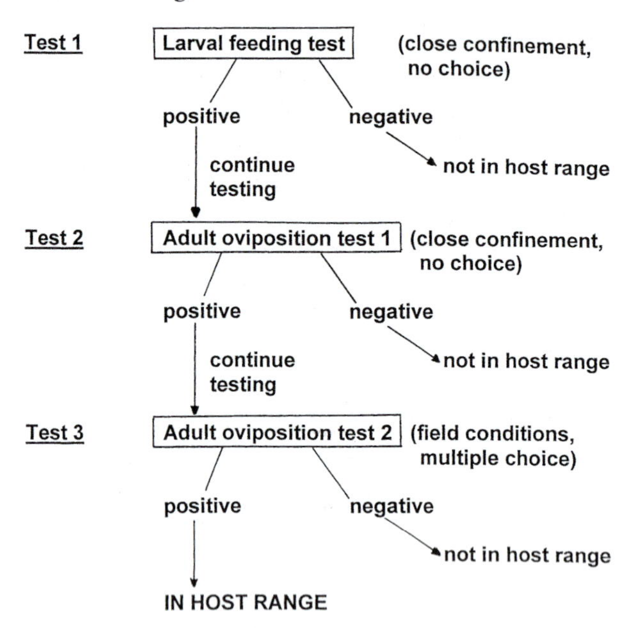

Figure 5.3. Reverse-order testing sequence for weed biological control agents (see text for explanation) (Wapshere, 1989).

This scheme recognises that, as for natural enemies of pest arthropods, the field 'ecological host range' may be much less than that indicated under the much more stressed conditions of laboratory cage experiments. No-choice tests on herbivores frequently reveal a considerably larger potential host range than is implied from choice tests alone. In such instances, emphasising the need for safety could lead to rejection of potentially valuable agents that pose no significant threat to non-target plants in nature (Briese, 1999), so that some workers advocate the use of open-field specificity tests to help overcome the numerous problems caused by confinement (Clement & Cristofaro, 1995). So-called 'continuation tests' can be undertaken in no-choice trials to determine whether non-target species can support the agent over successive generations. The basis for this is the implication that the more generations sustained, the greater is the likelihood that the plant species will be a true host (Day, 1999).

Additional background on the rationale, design and conduct of screening tests is included in Withers *et al.* (1999), van Driesche *et al.* (2000) and Gurr & Wratten (2000), among others.

Single or multiple agents?

Each agent species introduced may be regarded as a potential risk to the receiving environment, so that with greater numbers of agents the perceived risk increases. Many CBC practitioners favour the introduction of several agents against the same target pest, should they be available and be deemed suitable, in contrast to relying completely on the single 'best' agent species. On the one hand, a single agent lacks competition from other introduced consumers seeking the same resources and may thus expand to dominate the limited consumer guild. On the other hand, the anticipated outcome of a multiple introduction is that although different agents may compete with each other at times, they will produce a greater collective impact on the target pest than any single agent could inflict by itself.

Louda *et al.* (1998) suggested that the optimal introduction strategy would be to release 'the fewest and most effective agents with the lowest probability of non-target effects'. Another perspective likens CBC to a lottery, because of the difficulties of predicting how any agent will behave after introduction, with the outcome so dependent on chance that the best way forward is to import many species (after satisfactory screening tests) and leave them to sort out which, and in which combination, may be the most effective and capable of controlling the pest (McEvoy & Coombs, 2000).

McEvoy & Coombs (1999, 2000) identified two syndromes that have arisen under the 'lottery model' of CBC introductions (Fig. 5.4). First, the 'runaway importation rate' is indicated by the graph shown in Fig. 5.4a, where the number of agents introduced diverges strongly from the number of control target species (the example shown is from imported control agents against plant weeds in Oregon), which McEvoy & Coombs (2000) attributed to the increased pressures to pursue CBC, with the consequence that some CBC systems are becoming ever more complex in structure. Possible explanations, not mutually exclusive, include that (1) weed problems are becoming harder to solve, (2) scientific effort is not responding to meet increasing demands for agent safety, and/or (3) the learning curve may be getting worse. As McEvoy & Coombs put it, 'we may be practising the wrong things, speeding up the screening for new introductions for safety while neglecting to efficiently screen species for effectiveness'.

Second, the 'monitoring and evaluation gap' syndrome represents the heavy skewing of many CBC programmes to pre-release phases (including

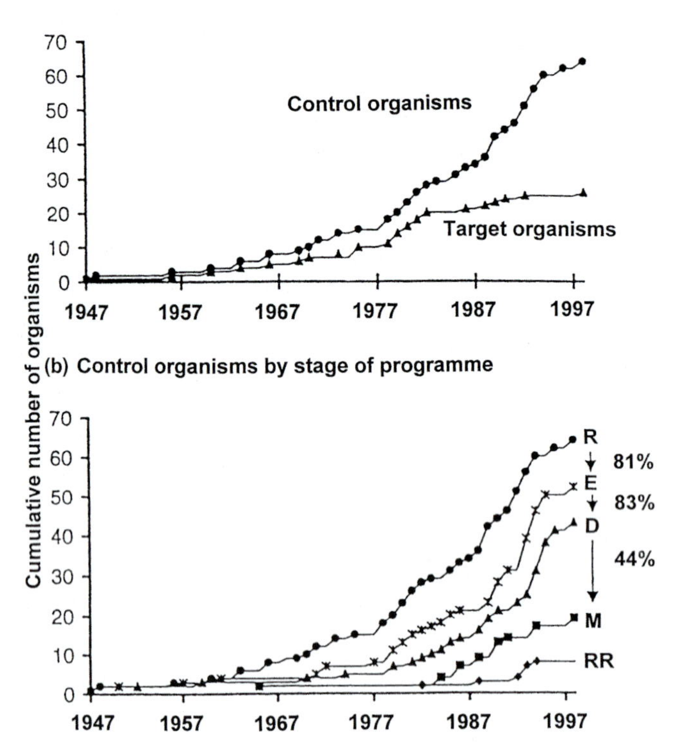

Figure 5.4. Some consequences of the lottery model of classical biological control (McEvoy & Coombes, 1999), with data on cumulative number of imported classical biological control agents for weeds in Oregon, USA (1947–98): (a) cumulative numbers of target species and control agents show strong divergence; (b) cumulative numbers of control agents progressing through sequential stages of release (R), establishment (E), redistribution (D), evaluation by monitoring (M) and re-release or repeated attempts (RR). Percentages at end of graph indicate success rates, thus 81% of released organisms became established, etc.

screening for safety) whilst paying much less attention to later effects on both target and non-target species. Assessing effectiveness of introduced agents goes hand in hand with reducing the number of different agents needed in any programme. However, a major deterrent to pursuing this properly is that full evaluation of any CBC agent (despite occasional spectacular early effects) may take about 10–20 years (McFadyen, 1998); funding for such long-term projects – especially if they do not show early promise – is rarely forthcoming. Thus, as McEvoy & Coombs (2000) also noted, many additional agents are being introduced before earlier introductions have been given adequate time to appraise their success. Unpredictable outcomes tend to favour the 'lottery approach', with the possibility of 'revenge effects' (see p. 180) arising from CBC and rendering it less effective and more risky. Screening for safety is indeed a key issue in agent selection, but moves toward selecting also for effectiveness are of fundamental importance in reducing non-target effects of concern for conservation.

Agents may act additively, synergistically or antagonistically, with little possibility of practitioners being able to predict the outcome of multi-species introductions except by direct field study (Roy & Pell, 2000). Additive interactions, apparently rare, assume non-interference between agents, so that their combined effect is the sum of the mortalities caused by each individual species. Synergistic effects imply a higher mortality than additive effects, facilitated by interactions between the agent species. Antagonistic effects imply the converse, with interference reducing the potential impacts of each or some agent species present. The last of these three options is often assumed and is the rationale often advanced for introducing a single-agent taxon. However, experimental validation of such effects is rare. One recent case (Rosenheim, 2001) found that one species (the green lacewing *Chrysoperla carnea*) acting alone amongst the numerous predatory taxa present in California cotton fields can suppress cotton aphid populations effectively. The full complex of predators acting together exerted only minimal suppression. Although this case deals mainly with native predators, and thus is not CBC, the conclusion that predator–predator interactions appeared to be the main factor restricting control is a salutary indication of unsuspected or undetected interagent effects that may be present in many biological control projects involving multiple species.

A slightly different alternative to the unrestricted lottery approach has been to select the 'best' members of different feeding guilds, or natural enemies of different target life stages, in the hope (or expectation) that these will show maximum complementarity with minimal interference effects. Thus, agents selected against a weed pest might include anticipated complementation of a foliage chewer by introducing taxa such as a root feeder, seed eater, flower-bud galler or stem-sap sucker; and for an arthropod pest, an egg parasitoid might be complemented by specialist larval and pupal parasitoids, should these be available. Each of these cases might avoid the kind of interference that could occur if, for example, three different foliage chewers or three different egg parasitoids were introduced together, by spreading the effect across different but more exclusive host parts or stages. However, the more different control agents that are introduced – in any CBC context – the greater is the challenge of monitoring their effects, 'increasing the likelihood that target *and non-target* effects go undetected' (McEvoy & Coombs, 2000; my emphasis).

The above discussion indicates the concerns that play a role in making rational decisions over which species to introduce and how many are needed. The experimental approach to CBC (i.e. trial and error) is not predictive and largely ignores ecological complexities (Ehler, 1990, 1999). But replacing this with a more predictive system (involving optimising which species or combination of species to introduce, and the impacts and wider effects of that/those species) remains a major challenge. As Ehler (1999) noted, an additional dilemma for practitioners is that the most specific (and, thereby, presumed safest) natural enemies are not necessarily those with highest impacts on the target pest. Multiple releases of such different species could 'clutter the environment with agents that might provide (at best) an intermediate level of control' (Ehler, 1999). The presence of such a complex might well hinder the potential success of better agents introduced at some later stage because of the potential they provide for interference and other impact-lowering interactions.

Native biological control agents

Emphasis on CBC, and the recent attention to neoclassical biological control (see p. 181), has stressed the values and roles of exotic natural enemies and the problems that these might cause. Another facet of biological control, in which the need for complex and unwieldy screening trials is removed, is to enhance the use of native natural enemies already present as residents in the regions where pest problems need attention. Such local species, mainly generalist predators and parasitoids, are almost invariably present, and many of them are amenable to manipulation in various ways – such as to attract and concentrate them on crops or to mass-rear them for release. Such manipulations anticipate the flexibility of the agent to be able to exploit the pest and to increase its abundance and co-occurrence with the target. Such agents can therefore obviate the need for sometimes demanding regulatory assurances in order to introduce exotic agents, together with the burden of proof for their safety. The topic of conservation biological control, involving habitat manipulations to conserve native natural enemies, is dealt with later (Chapter 6), but some of the general strategies and possible outcomes that might lead

Table 5.5. *Features that make some natural enemies more suitable than others for use in augmentation programmes (DeBach & Hagen, 1964)*

Development should be synchronous with that of the target pest, so that seasonal synchronisation for control does not need to be manipulated.
The species must be reared easily and economically in the production facility, perhaps in sufficient numbers and over extended periods to provide periodic releases.
The agent must thrive under the conditions in which it is to be released.
The agent should not attack other beneficial species (agents) operating in the same system or other agents likely to be introduced for the same purpose.
If non-pest species are present in the system, the agent should be specific to, or prefer, the target pest.

to risks in promoting native generalist agents are noted here to enhance the general theme of biological control and its wider ramifications.

Many native natural enemies, in addition to CBC agents, are enhanced routinely by deliberate augmentation, such as release of captive-bred stocks. Many species are produced commercially, with varying degrees of quality control, and are thus easily available for growers to distribute on their crops. They may thereby be introduced to particular areas where they are naturally sparse, or that constitute extensions to the natural ecological range, in substantial numbers. The aim of much augmentation is to provide superabundant natural enemies to impose rapid, short-term suppression of the pest, rather than to constitute a more permanent change in equilibrium; in the broader sense, 'augmentation' is any activity that has the primary purpose of increasing the impacts or abundance of natural enemies (DeBach, 1974; van Lenteren, 1986). The approach is justified in several different contexts (DeBach & Hagen, 1964), and particular natural enemies may be preferred over others because of any of a variety of biological or husbandry-facilitating features (Table 5.5). The contexts of particular appeal or suitability for augmentation are (DeBach & Hagen, 1964):

- Natural enemies are potentially effective but, in practice, ineffective because of environmental conditions that cannot be addressed easily in other ways.
- The pest is not controlled easily by any other means.
- The level of control needed may not be attained easily by any other means.

- Other methods of control, although feasible, are undesirable – for example, because of persistence of chemical residues.
- Treatment is necessary for only one or two members of a pest complex, and specific natural enemies are available to attack those key pests.

Such agents may be released in very large numbers, even by broadcasting them aerially over crops, and repeated releases (perhaps several times over a growing season) are not unusual. Such so-called 'inundative' releases are likened to applications of a biological insecticide (see p. 183), with the aim of suppressing the pest within the same generation and with little attention to any future control needs. The approach is particularly efficient for univoltine pests, for which optimal timing of a single release may be sufficient.

In contrast, smaller-scale 'inoculative releases' may comprise localised, spot releases from which agent populations develop and spread within the crop area over a longer period. They may thereby be effective over several pest generations within the same season. Some such agents may not survive periods in which the pest is absent and may need to be re-established in the following period of pest incidence. However, their natural biology predisposes them to revert to natural habits and survive on any available non-pest hosts in what is basically a natural community and environment for many of them.

The consequences of such augmentative releases are (sometimes massive) increases in the local abundance of natural enemies – these often being mass-reared under closely regulated standardised conditions – with the potential to attack other, non-target species should opportunity to do so exist. The very large numbers involved clearly pose the possibility of increased non-target impacts. However, Howarth (2000) suggested that in most cases, uses of native agents would not lead to changes in populations of native non-target species because the system should return to pre-treatment levels after management has ceased. Notwithstanding this supposition, inundative releases in particular can result in highly elevated populations of local agents, and these should be monitored for non-target effects.

'Quality control' in mass-rearing of agents has two important components: maintenance of biological capability and assurance of taxonomic integrity. Commercial rearing facilities, of *Trichogramma* wasps for example, may produce several hundred million individuals a week (Smith, 1996) but range in scale from large centralised facilities to small individual operations. Rearing conditions may involve unnatural standardised climatic regimes to avoid onset of diapause, use of artificial or semi-artificial diets, use of prey reared on unnatural host plants, and many other components changed from the animal's former wild lifestyle as a result of needs for logistic or commercial expediency and enhanced rearing success. Each of these numerous factors may influence the agent's performance or field capabilities or preferences in ways not reflected in lessening of numbers produced. Routine measurements for quality control in commercial production units tend to be 'production-oriented' (King *et al.*, 1985), involving maintenance of numbers produced per unit cost (so involving measures such as fecundity or emergence rates) rather than 'behaviour-oriented' or 'genetic', perhaps more relevant to how an agent may perform after release. A biologically labile agent reared in captivity over many generations may not have wholly the same attributes as those in the original wild parental stock. At the least, there may be some suspicion that some changes in host preference might occur and affect its use against the target species.

Occasionally, agents are misidentified, so that mass-reared individuals purported to be a particular native species may in fact be another species or even an exotic one. The green lacewing complex *Chrysoperla carnea* (see p. 161) comprises a number of sibling species structurally very similar and difficult for a non-specialist to differentiate. Their variety and wide use for augmentative control has fostered widely the idea that 'green lacewings' are all good biological control agents. The different distribution of sibling members of the complex in Europe, where relatively free exchange of agents between countries occurs, has led, for example, to taxa from Italy being introduced into parts of northern Europe where they do not occur naturally. Interpretation of biocontrol outcomes based, however inadvertently, on the wrong agent species may (at the least) be misleading as a predictive precedent. A recent survey (O'Neill *et al.*, 1998) of reared samples of '*C. carnea*' purchased from three commercial insectaries in the USA showed all to be *C. rufilabris*, a species not included in the *carnea*-complex and biologically very distinct from the species in that group. As another example, misidentification of the parasitoid wasps *Aphytis lingnamensis* and *A. melinus* as *A. chrysomphali* thwarted early attempts to control red scale on citrus in California (Rosen & DeBach, 1979). Such cases emphasise the importance of sound systematic appraisal to underpin selection and use of biological control agents and to help predict possible risks of their use.

Pre-emptive classical biological control

This practice involves the introduction and establishment of agents to control important key pests that are not yet present in the area of agent introduction but that are anticipated to arrive. It thus involves the deliberate introduction of exotic agents that are expected to form self-sustaining populations on alternative hosts or prey and 'be ready' to control the pest, should it arrive at some future time. This exercise has been promoted or undertaken at two different levels. First, Ehler (1998) suggested the introduction of selected generalist predator species at likely points of entry and other high risk-areas as a deterrent to establishment of pests that might arrive. This strategy was acknowledged as having environmental risk.

Second, more precisely targeted purposes are exemplified by Hughes *et al.* (1994, p. 144), who introduced the parasitoid *Aphelinus varipes* to Australia in the anticipation of subsequent arrival of its preferred host (Russian wheat aphid, *Diuraphis noxia*) as a potentially highly damaging pest to cereal crops. In this case, the wasp was considered highly unlikely to attack the small native Australian aphid fauna by invading natural ecosystems.

Documenting and appraising the risks of classical biological control

CBC involves the permanent addition of new species to the receiving environment. As Holt & Hochberg (2001) have emphasised, this process is 'deliberate community assembly' and subject to any rules of natural community assembly involving establishment, impact and the wider landscape context of the introduction. Each of these facets may need to be considered in attempting to predict the risk that an introduced species may provide.

Ehler (1999) believed that a successful CBC agent should have three main attributes:

- colonising ability, allowing it to keep pace with disruption of the habitat, for example from periodical harvesting or tillage;
- temporal persistence, so that the agent can maintain its position even when the target pest is scarce or absent;
- opportunistic foraging, enabling the agent to rapidly exploit a pest population.

These very features cause concern, and they imply needs for an agent to have some 'ecological flexibility', sufficient to attack non-target taxa and to stray from the immediate arena of interaction with the target. Risks may arise directly from such flexibility. Holt & Hochberg (2001) delimited some of the interactions of concern as 'community modules', illustrating various ways in which an agent may participate in trophic interactions with non-target species. These modules extend the more common single feeding links of a conceptual food chain to other species, whilst avoiding the massive complexity brought about by studying whole, complex food webs. The authors pointed out that such modules are 'building blocks', but ones that can be studied far more easily in laboratory or field – both contexts in which focus on a limited variety of non-target species may be predominant. The major modules of concern in assessing indirect effects of CBC (that is, those effects on species other than the intended target) are summarised in Fig. 5.5 to illustrate the variety of those concerns.

In practice, the 'shared predation' module emphasises structures in which the agent attacks (or can attack) two or more prey species. Extending this conceptually makes this relevant also to contexts such as (1) a herbivore attacking two or more plant species, (2) a parasitoid attacking two or more host species, (3) a hyperparasitoid sustained by two primary parasitoid species, (4) a pathogen (see p. 183) capable of attacking two or more hosts, or (5) a predatory arthropod attacking several prey arthropod species. Generalising to 'shared natural enemies', as above, reveals three possible situations where risks may arise in CBC. These (Holt & Hochberg, 2001) are:

- An introduced specialist agent may be consumed by resident generalist natural enemies. Although this is not specifically clear from Fig. 5.5, it emphasises that indirect interactions between species at one trophic level can be mediated through responses by a higher trophic level. An introduced agent sustained at higher levels of abundance on the target species could indirectly enhance the diet of resident natural enemies, which in turn attack other, resident species more effectively. As specific examples (both discussed by Holt & Hochberg, 2001): (1) experimental introductions to southern Finland of a braconid parasitoid of pierid butterflies were made into populations of a congeneric wasp that attacks a nymphalid butterfly (*Melitaea cinxia*), leading to greatly increased levels of attack by a generalist ichneumonid hyperparasitoid, so increasing attack rates on the native parasitoid up to the point of local extinction

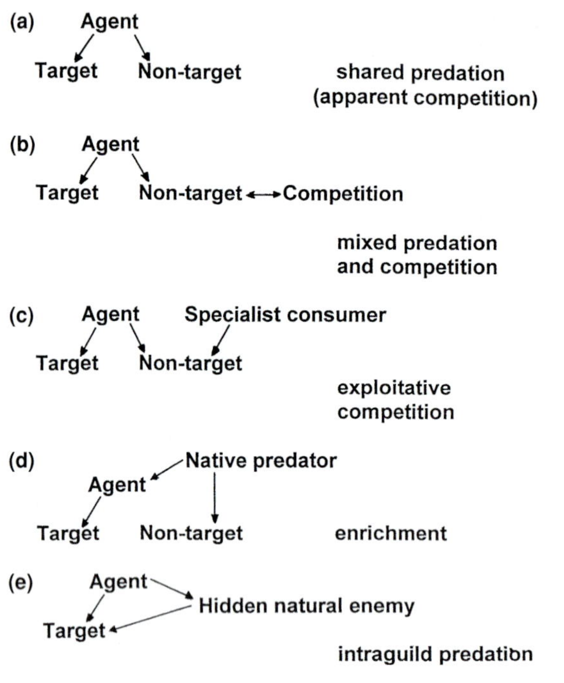

Figure 5.5. Approach of community modules for assessing impacts of classical biological control effects (Holt & Hochberg, 2001). In this sense, a 'module' represents the interactions among a small number of species, and lines indicate that two species interact. (a) Impacts on non-targets reflect interactions between agent and target; (b) impacts on non-target aggravated in presence of competing species; (c) agent exploits non-target species that is required by another non-target consumer; (d) introduction of agent enriches diet of a native predator, with impacts on non-target prey; (e) agent both competes with and attacks a non-target natural enemy.

(van Nouhuys & Hanski, 2000); (2) a possible additional risk of an introduced herbivore on weeds is that the abundance of native inquilines and parasitoids can be enhanced and place an array of native hosts at risk (Schonrogge *et al.*, 1996).

- An introduced agent may attack one or more species other than its putative target, so providing the most immediate concerns over its effects.
- The risks of shared predation can transcend habitat boundaries, so that a predominant habitat (such as an agroecosystem) in which the agent interacts with its target can provide sources of individuals that may invade other habitats in which only non-target species are present.

Effects of the latter two situations are linked with the population dynamics of the agent–target interaction and the whole gamut of laboratory and field predator–prey studies. The theoretical implications that arise from these may thus all be relevant in predicting possible outcomes. If other factors are equal, then non-target species may be most at risk when (1) the non-target has a low intrinsic population growth rate, so that replenishment of losses is slow; (2) the natural enemy has a high attack rate, even when the non-target is rare; and (3) the target species can sustain high population levels of the natural enemy. Empirical studies have revealed a wide variety of outcomes (see Holt & Hochberg (2001) for discussion), with modelling exercises sometimes showing features such as transient risks of extinction of non-targets during periods of agent establishment when the species have not yet reached equilibrium. The theoretical exercises discussed by Holt & Hochberg led them to suggest several rules of thumb that should be considered when evaluating risks from CBC agent (Table 5.6).

Modelling exercises have considerable appeal in attempting to predict outcomes of CBC introductions, but there are obvious limitations (based on incomplete biological knowledge) in the extent to which they can be used in the absence of *post hoc* demonstration of the results. Gurr *et al.* (2000) noted that most modelling has emphasised likely effects on the target species rather than on non-targets, in large part reflecting the far greater knowledge available on the pest target and the relative lack of focused demand for other approaches, especially until recently. Published models of CBC systems have indeed been important in augmenting knowledge of those particular systems and, more generally, in adding understanding to the wider discipline.

Although the potential long-term non-target effects of CBC agents are the more frequently expressed concern, Lynch *et al.* (2002) have pointed out that they may also cause transient impacts more akin to 'pulse' disturbances and resulting in severe effects over a short period following their introduction. This situation might occur even for non-target species that are only weakly accepted by the agent, so that pre-release laboratory testing might imply its safety for release. However, Lynch *et al.* (2002) noted that large impacts on the non-target might occur soon after release for the simple reason that the agent population is likely to increase substantially in response to a very high density of its target, and this high density might dominate its low acceptance rate to cause strong declines or local extirpation of non-target species. This supposition was supported by models of the host-parasitoid dynamics,

Table 5.6. *Possible rules of thumb arising from theoretical studies of shared predation and that merit consideration in evaluating possible risks from classical biological control (CBC) agents (Holt & Hochberg, 2001)*

1. A control agent only moderately effective against the target may be more abundant than an effective agent and so pose a greater risk of incidental attacks on non-targets.
2. Risk of extinction of a non-target is enhanced if it has a lower intrinsic growth rate or suffers a higher attack rate from a shared predator than the target species Measuring magnitude of attack does not fully characterise risk to non-targets.
3. Risk to non-targets is greater if the agent experiences little density-dependence due to factors other than its own resource availability. Absence of such factors increases agent numbers sustained by the target species, increasing risk of exposure to non-targets.
4. Risk to non-targets is greater if predators have high maximum attack rates at high prey abundance, inflating indirect effects of target species on non-targets, mediated through shared natural enemy.
5. If points 3 and 4 above both occur, then greatest risk to non-targets arises if target can sustain high production with predation.
6. If points 3 and 4 above do not hold, then risk to non-targets may be greater at intermediate abundances of alternative prey.
7. Non-target species that can coexist with the agent at equilibrium may risk extinction during agent establishment phase.
8. Even if target and non-target do not co-occur in same habitats, risks are present. This is particularly serious for non-target species in habitat fragments intermingled with habitats occupied by the target species, when the target is highly productive and the agent is only moderately successful in controlling it.
9. Use of specialist natural enemies does not necessarily preclude all negative effects, because the specialists can themselves be prey for resident generalist enemies, whose numbers and wider impacts may then increase.

but models also indicated that this risk might be reduced considerably. Non-target hosts cannot, in these cases, support the agent population for long but may suffer a severe transient spill-over shortly after agent introduction. The sheer numbers involved, even with low propensity to attack, may be sufficient to cause extirpation.

'Acceptability of risk is a societal issue, which may evolve with changing societal values and scientific knowledge' (Hopper, 2001). Similar realisations on the needs for effective compromise (e.g. Samways, 1997) lead to hard questions as to what risks are acceptable, with the actual risks intangible or predicted from what are essentially little more than crystal-ball-gazing exercises, despite continuing attempts to improve on this situation. The most common assessment of a CBC project is simply 'it worked' or 'it didn't work' (Prinsloo & Samways, 2001), rather than any wider appraisal. Prinsloo & Samways made this comment in the context of assessing host specificity of chalcidoid wasp parasitoids introduced to South Africa as CBC agents. Their survey implied that only 7 of 45 species are apparently monophagous – 'apparently' because research has not been comprehensive, whilst the remaining 38 species are known to have wider host ranges. The eight species of *Trichogramma* amongst these are known to attack collectively at least 60 species of Lepidoptera, including members of all five main butterfly families. However, the only direct evidence of introduced chalcidoids adapting to non-target native hosts in South Africa involves three species of Aphelinidae (*Aphytis*) attacking five endemic diaspidid scale insects (Prinsloo & Neser, 1994) – a case nevertheless sufficient to establish the reality of host-switching in the receiving environment.

The major general conclusion from the discussion above is that generalised risks from CBC can be proposed easily as a consequence of adding exotic species to a new, favourable environment and fostering their wellbeing and increase for pest suppression. Wider 'community effects' (as illustrated in the various modules in Fig. 5.5) are perhaps almost inevitable, may be widespread, and inherently suggest possibilities of 'risk' (however the term may be interpreted) to the receiving environment and the native species there. The actual nature of this risk arising from the permanent establishment of CBC agents is much harder to appraise, and it is pertinent to focus more closely from this very broad scenario by examining a number of the cases that have caused specific concern to conservationists. These arise directly from the spread of agents beyond the crop environments or other areas in which the target pests are present and the documented or inferred detriment to native taxa that has, or may have, resulted. Substantial numbers of non-target effects may occur (Table 5.7), but many of these may be transient or unimportant.

Howarth (2000) tabulated some 60 titles of accounts relating to non-target effects of biological control-agent introductions. Some are very general, but many present evidence and discussion pointing to incidences of population declines, actual or potential local extinctions, or global extinction and wider community effects. Many of the most severe effects have resulted from introductions of CBC agents to remote island ecosystems, particularly in the Pacific region, as especially vulnerable habitats to invasion. Continental areas have often been regarded as more resilient to invasions, although this is by no means universal. Continental systems

Table 5.7. *Indication of incidences of non-target attacks by biological control agents (after Stiling & Simberloff, 2000)*

Agent category	Number of species introduced	Attacking non-targets		Area	Reference
		Number	%		
Parasitoids	313	50	16	USA	Hawkins & Marino (1997)
Arthropod herbivores	33	7	21.2	Worldwide	Turner (1985)
Parasitoids and predators	40	15	37.5	Canada	McLeod *et al.* (1962)
All agents	243	33	13.6	Hawaii	Funasaki *et al.* (1988a)

may be biologically resilient, or the balance of the outcomes may simply reflect the very large number of CBC introductions to islands. Or, as Samways (1997) noted, the more complex continental environments may simply render detection of specific effects much more difficult, even when they do occur.

Much of the impetus for recent concerns over CBC arose from Howarth's (1983) formulations of concerns for the Hawaiian islands (see also Howarth (1985), Howarth (1991) and Gagné & Howarth (1985) for further expressions of this concern over non-target effects), which catalysed intensive (and sometimes emotional) debate that continues to the present day and involves the polarised views summarised in the opening section of this chapter with comments such as 'there appear to be no rigorous studies showing that arthropods intentionally introduced for biological control have reduced abundances, let alone caused extinction, of native non-target species' (Hopper, 1998). The conservationist's dilemma is thus to appraise the conflicting claims and counterclaims (Lockwood, 1996), opinions and extent of evidence (balanced or biased) presented as realistically as possible, with the realisation that 'the conflicts of interest concerning the environmental impacts of pest control will continue to intensify' (Howarth, 2001). Concerns are exacerbated by continuation of the short-term approach of introducing any or all agents and leaving them to interact in the new environment, particularly if these are generalists, rather than pursuing effective (even if expensive) screening tests before release.

The broad approach is exemplified by Walker *et al.* (1997), who introduced about seven species of generalist egg parasitoid wasps from Asia for control of sugar-beet leafhopper (*Circulifer tenellus*) in California.

Howarth (2000) noted the contrast between (1) risks from individual introductions resulting from neglect (such as lack of screening, as above) and (2) risks from introductions in which the risks have been evaluated and accepted. 'Our society has invested very little in the science of understanding collateral damage to native insects caused by biological control agents' (Strong & Pemberton, 2001).

'Risk assessment' was considered by the National Academy of Sciences (NAS, 1993) (formulated primarily for detecting risks of chemical exposure to humans) as a four-stage process, and that approach was developed in the contest of biological control agents for weeds by Lonsdale *et al.* (2001). The NAS formula has thus been the basis of much wider context evaluation, as follows:

- Hazard identification: identification of the kinds of danger/risk that might occur, for example the feeding of an agent on non-target host or prey.
- Dose–response assessment: the relationships between exposure and the harm that results. This parameter involves estimating the degree of damage based on perceived input from pre-release studies. It takes the form of 'If the agent attacks, what will be the result?' rather than the generalised scale of 'What damage will a given density of the agent cause?'(Lonsdale *et al.*, 2001).
- Exposure assessment: estimation of how much of the non-target(s) will be exposed to the agent, for example in weed control by climatic modelling to determine exposure through overlap of range.
- Risk characterisation: combining information from the above three stages to estimate overall amount of damage that will be caused.

As Lonsdale *et al.* (2001) stated, assessment (even in the more ecologically focused form of NAS (1993), as discussed by Lipton *et al.* (1996); see Table 5.8) depends on predictive capability, which is inherently difficult. Problems of ecological risk assessment remain also because it is very difficult to be explicit about what might really be predictable (Louda *et al.*, 2003a, 2003b). For example, Louda *et al.* (2003a) suggested that host specificity and individual behaviour cannot predict ecological host range and impact on native populations under natural conditions.

Table 5.8. *An ecological risk-assessment framework for biological control agents, building on the format of the National Academy of Sciences (NAS, 1993) (see text) (Lipton* et al. *1996, as modified by Lonsdale* et al. *2001)*

Process	Comments
1. Receptor identification	Characterise components of system and determine those most vulnerable to disturbance of the system. Modelling (such as by a flow chart) may be useful in tracing non-targets and trophic levels potentially affected.
2. Hazard identification	Identify hazard for species identified in stage 1, and the system level (individual, population, community) at which hazards operate.
3. Endpoint selection	Choose species/system levels sensitive to hazard to use as indicators for monitoring.
4. Relationship assessment	From ecological knowledge of the system, evaluate feedback loops to identify secondary/tertiary effects/risks. If new receptors identified, necessary to repeat stages 2 and 3; stages 2–4 together are an iterative process to identify potential risks.
5. Response assessment	Explore consequences of low, medium or high density of agent (original NAS indicated use of dose–response curves, but equivalent data not usually available for CBC agents).
6. Exposure assessment	Attempt to predict range of CBC agent, and overlap (space, time) with receptors.
7. Risk assessment/uncertainty analysis	Combine all of the above to evaluate/ encapsulate degree of risks posed to ecological endpoints.

CBC, classical biological control.

Usually, information on the population consequences is not available, or the likely interactions are not understood. Louda *et al.* implied the need for further information on a variety of difficult-to-measure parameters, such as (1) dependability of host specificity test outcomes, (2) host choice with various combinations of resource availability, (3) population responses of the agent after dispersal into areas lacking preferred target species, and (4) responses to changed environmental conditions and phenology/seasonal variability of resources, among others. In general, native relatives of target species are most vulnerable to non-target attacks, but the 'expert opinion' on the risks suggested by specificity tests is sometimes insufficiently informed.

A major problem with elucidating the roles of CBC agents in events such as non-target extinction (or extirpation) is that most of these losses have not been monitored, and detailed studies of such cases have generally commenced only after the species has disappeared or declined substantially in either or both of range and abundance. Many severe impacts have simply not been recorded (Simberloff, 1992), but Howarth (2000) claimed that more than 100 documented extinctions of non-target species have been caused by biological control introductions. Many of these are not strictly agricultural in orientation, but cases such as the losses of land snails on Pacific islands such as Moorea (*Partula*) and Hawaii (where more than two-thirds of some 750 species are extinct or endangered), with these losses attributed to the introduced predatory snail *Euglandina rosea* (e.g. Clarke *et al.*, 1984; Murray *et al.*, 1988; Hadfield *et al.*, 1993; Cowie, 1992), are a salutary indication of the unexpected effects of exotic generalist invasive species, as, soberingly, introductions of *Euglandina* to new areas apparently continued well after its effects were understood (Howarth, 2001).

Exotic parasitoids in the Hawaiian islands

According to Gagné & Howarth (1985) (following Howarth, 1983), parasitoids introduced deliberately to the Hawaiian islands have contributed in a major way to the decline and extinction of several species of endemic Lepidoptera. Surveys by Asquith & Miramontes (2001) endorsed this viewpoint, which has prompted fervent debate and is perhaps the single most important accusation casting doubt on the safety of CBC in recent years. Asquith & Miramontes (2001) used Malaise traps to capture Ichneumonidae and Braconidae in mesic/wet forest on Kauai. Both of these wasp families are represented poorly in the native biota, and a high proportion of the two-year catch of 2017 individuals from natural environments was composed of exotic species. In the total of 17 species (one braconid, 16 ichneumonids), two CBC agent species comprised 45% of the total wasps. These are the braconid *Meteorus laphygmae* (introduced in 1942) and the ichneumonid *Eriborus sinicus* (1928). Seven adventive species (below) were also captured, with *Pimpla punicipes* comprising a further 34.9% (n = 703) of the wasps. These catches illustrate well the expansion of the Hawaiian fauna by exotic species, as illustrated by the increase in Braconidae from only two native species to a mid-1990s total of 76 species in 42 genera. Many of these are categorised as 'adventive', and several workers (such as Funasaki *et al.*, 1988a) have argued that it is these exotic taxa rather than the relatively few known CBC agent species that have had severe impacts on native insects. All alien species captured by Asquith & Miramontes (other than two undetermined taxa) have been reported from native Lepidoptera hosts, with their Malaise trap data supporting strongly that both CBC agents and other adventive parasitoids have invaded native forests and attacked native hosts. The general richness of these groups of parasitoids has been doubled by these additions, and they may outcompete native parasitoids for hosts – although there is as yet no direct evidence of loss of native parasitoids as a consequence of this.

Asquith & Miramontes (2001) demonstrated that previous assumptions (Funasaki *et al.*, 1988a) on the high host specificity of most introduced wasp agents were misleading. Additional hosts were documented for 7 of 32 species – nominally a rather small proportion, but still nearly 25% of agent species of these groups, as the most conservative evaluation of host range. The braconid *M. laphygmae* was reported from 13 native host moths, and *E. sinicus* from only three; in contrast, the adventive ichneumonid *Trathala flavoorbitalis*, a specialist on Pyraloidea, was listed from 34 host species. The recent historical practice of CBC in the islands is illustrated by the release of three parasitoids in the late 1980s for control of diamondback moth (*Plutella xylostella*) (Funasaki *et al.*, 1988b), despite the sympatric occurrence of the closely related native species *P. capparidis* (for which there is no record of screening the wasps for safety). *P. capparidis* now appears to be extinct, at least at its type locality and probably from the entire island of Oahu.

The category 'adventive' in the above discussion merits comment. The term implies simply that the species has 'arrived' rather than been introduced deliberately, so such species are regarded as 'natural arrivals'. Howarth (2001) claimed that many species considered by other workers as inadvertent introductions were, in fact, undocumented purposeful introductions for CBC; Howarth stated that early workers documented only the introductions that they felt were successful. Notwithstanding the validity of this interpretation, and that much of the evidence of harmful effects of CBC agents in Hawaii will remain somewhat controversial, Asquith & Miramontes (2001) confirmed the high 'alien' component of Ichneumonidae in native forests and assumed the 'strong probability' of severe ecological effects (together with current inability to predict future effects) as sufficient to recommend that 'the purposeful introduction of parasitoids and predatory insects into Hawaii should be discontinued'.

Not surprisingly, this view is not accepted universally by CBC practitioners. The reality of parasitoid drift to other hosts remains unclear for many taxa, but some calls for increased intensity of screening provide disincentives to CBC (Messing, 2000). Messing discussed a case of screening for the aphidiid wasp *Lysiphlebus ambiguus* introduced to Hawaii for control of a sugarcane aphid (*Sipha flava*), in which testing against native Psyllidae was considered necessary, despite no aphidiid (a well-studied group of parasitoids in CBC) ever having been recorded as attacking a psyllid. He noted: 'The waste of time, money and quarantine space, and the drop of research morale that this engenders, is detrimental to the practice of biological control in a state where many exotic pests threaten both native and agricultural ecosystems.'

Two further recent Hawaiian studies are also relevant to this summary. First, Follett *et al.* (1999) appraised CBC attempts to control green vegetable bug (*Nezara viridula*) by introduced parasitoids. The egg parasitoid *Trissolcus basalis* (Scelionidae) attacks the native koa bug (*Coleotichus blackburniae*) to produce high levels of parasitisation in the laboratory, but such effects were not considered important at the time of parasitoid release. The thoughtful retrospective analyses of this case provided by Follett and colleagues demonstrate the uncertainty of the conclusions made by some workers that Hawaiian pentatomoid bugs have declined because of these CBC introductions. Koa bug now occurs mainly at higher elevations, reflecting (at least in part) degradation of much formerly suitable lowland habitat, where *T. basalis* is apparently absent and the other parasitoid (*Trichopoda pilipes*, Tachinidae, which attacks late instar bug nymphs and adults) attacks only to a very low extent. Elevation gradients may thus provide refuges from attack. Both parasitoids are well established in lowland koa bug habitats, which are largely distinct from *N. viridula* habitats, but their quantitative impacts in relation to those of alien ants and continuing habitat loss remain unclear.

Second, Duan & Messing (2000) summarised the extensive programme on evaluation of opiine braconid wasps to control pest fruit flies (Tephritidae) in Hawaii. Some of the wasps had been introduced early in the twentieth century and, in all, more than 30 tephritid parasitoids (from six families) were imported from many parts of the world. With later delineation of the endemic non-pest Hawaiian Tephritidae, and other exotic flies introduced as CBC agents for weed control, side effects of such agents assumed both biological and practical importance.

There is little evidence for harmful effects by the tephritid parasitoids already introduced, and Duan & Messing (2000) emphasised the need to evaluate the risk of any future introductions contemplated to counter pestiferous Tephritidae.

The Levuana moth in Fiji

The eradication of the coconut moth, *Levuana iridescens*, an endemic Fijian zygaenid pest of coconuts, is lauded widely as one of the most successful early cases of CBC. The campaign was documented in detail by Tothill *et al.* (1930), and Howarth (2001) noted it as 'possibly the best documented extinction of a native insect' by a CBC agent. In only two years, the moth went from being a serious agricultural pest to being endangered, and no authentic record of the moth has been made since 1929 – although there are indications that it might have survived in small numbers at least until the 1950s.

Tothill *et al.*'s account, somewhat unusually for the time, includes full details of the intensive post-release monitoring that occurred during the programme. However, Sands (1997) noted the report (Paine, 1994) of a later outbreak of *Levuana* in Fiji in 1956, and suggested that it might well not be extinct but surviving in refuges. He also queried whether the moth was truly endemic to Fiji or possibly exotic and present elsewhere in South-East Asia.

The agent implicated in the loss of *L. iridescens* was the tachinid fly *Bessa* (formerly *Ptychomyia*) *remota*, introduced from Malaysia. Because *Levuana* was colonial, a single fly could exploit and kill all or most members of a colony, once encountered, and the tachinid was sustained on alternative hosts during periods of levuana scarcity. Indeed, the fly still occurs on Fiji so (if *Levuana* is indeed extinct) has now been sustained for up to 70 years on only non-target hosts. One non-target species, another zygaenid (*Heteropan dolens*), was extirpated from Fiji around the same time as levuana.

Native butterflies on Guam

Nafus (1992, 1993, 1994) examined the parasitoid spectrum attacking the closely related nymphalid butterflies *Hypolimnas bolina* and *H. anomala*, whose caterpillars fed on *Pipturus* on Guam. His aim was to determine the extent to which CBC agent parasitoids exploited these non-target hosts, with the wider view of determining any possible role of CBC in the extinction of Guam's butterfly fauna. Four of the 20 resident species of the island appeared to have become extinct there. *Papilio xuthus* was once abundant but has not been seen since 1968; *Appias paulina* and *Euploea eleutho* have not been collected since the 1940s, but *Vagrans egestina* was found as recently as 1978. Nine species of pest Lepidoptera on Guam (including two butterflies, *Papilio polytes* and the banana skipper *Erionota thrax*) had been targeted for CBC in the past, and 27 species of parasitoids and predators introduced against these. Two Braconidae (*Apanteles* (now *Cotesia*) *papilionis* and *A. erionotae*) and a pteromalid wasp (*Pteromalus luzonensis*) were specifically imported to control butterflies, and the others were imported to suppress moths (Noctuidae, Pyralidae). Both the butterfly parasitoids are polyphagous species, with potential to attack native species (Nafus, 1993). Thus, 50–100% of eggs of the sphingid moth *Agrias convolvuli* are attacked by *T. chilonis* (Nafus & Schreiner, 1986).

Nafus (1994) reported 18 predators/parasitoids (including the cane toad, *Bufo marinus*) consuming *Hypolimnas*, including two CBC agents, *Trichogramma chilonis* and *Brachymeria lasus*. Native parasitoids, together with ants and food limitation, were the major causes of mortality in *H. bolina*. However, *B. lasus* parasitised an average 25% of the pupae, so that *H.bolina* was occasionally affected adversely by this agent. In contrast, no agent appeared to cause significant mortality of *H. anomala*. The host range of *B. lasus* is very broad, and Nafus (1994) believed that it could 'easily contribute to the extinction of a rare species' because of the inverse density-dependent parasitisation facilitated by exploiting a variety of alternative hosts. However, there was no evidence that any CBC agent contributed significantly to the earlier butterfly extinctions on Guam.

On both Guam and Hawaii, introduced 'tramp' ants (such as Argentine ant, *Linepithima humile*, which has been implicated in reducing populations of native moths in Hawaii (Loope *et al.*, 1988) may be important endangering organisms, not least because the paucity or absence of native ant species has deprived the native fauna of developing any strong 'counter-ant strategies' to help thwart such invasives.

Native Saturniidae in North America

Although primarily a forest rather than agricultural pest-control programme, the long-term biological control campaign against two lymantriid moths (gypsy moth, *Lymantria dispar*, and browntail moth, *Euproctis chrysorhoea*) has attracted considerable recent comment in relation to non-target effects and merits notice here. The moths are important defoliators of forests in North America.

The tachinid fly *Compsilura concinnata* was introduced against gypsy moth, within a range of 13 target Lepidoptera in North America, initially in 1906 but then repeatedly over the next 80 years or so. Despite its evident polyphagy, general lack of perceived extinctions allowed Coulson *et al.* (1991) to claim that 'there is no evidence that it has had a profound effect on any native species'. However, by that time, it had been reported as attacking more than 180 species of native Lepidoptera, Coleoptera and Symphyta in North America. Indeed, as early as 1919, decline in the saturniid *Callosamia promethea* had been observed in areas of the fly's activity (Culver, 1919).

The agent depends on access to non-target hosts. The primary target species (*L. dispar*) is univoltine and may occur in enormous numbers, so allowing massive tachinid populations to develop rapidly. *C. concinnata*, though, has three or four generations each year, so that up to three generations (depending on the locality) must be passed on alternative host species.

The implications of this need for non-target hosts are serious. Consider the data of Gould *et al.* (1990), who created 'artificial outbreaks' of gypsy moth in Massachusetts and confirmed that *C. concinnata* has a density-dependent response to host increases – at 'moderately high levels' (a million hosts per hectare!), parasitisation rate exceeded 90%, with each attacked caterpillar contributing an average of one female fly. The consequent million or so flies per hectare would thus need to search for 50–100 million non-target host larvae per hectare of forest in (as Howarth (2000) emphasised) conditions where no target hosts are present, because of phenological disparity.

The documented host range of *Compsilura* in North America included 12 species of silkmoths, Saturniidae, several of which have declined substantially in recent years. Boettner *et al.* (2000) undertook field trials to determine the effect of *C. concinnata* on populations of three such non-target species of Saturniidae. Two main experimental species, *Hyalophora cecropia* and *C. promethea* (as above), were exposed to the fly in Massachusetts forests in which the parasitoid was known to be abundant, as follows:

- Hatchling larvae of *H. cecropia* were placed in groups of five ($n = 500$ caterpillars) per tree. No larvae survived beyond the fifth instar.
- At the same time, groups (100 caterpillars each instar) of each of the first three instars were deployed. *C. concinnata* caused 81% mortality of these stages.

- *C. promethea* larvae were deployed in the field in aggregations of 1–100 caterpillars. Parasitisation by *C. concinnata* was 69.8% for caterpillars exposed for six days (on 135 trees), and 68.6% in a second trial of caterpillars exposed for eight days (on 115 trees). A total of 1407 caterpillars were used.

Fortuitous discovery of an egg mass of a third saturniid, the state-threatened *Hemileuca maia maia*, allowed monitoring of parasitoid incidence; *C. concinnata* was responsible for 36% (n = 50 caterpillars) mortality of this relatively small sample (one, nevertheless, that is far larger than is commonly available for any threatened invertebrate species, which by their very nature are difficult to evaluate quantitatively).

The results of these trials suggested strongly that *C. concinnata* has played a role in the decline of these saturniids and, by implication, probably also in that of other non-target hosts. The relatively short exposure time in the trials – at the most, only eight days of a developmental period normally of around two months – is also likely to have given only a minimal estimate of overall attack rates. In addition, several flies can develop in such large hosts: Boettner *et al.* (2000) reported one *C. promethea* caterpillar yielding 14 adult *C. concinnata*.

It is important to appreciate that the above cases, and others with similar implications for the decline of non-target invertebrates, reflect the outcomes of introductions made up to many decades ago, when the kinds of screening protocols discussed earlier (including tests of herbivores against plant weeds) to investigate agent safety were in their infancy or were non-existent. Sands (1997) made the important point that *Bessa* would not now qualify for adoption as a CBC agent against *Levuana*, because it is not specific even to the family containing the target moth. Similar restrictions are likely for the other example agents noted above, but the zeal of practitioners, often with the best motives for pest suppression, can lead easily to clandestine introductions of inadequately screened agents. These might include ambiguous adventives of the kind noted above in the Hawaiian fauna. Others will remain undocumented, and there is clear potential for undesirable consequences. Howarth (2000) claimed that 'clandestine programmes by private individuals are also being conducted', and included a number of references to indicate that 'unsanctioned and smuggled biological control agents may be a significant problem'. The progressive development of regulatory controls (Chapter 10) is only a partial stricture on such operations, which may always have the potential to 'evade the system'.

It is easy to exaggerate the inadequate practices of historical CBC campaigns such as those exemplified above, and to extend their limitations to the discipline as a whole. In general, they were undertaken by the highest contemporary standards, and perspective on non-target effects has changed dramatically since they were undertaken, because of increased concerns for conservation extending beyond the vertebrate world. Risk assessment is now undertaken more widely as a central part of CBC. An agent valuable in one part of the world may have serious non-target effects elsewhere, emphasising the need for local or regional evaluation. The pyralid moth *Cactoblastis cactorum* successfully controlled massive infestations of prickly pear cactus (*Opuntia*) in Australia as one of the widely heralded 'classic successes' of weed CBC (Dodd, 1940). Other than for some transient damage to adjacent fruit crops during initial population explosions of *Cactoblastis*, no non-target concern was evinced. The same moth in Florida threatened the last-known individuals of *Opuntia spinosissima* (Bennet & Habeck, 1995), with Howarth (2000) commenting that if this threat had not been noticed by the serendipitous presence of biologists, then the cactus might well have been lost and its extinction attributed simply to 'habitat destruction'. Johnson & Stiling (1996) note that the last few individuals of the *Opuntia* are now caged to exclude herbivore attack, and there are concerns that *Cactoblastis* could spread in south-western North America, where other cacti may be vulnerable to attack.

Nevertheless, as Hopper (2001) noted, retrospective analyses of past CBC campaigns and their impacts can 'provide the testing ground for predictions about non-target effects', together with evaluating the methods used for pre-release agent selection and the predictions that followed. However, even for some recent CBC programmes, expediency has taken precedence over careful selection of agents. For example, the North American campaign to use CBC against Russian wheat aphid (*Diuraphis noxia*, c.f. p. 144) led to introduction of 29 exotic predators and parasitoids (and two fungal pathogens) from 1986 to 1993. The releases included more than a million individuals of 12 species of ladybird beetles (Coccinellidae) new to North America (Quisenberry & Peairs, 1998), with the emphasis on collecting and releasing, in as many sites as possible, as many species of potential agents as could be found, even though '. . . few sound criteria and techniques were available for making such choice' (Hopper *et al.*, 1998; Strong & Pemberton, 2001). The possible non-target effects of Coccinellidae introduced to North America

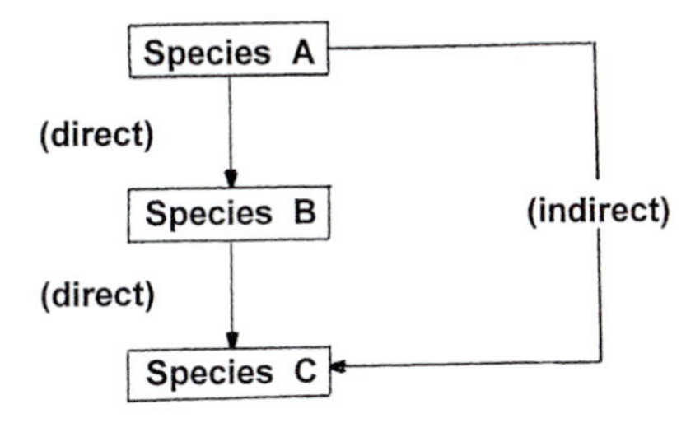

Figure 5.6. Relationships between direct and indirect ecological effects of biological control agents (Lonsdale *et al.*, 2001). Arrows denote interactions between species. In this example – of an insect herbivore (A) feeding on a weed (B) – the direct effect leads to reduction of B and drives down populations of another consumer (C), also feeding on B, by an indirect effect. If A was a biological control agent and B a weed, then this indirect effect on C might be condoned; if, on the other hand, B was a native plant and C a rare specialist native insect herbivore, then both the direct effect of A on B and the indirect effect of A on C would be likely to be deemed undersirable.

were summarised by Obrycki *et al.* (1999). Competition and intraguild predation with imported ladybirds have been implicated in decreased densities of native Coccinellidae there and, up to 1985 about 179 coccinellid species had been imported to North America as CBC agents (Strong & Pemberton, 2001).

Wider ecological effects

The foregoing summaries have emphasised the possible direct non-target effects of biological control agents, particularly exotic species. Wider ecological implications also occur from changes to the receiving environment and its biota, through changes in 'ecological balance' influencing the abundance and wellbeing of members of the community. These effects are indeed hard to define, and many constitute parts of the generalised fears that conservationists often voice for the presence of invasive species or other abundant exotic taxa. The equivalent syndrome was termed 'diffuse dread of exotics' by Messing (2000). Indirect effects occur when other species, not attacked directly by the agent, are affected by the agent's abundance or by its feeding on a species (Lonsdale *et al.*, 2001) (Fig. 5.6). The bottom line appears to be that there are potential major ecological risks involved in releasing exotic biological control agents, so that a conservation perspective argues for more caution and more rigorous evaluation of the benefits and potential costs involved (Louda *et al.*, 2003a, 2003b). The impacts that may need to be considered, and for which protocols should, ideally, be formulated, extend beyond immediate impacts on non-target species to include topics such as (1) changes in ecosystem processes, (2) reduction in diversity in the receiving community and environment, and (3) effects on populations participating in food webs into which the agent is released. Such themes are commonly intangible in practical terms, and the requisite baseline information is very difficult to obtain without considerably more resources than are usually available or deemed worthwhile. As Louda *et al.* (2003a, 2003b) advocated, whilst predicting community interactions is indeed complex, realistic ecological risk of introductions cannot be evaluated without it – the few studies undertaken, such as Louda's extensive work on impacts of *Rhinocyllus conicus* on thistles in North America (Louda *et al.*, 1997), suggest that at times such effects can be considerably more complex than usually suspected.

Fears over the consequences of CBC will inevitably continue, with the needs for risk analysis embedded in future screening and selection programmes. There are relatively few cases in which CBC agents have been documented attacking species of known conservation concern (for example, those listed in Red Data Books and schedules of protected species) and, for some, the relative importance of CBC agent attack and other factors influencing decline through habitat loss or range reduction remains unclear. The limitations of screening tests render it unlikely that all such cases can ever be predicted before agent release, but the implications of such tests remain of massive importance because, at the least, they prevent the releases of the more obviously hazardous species. Phenomena such as host-switching by agents may still occur, but laboratory inferences (that is, in relatively artificial environments) may not always carry to field situations. Thus, the ladybird *Coccinella septempunctata* was tested in the laboratory to see whether it would eat eggs and caterpillars of a threatened North American lycaenid (*Erynnis comyntas*) when aphid food was not available. It did so (Horn, 1991), but such feeding has not been observed in the field.

McEvoy & Coombs (1999, 2000) discussed the relevance of 'revenge effects' arising from CBC, indicating that the proliferation of agents has the potential to render the practice less effective and more risky, so affecting its practicality. There are several origins of revenge effects, in addition to the direct effects of attack on non-target species:

- By concentrating on some agents (such as 'new' agents), neglect of others may occur. Allied to this, reliance on exotic agents as a panacea can lead to overlooking native species already in place; and by concentrating on mitigating harm caused by established invaders (the pests themselves), new invaders might not be predicted and prevented. Quarantine needs must not be overlooked in embracing management of invasive taxa.
- One control agent undermines another, more effective agent and so leads to an increase in pest density. The underlying principle is that proliferation of an agent species may increase the chances of antagonistic interactions between the different agents, such as when (1) one agent preys on another, (2) an agent interferes with another's access resources shared by both species, and/or (3) an agent reduces resources needed by the other species. Such interactions may provide difficulties in evaluating the relative benefits or negative effects of multiple agent introductions.
- The target pest is replaced by another pest that may be even more difficult to control. This has been particularly evident in some weed-control programmes, in which control of a target weed has occasionally led to its replacement by another weed not susceptible to any agent already in place in the local environment.

Neoclassical biological control

The theme of neoclassical biological control was brought effectively to the notice of conservationists through proposals to introduce exotic parasitoid wasps to attack native rangeland grasshoppers in North America (Lockwood, 1993a, 1993b; Carruthers & Onsager, 1993). This practice thereby has a very different ecological basis from CBC. In the latter, historical associations interrupted by displacement of the pest are re-established by introducing natural enemies previously associated with the pest. In neoclassical biological control, however, new ecological linkages are sought and promoted, so that the agents being introduced demonstrably have sufficient ecological flexibility to adopt novel hosts or prey species. Although the practice has been supported by some ecologists to complement more traditional CBC strategies by 'trying to take advantage of the lack of evolved commensalism common to many exploiter–victim systems' (Hokkanen & Pimentel, 1984, 1989), the deliberate introduction of novel generalist taxa as permanent residents into other faunas is clearly of concern. The perceived necessity of these being preadapted to change hosts as a selection criterion for their use also creates concerns that they will move to additional native taxa, including those of conservation interest. As Lockwood (1993b) put it, 'Host ranges of the exotic organisms are essentially unknown . . . so ecological safety assurances are unfounded.'

North American rangeland grasshopper pests represent a large taxonomic group not dominated primarily by pest species. Perhaps fewer than 10% of acridid grasshoppers on any continent have serious pest potential (Lockwood, 1998), and pests and taxa of conservation concern may be congeneric (Samways & Lockwood, 1998).

Neoclassical biological control agents have potential to expand into new areas (Lockwood & Ewen, 1997), and the rangeland grasshopper example of introducing a scelionid wasp (as an egg parasitoid) and a fungal pathogen has led to calls for a very high level of scrutiny and care. In such contexts, the same properties of host populations that facilitate spread and establishment of biological control agents on specific targets are of concern in spreading agents more widely into natural faunas. Other management options for the grasshoppers may be preferable (Lockwood, 1998; p. 279). The debate over release of the fungus *Entomophaga praxibuli* and the wasp *Scelio parvicornis*, both native to Australia, involved a variety of issues in addition to purely conservation concerns. These include possible adverse interactions between native and exotic biological control agents, negative impacts on native grasshoppers affecting weed control, and various ecological processes. Both organisms were to be released without thorough host range testing and without plans to systematically monitor the agents, their hosts or non-target taxa following release (Lockwood & Ewen, 1997). Application to release the scelionid was denied in 1994, with the wasp being designated an 'indirect plant pest' under the US Federal Plant Pest Act, because of its likely suppression of several native grasshopper species that can control rangeland weeds and their potential to spread within the rangelands.

A second example of neoclassical biological control involves the European braconid wasp *Peristenus digoneutis*, also imported into North America to control native mirid bugs (predominantly *Lygus lineolaris*, the tarnished plant bug, a serious pest on various fruit and vegetable crops). It was especially effective in alfalfa, but the wasp also attacked other mirids, although *L. lineolaris* was clearly a 'preferred host' (Day, 1996). For this species, likely side effects appeared not to be significant, and the major secondary host species (the European bug *Adelphocoris lineolatus*) gave *Peristenus* a secondary role as a CBC agent on alfalfa.

Insect pathogens and pest management

The use of pathogens in agricultural and forest pest management has been dominated by a single complex taxon, *Bacillus thuringiensis* ('*Bt*'), because it has been believed widely to be sufficiently host-specific for relatively precise targeted applications, particularly against Lepidoptera. *Bt* occurs naturally in the environment, in a variety of sites, including soil, the phylloplane microflora, and various plant and insect materials. Despite this, the future of commercial formulations of *Bt* in the wider environment is understood only in rather general terms, so that concerns occur through fears of its accumulation and persistence. Numerous isolates of *Bt* have been developed since it was used initially in France in 1938 (with its potential recognised much earlier, in Japan in 1901), and it now comprises more than 90% of all pathogen-based biopesticides. The insecticidal properties of this Gram-positive bacterium are associated with endotoxins from the crystalline protein inclusions and have been designated by 'Cry' categories to reflect their variety. The predominant activity is caused by delta-endotoxins produced by each serotype or strain, but some *Bt* strains produce toxic exotoxins and other active principles. As Glare & O'Callaghan (2000) emphasised, the simple epithet '*Bt*' gives little indication of the variety of toxins, spores, crystals and formulations that have been used in any particular study, so that summarising and appraising its effects is difficult. Formally, segregates regarded previously as varieties are now treated as subspecies, and some 170 delta-endotoxin genes are now known (Crickmore *et al.*, 1998, 2000).

Bt has been used widely against pest Lepidoptera in forests, agriculture and horticulture, most commonly as *Bt. kurstaki*. In Australia, for example, it has been an integral component of IPM for *Helicoverpa* on cotton and other crops, *Pieris rapae* and *Plutella xylostella* on brassica crops, and *Epiphyas postvittana* and others on fruit trees. Promotion of *Bt* in such contexts has reflected the high level of product specificity, particularly in minimising non-target effects on natural enemies. However, some recent concerns have arisen over its continued widespread use. These concerns range from possible effects on people (through broadcast of *Bt* in urban areas) to those over specific and community level effects on non-target invertebrates. Thus, there is perceived potential for *Bt* (in formulations such as *Bt. israelensis*, used against larvae of mosquitos and blackflies) to

Table 5.9. *Effects of microbial pesticides on non-target beneficial arthropods (Flexner* et al., *1986)*

Effects from indirect mortality of natural enemies probably more significant across all microbial types than are direct mortalities from microbial pesticides
Significant levels of direct mortality of beneficial insects can be caused by bacterial and protozoan microbial pesticides
No direct mortality from viruses of insect pests have been documented for arthropod natural enemies
Direct effects of fungi on arthropod natural enemies not well studied; probably underestimated until now
Unique standard methods need to be developed for microbial assay and for explaining more exactly the dosage administered

distribute widely in waterways, with possible wider influences on aquatic communities, so that concerns for non-target taxa occur in both terrestrial and freshwater environments. Very recently, additional concerns have arisen through incorporation of *Bt* into genetically modified plants, so that the major problems may arise from (1) direct application of *Bt* and (2) genetically modified plants facilitating *Bt* transfer. The bulk of available information was summarised by Glare & O'Callaghan (2000), and the following is largely a precis of the non-target effects that they discussed. Possible effects on endangered invertebrate species add a further dimension.

General risks of microbial pesticides to non-target organisms are evaluated in Table 5.9.

Direct applications of pathogens

Effects of *Bt* on many non-target taxa appear to be negligible or non-existent, but, reflecting its wide variety of applications (so that, for example, more than 300 taxa of Lepidoptera have been 'controlled' by *Bt*; and *Bt. kurstaki* has been claimed to be effective against members of 318 genera of insects – although, for fuller perspective, insects from 166 genera have been reported to be non-susceptible), safety concerns have naturally arisen. As Glare & O'Callaghan (2000) emphasised, effects are related to toxicity, so that inferences from many laboratory screening studies using doses far higher than those used conventionally in the field may be misleading. Other factors influencing non-target impacts include targeting of applications, environmental factors influencing persistence and insect feeding behaviour, each of which affects the amount of *Bt* to which the organisms may be effectively exposed. Very broadly, from the compendious literature listed by Glare & O'Callaghan, phytotoxic effects and harm to vertebrates are very rare indeed. Within the invertebrates, low doses of several toxins did not affect two Collembola species, molluscs and crustaceans are generally insensitive to *Bt* at field application rates, entomopathogenic nematodes are generally non-susceptible (but some purported nematicidal *Bt* strains have recently been developed), and several studies have failed to reveal adverse effects on spiders. A few mites have been susceptible to relatively high dosages in the laboratory. Several workers have inferred that there need be little concern for non-target effects on aquatic invertebrates, other than Diptera related to targets of *Bt. israelensis*; more typical aquatic orders such as Ephemeroptera, Plecoptera and Trichoptera (together used widely to infer water quality, as the 'EPT index') appear to be largely non-susceptible at normal field doses, although a few studies on stoneflies in Canada (Kreutzweiser *et al.*, 1992, 1994) revealed high mortality at high doses.

Most concerns have been expressed for Lepidoptera or, pragmatically, for natural enemies such as parasitoid wasps and predatory insects as other components of IPM. Intrageneric variability in susceptibility in some moths leads to concerns over non-target effects of sprays for gypsy moth (*Lymantria dispar*) and spruce budworm (*Choristoneura fumiferana*) in North American forest systems with naturally high species diversity. In West Virginia, *Bt. kurstaki* used against *L. dispar* caused significant decline in other macrolepidoptera (Butler *et al.*, 1995), but with no significant effects on species present after the spraying period and none on non-Lepidoptera, the latter finding being corroborated by other studies on similar systems. More generally, *Bt* applications against forest Lepidoptera have sometimes been associated with decreased abundance of larval and adult moths, these sometimes extending to the year following application. Most affected species recovered in the first year after spraying. In Oregon oak forests, Miller (1990) monitored 35 species of Lepidoptera for three years after spraying with *Bt. kurstaki*. Two years after spraying, no significant effects remained on numbers, but species diversity was still reduced. Aerial drift may occur, as with any other aerially applied pesticide, necessitating consideration of non-target effects beyond the target areas. And, as for other pesticides, such uses should, ideally, be avoided if nearby areas support taxa of conservation significance, such as endangered species. Understandably, few such species have been tested specifically for susceptibility to *Bt*, but the Karner blue butterfly (*Lycaeides melissa samuelis*) was susceptible to *Bt. kurstaki* at high laboratory doses (Herms *et al.*, 1997). In contrast, the monarch (or wanderer, *Danaus plexippus*) showed only low susceptibility in laboratory tests (Leong *et al.*, 1992).

Local extinctions of non-target taxa of Lepidoptera have been inferred to occur with multiple applications of *Bt*, and timing may be particularly critical in the case of univoltine vulnerable species. Whaley *et al.* (1998) suggested that such effects may be usual and that they should be expected. In areas where sensitive species occur, Whaley and colleagues suggested the possible use of exclusion plots in situations where alternatives to *Bt* were not available and drift is likely. They found significant mortality of caterpillars of two Lycaenidae (*Incisalia fotis*, *Callophrys sheridanii*) fed on host plants gathered up to 3 km from the down-canyon edge of application sites in Utah. Exclusion plots, they suggested, could be covered temporarily with plastic sheets, which are removed once the drift has passed.

Impacts of *Bt* on natural enemies have been of wide relevance in developing its use in IPM, and *Bt* has often been promoted as compatible with most conventional biological control programmes. Numerous trials, involving many of the numerous species of arthropod predators and parasitoids, have largely endorsed this view. *Bt* has only rarely been found to be directly toxic to predatory insects, although a variety of adverse effects (such as declining growth rate and fecundity, reduced fertility, and delayed mortality in the ensuing predator generation) have been found in some cases. Direct toxic effects were reported for the predatory mite *Metaseiulus occidentalis* (Chapman & Hoy, 1991), and some trials on green lacewings (*Chrysoperla* spp.) also yielded ambivalent results. For parasitoid wasps, no effects have been documented in attacking non-susceptible hosts. In other instances, a variety of indirect effects (as for predators, above) have been reported, although no impacts have been found in many instances.

For other beneficial arthropods, numerous studies have implied safety of *Bt* against honeybees, even when used to control wax moth (*Galleria*) in commercial hives. Effects on most earthworms also appear to be small.

Bacillus thuringiensis in genetically modified plants

Use of *Bt*-endotoxin genes in genetically modified plants is now widespread, and the practice has engendered some concerns (see p. 303). In particular, these were fostered through a study showing that transgenic pollen from corn plants was toxic to caterpillars of the butterfly *Danaus plexippus* (Losey *et al.*, 1999) through contaminating the surface of milkweeds (*Asclepias*) on which the caterpillars feed. About half the caterpillars died after four days of feeding. Although that study has been criticised for lack of detail, it did raise important issues that demonstrate some potential problems of uncritical *Bt* use in the wider environment. The possibility of harm to *D. plexippus* from *Bt* corn-pollen deposition has led to considerable recent debate on the reality of risks. Need for risk assessment is recognised widely for this and similar cases (Wolt *et al.*, 2003). Female butterflies do not avoid *Bt* pollen-contaminated plants, and conservation of larval habitats is of wide relevance in countering possible adverse effects of such *Bt* use (Jesse & Obrycki, 2003).

In their encyclopaedic review of *Bt* safety concerns, Glare & O'Callaghan (2000) gave a qualified verdict of 'safe to use', in that they considered it sufficiently specific to not eradicate even the most 'at-risk' non-target insect populations. They also noted it prudent to continue to monitor *Bt* applications closely for negative impacts and to continue to improve techniques for such monitoring.

Other pathogens may also be relatively safe in the wider environment. 'Extensive laboratory tests' of the New Zealand endemic bacterium *Serratia entomophila* on non-target organisms indicated that it was highly specific to a target scarabaeid beetle (grass grub, *Costelytra zealandica*). Closely related beetles could not be infected (Jackson *et al.*, 1993).

Entomopathogenic nematodes

Very few studies have been made on non-target impacts of entomopathogenic nematode worms, and interest in this group as control agents has increased considerably, in large part because they have been believed widely to be largely restricted to the areas of release or inoculation. Two of the main genera employed, *Steinernema* and *Heterorhabditis*, spread naturally in the environment only very slowly, although movement of infected hosts can sometimes change this scenario considerably. Indeed, individuals from an inoculative release of *S. scapteriscus* used to suppress pest mole crickets on golf courses were dispersed more than 20 km by their hosts (Parkman *et al.*, 1994). Parallel long-distance dispersal of some other exotic nematodes in such ways can not be ruled out. However, Kerry (1995) and others have enumerated a wide spectrum of natural enemies important in countering population increases of these nematodes. Bathon (1996) therefore believed that effects of nematodes will generally be restricted to treatment plots and nearby sites and emphasised that there is no report of a release of nematodes (whether native or exotic taxa) severely affecting the non-target biota in a release area.

Nevertheless, a cautious approach to introductions of exotic nematodes was advocated, and Australia, New Zealand (Bedding *et al.*, 1996) and some other countries impose quarantine or other regulatory measures to increase safety assurances. Thus, for Australia, the necessary information by which to assess whether particular nematodes may be a problem if released must be obtained. This information is mandatory both for exotic species proposed for introduction and for exotic strains of native species. Nematodes thus parallel many other taxa for which risk assessment is required for introductions. Intriguingly, many entomopathogenic nematodes are extremely polyphagous in laboratory trials (Georgis *et al.*, 1991), so that impacts on non-target taxa would appear to be likely, even if not confirmed in practice.

Entomopathogenic fungi

Aerial ULV oil formulations of a strain of *Metarhizium anisopliae* var. *acridum* have been used successfully in trials against nymphal bands and adult swarms of *C. terminifera*. Persistence of the fungus is of some concern to environmentalists, as about half the locusts fed on vegetation collected from treated areas seven days after spraying died. There was also high mortality of hopper bands invading from nearby untreated areas. Nearly 2500 ha of bands and swarms were treated in the Cannon Range National Park, South Australia. *Metarhizium* is seen as the most promising alternative control agent to replace fenitrothion (see p. 114), and current formulations listed in Australia have been approved for use both in sensitive areas and on organic farms as a preferred option (Hunter *et al.*, 2001).

6 · *Cultural aspects of pest management*

Cultural control emphasises making environmental manipulations within and near agroecosystems to hinder the pest or to promote local wellbeing of native natural enemies. This practice is motivated largely by needs to promote conservation biological control as a means of overcoming some of the fears and practical problems that arise over use of exotic control agents. Habitat modifications include a wide variety of changes in agricultural practices in space and time, many involving lessening the intensity of human interventions. The wider benefits of such changes can be considerable and represent an active current endeavour to enhance invertebrate conservation in agroecosystems.

Introduction

This third chapter on the implications of the major components of modern crop protection for conservation introduces the various aspects of 'cultural controls', in which the crop environment and its surroundings are manipulated to help in pest management. Such methods constitute a third major strand to IPM, together with pesticides and biological control. The major principle is that availability of resources needed for the wellbeing of pests or of their natural enemies can be influenced in space and time by agricultural practices, so that manipulations to the crop environment may decrease the incidence and impacts of pests and increase that of natural enemies. Such modifications may therefore be important in pest management but may also harm or benefit natural biota, including the numerous invertebrates associated with agroecosystems and nearby areas. Both intra-crop environments and wider environments such as crop boundaries and surrounds are relevant considerations, with the latter having massive practical relevance as reservoir habitats for natural enemies, from where they can disperse (either naturally or by providing inducements such as attractant baits) into the crops as control agents. Many such modifications in management draw on ecological knowledge, particularly aspects of invertebrate–plant and of invertebrate–invertebrate interactions. Conditions can be changed to make the local environments (1) less favourable for pests or (2) more favourable for natural enemies, thereby anticipating more effective pest control and involving factors relevant to native and/or exotic agents. Such strategies may involve manipulation of the entire habitat (as in some of the cultural methods noted below) but may also include changing aspects of the 'functional environment' – that is, 'the physical and biotic elements that directly or indirectly impact survival, migration, reproduction, feeding, and behaviours associated with these processes' (Letourneau & Altieri, 1999). The major impetus for conservation biological control (see below) has invoked many of these aspects and their enhancement for natural enemies. As several commentators have observed, increased potential for biological control has been the single most important driver for development of management to increase arthropod diversity on farms.

Indeed, Ehler (1998) noted that conservation of natural enemies is probably the oldest form of biological control of insect pests; however, its recent modifications include more formal considerations of tritrophic systems (whereby features of the pest's host plant play a role in the interaction between the pest and higher level consumer agents) and thus extend the simpler predator–prey and herbivore–plant interactions into a wider community context of increased relevance for conservation, by its potential for wider habitat and resource considerations. Habitat management to improve availability of resources needed by natural enemies has thus been important in promoting conservation, not least because it draws on ecological knowledge of the agents and the qualities of the environments in which they thrive. In addition, the implications of natural habitat loss to agriculture have emphasised increasingly the importance of natural remnant habitat patches and the qualities of borders around cultivated areas, with buffers (such as conservation headlands, see p. 221) now an integral management component of many agroecosystems. These themes are considered more thoroughly in Chapter 7.

Much of cultural control is inherently sympathetic to conservation, most fundamentally because the practices involve increasing diversity within the simplified cultivated systems, be they for annual or perennial crops. They thus (1) help to overcome the limitations imposed by monoculture-dominated environments and (2) may help to avoid the 'reversion to successional zero' typified by harvesting of crops and replanting the following season after a period of fallow. And, importantly, such environmental management is linked with reduced pesticide inputs

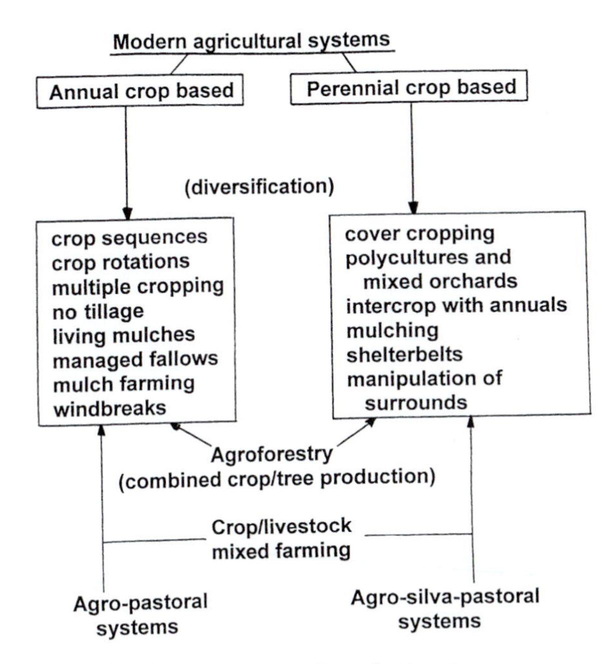

Figure 6.1. Component practices of cultural pest management. Strategies for plant diversification in annual and perennial crops, with the major components of each listed. Agroforestry (see Chapter 10) is included (after Letourneau & Altieri, 1999).

and, often, lessened intensification in other ways. The range of component practices is considerable (Fig. 6.1). Likewise, the range of benefits from such enhancement to agricultural biodiversity and its functions is also very diverse (Fig. 6.2) (Altieri, 1991; Altieri & Nicholls, 1999)

Three main axes of agroecosystem or crop management involve cultural practices:

- Temporal dimensions involve modifications to the seasonality of planting or harvesting regimes, influences of different successional sequences and rotations, and management inputs.
- Spatial dimensions involve parameters such as the area of cropping, the use of monocultures of multiple crops in an area, and the mosaic arrangement of cropping areas in the wider landscape.
- Biological dimensions include variation in the crop variety, age, susceptibility and exposure, and apparency of the crop(s) in relation to the biological features of the key pests present and their natural enemies.

This very wide scope involves mechanical aspects of management (such as cultivation/tillage, harvesting regimes), chemical aspects (in addition to

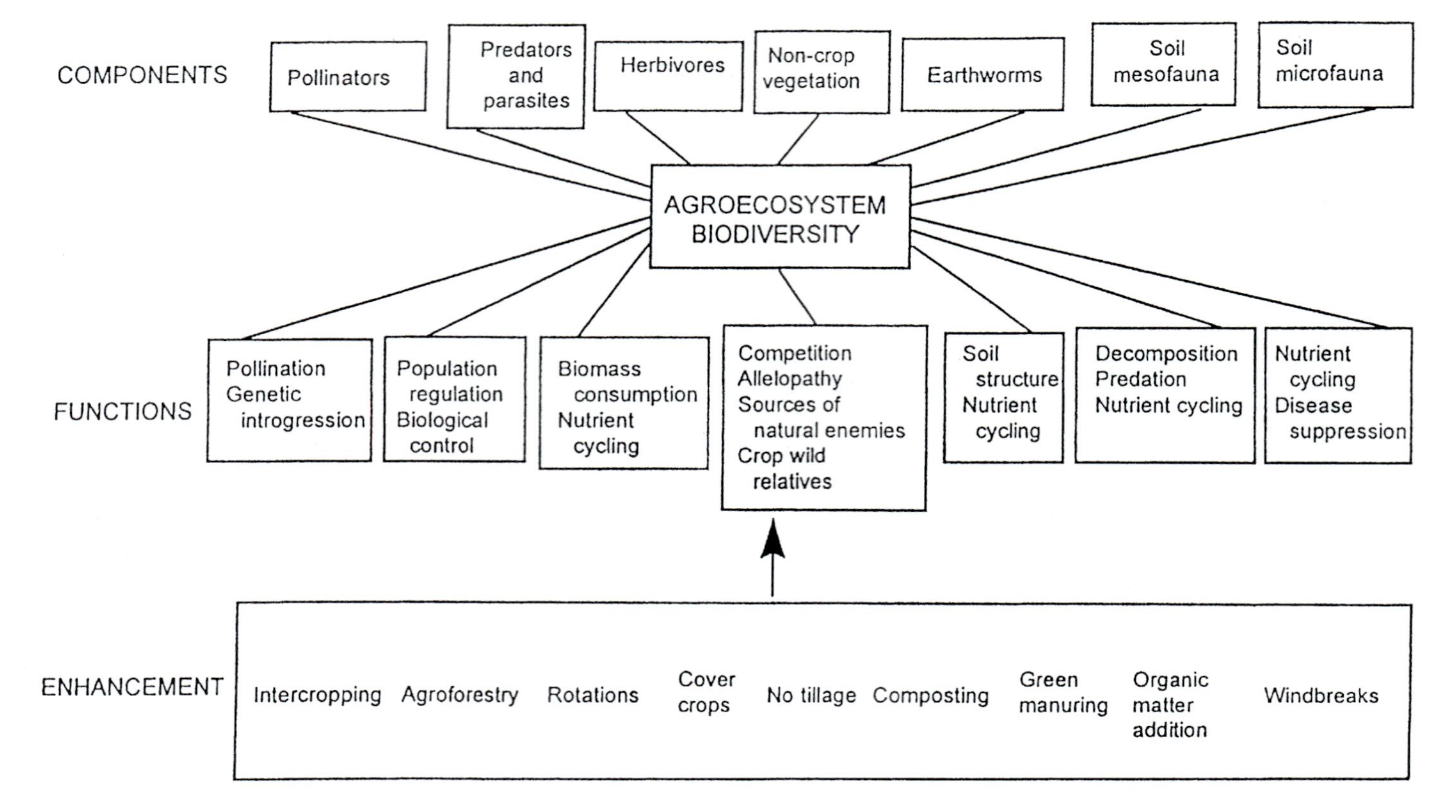

Figure 6.2. Components, functions and enhancement strategies for biodiversity in agroecosystems (after Altieri & Nicholls, 1999).

Components of environmental diversity

Temporal

- monoculture
- rotation
- sequential crops
- continuous crops
- fallow
- annual/perennial crop
- early/late planting
- asynchronous
- harvesting date
- time of weeding
- long/short maturing crop
- good/poor season
- timing of climatic events

Spatial

- species-pure crop
- multiple cropping
- agroforestry
- large/small field
- grouped/scattered fields
- simple/complex surrounds
- high/low density
- weedy/weed-free
- uniform/mixed varieties
- local or regional crop mosaics
- trap crops

Biological

- tolerant or susceptible variety
- resistant/pure line
- uniform/mixed age
- monophagous/polyphagous pest
- large/small pest complex plus natural enemies
- multivoltine/univoltine pests
- apparent/non-apparent crop
- high/low chemical defences in crop
- simple/complex food webs
- native/exotic pest(s) and natural enemies

Figure 6.3. Cultural measures. The three major dimensions of components involved in the diversification of agroecosystems (after Letourneau & Altieri, 1999).

pesticides (Chapter 4), applications of fertilisers, water and other chemicals may affect biota), and biological aspects extending well beyond the scope of conventional biological control (Chapter 5). They draw, very broadly, on the need to reduce intensification, overall disturbance and habitat loss or fragmentation in production systems, and one recent summary of the components of cultural management is given in Fig. 6.3. As Tscharntke & Kruess (1999) (see also Kruess & Tscharntke, 2000) remarked, habitat fragmentation (both in space and time) 'can be manipulated in such a way that populations of beneficial arthropods increase relative to pests' (the conventional basis of conservation biological control), and this is an important component in facilitating wider conservation and biodiversity increase in landscapes by also increasing connectivity of habitat patches and so reducing the drastic effects wrought by fragmentation.

Much of this endeavour devolves on vegetation management, with the twin primary motives of (1) lessening pest incidence and (2) its corollary of maximising natural enemy numbers and wellbeing, both sought by reducing disturbances and maintaining or enhancing critical resources, including refugia. Reducing disturbance also involves two axes – the level and the frequency of disturbance with the general principle that low-level infrequent disturbance might be preferable to high-level frequent disturbance for conservation.

Vegetation management is a diverse discipline, and involves the following aspects (after Letourneau & Altieri, 1999):

- Managing the crop in time and space.
- Managing the composition and abundance of non-crop vegetation in and around cropping areas.
- Soil type and condition.
- Managing the surrounding environment in the wider habitat matrix.
- Optimising the type and intensity of management.

Much of the impetus for such management is to reduce the potential for pests to cause damage to crops and, whereas the individual vagaries of invertebrate population dynamics are commonly difficult to predict, several generalisations over the structural characteristics of agroecosystems that reduce pest incidence are gradually becoming clearer. These, as listed by Letourneau & Altieri (1999), provide a useful framework for developing cultural management principles, as follows:

- Low-pest-potential systems have heightened crop diversity through mixtures in time and space, as opposed to uniform monocultures.
- If primarily monocultures, these are rendered discontinuous through rotations of different crops, use of crop-free periods at times of pest incidence, and use of rapidly maturing crop varieties, all of which can help to thwart continuity of pest population increase and attack.
- Rather than vast areas of monoculture crops, pest incidence is lowered in small mosaic-arrangement fields within a landscape also containing uncultivated land, potentially providing shelter and alternative food for natural enemies.
- A predominant perennial crop component. Thus, orchards are generally considered more stable than annual field crops, because of reduced levels of disturbance and the opportunities they have to develop greater structural and biological diversity.
- High crop densities in the presence of tolerable levels of weeds as 'background', again increasing structural diversity and acting as possible refuges and alternative habitats for natural enemies.
- Higher genetic diversity, brought about by use of several lines or varieties of the same crop in an area.

The agroecosystem gradients of disturbance and diversity may thus range from (1) a monoculture of a single crop variety grown over a large area and throughout all or most of the suitable season every year to (2) a sequence of plantings and harvests of the same or different crops, each on small scales and intermingled with others, and also from (1) short-term crops, where disturbance may occur up to several times each growing season with harvest/tillage/planting and allied processes to (2) long-term perennial orchard or agroforestry crops where disturbance is infrequent and, often, minimal. Plantations of the exotic *Pinus radiata* grown for softwood in Australia, for example, may be largely unmanaged except for thinning on one to three occasions over their 25–30-year pre-harvest development.

This chapter outlines the various forms of mechanical disturbance included as facets of 'cultural control' for pest management in agroecosystems, and the emphasis on 'conservation biological control' that has been a major impetus for their development is noted first to introduce these topics. It is important not to be waylaid by semantics in defining the topics covered here. Some workers use the term 'cultural control' to mean the forms of habitat diversification used to reduce pest dispersal, colonisation or reproduction on target crops – for example, by use of trap or decoy crops, visual or olfactory confusion of pests to hinder their colonisation, and similar tactics. By contrast, if the main aim of affecting structure of the system is to enhance performance of natural enemies, then this is conservation biological control as a distinct topic. Both these themes may operate together and be conjoint strategies. They may thus be embraced as the wider 'cultural control', with conservation biological control then being a particular subset of this discipline.

Conservation biological control

Much environmental management in agroecosystems has been developed partially or primarily to enhance the efficiency of biological control – either by improving conditions for the wellbeing of CBC agents or helping to sustain and increase the abundance and effectiveness of native agents. Such modifications are clearly driven initially by economics but are important in helping to lessen aspects of agricultural intensification (such as pesticide use) as part of an overall drive to improve implementation of IPM. Much of the topic, then, is involved directly with conservation, with the principles of habitat enhancement for natural enemies having much wider importance in the maintenance of native biota in the agricultural landscape mosaic. In some cases, the crops may not be the most important component of the ecosystem for native natural enemies (see Ferro & McNeil, 1998). However, the plants on which invertebrates live, feed and reproduce are often the most significant components of their habitats, not only for herbivores but also (as noted above) for the various physical and chemical influences of plants on predators and parasitoids, affecting their capability as consumers and thus as control agents (Powell, 1986).

'Conservation biological control involves the use of tactics and approaches that involve the manipulation of the environment (i.e. the habitat) of natural enemies so as to enhance their survival, and/or physiological and behavioural performance, and resulting in enhanced effectiveness' (Barbosa, 1998). It is a topic of considerable recent interest to both pest managers and conservationists. Until recently, much of the research on conservation biological control had been largely theoretical, but emphasis on its implementation has grown considerably, with increasingly sophisticated applications of ecological knowledge to assure its effectiveness. Habitat management can involve virtually any aspect of critical resources, including provision of supplementary foods (such as alternative hosts or prey, often present in natural communities), complementary foods (such as honeydew, pollen and nectar, again often furnished by natural vegetation), modified microclimates (such as by provision of windbreaks for shelter and shade) and refuges (such as overwintering sites or other places where agents can survive during periods when the crop (and pests) are absent). Many workers, following Ehler, have stressed that the practice is by no means new, with one of the earliest documented cases being of Chinese citrus growers placing bamboo poles between trees to facilitate dispersal of the predatory tree ant (*Oecophylla smaragdina*), but it has recently received renewed and innovative focus within IPM. Olkowski & Zhang (1998) noted that nests of *Oecophylla* are still sold in markets in south-eastern China, with the main role of the ants being the killing of larger pests without interfering with the smaller natural enemies of scale insects. Orchards with ants have about 62% less fruit damage than those in which chemical pesticides alone are employed, and – in addition to bamboo or hessian walkways between trees – the tree bases are surrounded by water-filled moats to prevent ants escaping when they reach the ground (Fig. 6.4).

Habitat modification can affect the crops, their pests and beneficial organisms, and the variety of modifications is indeed wide (Bugg & Pickett, 1998). The allied topic of 'farmscaping' (Chapter 7) is the broader manipulation of agricultural environments, with the wide range

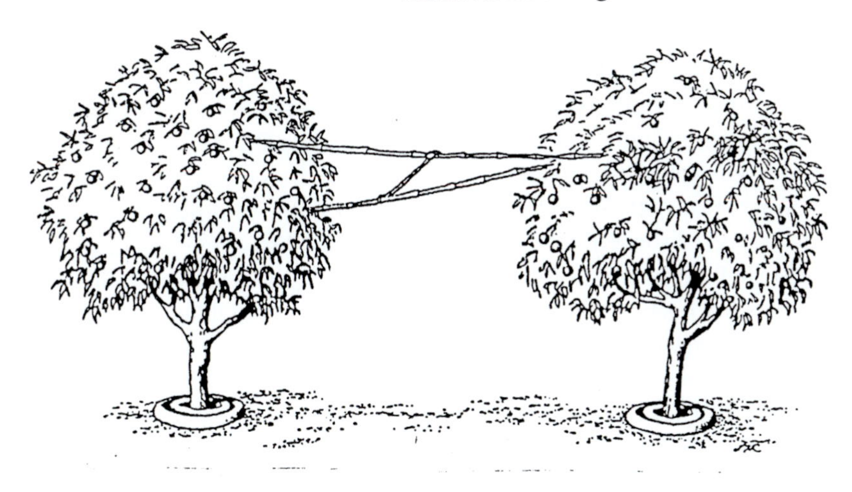

Figure 6.4. Control of pests in Chinese citrus orchards by promoting movements of tree ants, *Oecophylla*, between trees. Bamboo poles or hemp ropes are stretched between trees to allow passage of ants through the canopies, and water moats (made of clay) around the trunks prevent the ants from leaving the system (Olkowski & Zhang, 1998).

of these indicated in cultural controls (embracing conservation biological control) to enhance more sustainable, ecologically based farming systems – albeit with a strong emphasis on plantings and modification of vegetation (Bugg *et al.*, 1998) and drawing on the wider principles of landscape ecology and restoration (Chapters 7 and 8).

More particularly, here, numerous aspects of plant structure, community composition and variety affect the behaviour and wider wellbeing of natural enemies and influence the ways in which plants may mediate the interactions between herbivores (pests) and their predators and parasitoids. For example, a summary of plant effects on invertebrate predators (Barbosa & Wratten, 1998) included the following broad options, which can overlap in various combinations:

- changing the morphology and chemistry of crop varieties;
- changing the growth form of the crops (plant architecture, canopy structure);
- diversifying the vegetation within and near cropping areas;
- changing the spatial and temporal relationships and distribution of crop patches.

As with parasitoids (Barbosa & Benrey, 1998; Marino & Landis, 1996), a suite of factors may operate synergistically or antagonistically to affect performance of predators, by influencing sensory communication and searching efficiency, ability to move freely, fecundity and fertility, and many other features. Agricultural systems subject to frequent disturbance are recognised widely as environments that are difficult for many parasitoids to exploit effectively, with the most extreme cases being annual monoculture systems (Landis & Menalled, 1998). A number of workers have reported these as the most inimicable to colonisation and establishment by both parasitoids and predators. Thus, parasitoid failure in such systems has been attributed to indirect effects of pesticides, lack of adequate food resources (nectar, pollen), paucity of alternative hosts, lack of shelter, and tillage and cultivation regimes. Direct effects (such as killing by pesticides) can remove large numbers of natural enemies, but indirect effects (such as herbicidal killing of weeds) can also have dramatic effects – for example, if those weeds furnish the major nectar or pollen foods needed by adult parasitoids. Weed control may also affect host availability by provision of alternative hosts and by constituting reservoirs for the natural enemies. The physical structure of a crop stand influences movement and attack rates of parasitoids. In one experimental trial (Coll & Bottrell, 1996), the movement of the wasp *Pediobius foveolatus* (Eulophidae) was compared in four crop habitats: beans planted at (1) high density and (2) low density, and beans interplanted with (3) short maize or (4) tall maize. Maize height was a major influence on wasp movement, with lowered immigration/emigration rates within the bean–tall maize combination. Lower emigration rate was interpreted as a response to high shade levels within the crop, but wasps accumulated in these mixed-species, structurally complex cropping habitats, where parasitisation rates on Mexican bean beetle (*Epilachna varivestis*) were higher than in monoculture beans.

The basic rationale of some interactions between farm production practices and parasitoid population dynamics is indicated in Fig. 6.5 (Landis & Menalled, 1998), with the emphasis that production systems can influence (1) complexity of the environment, (2) the spatial distribution of crops and refuge habitats and (3) the disturbance regime, and that a diverse, abundant local parasitoid assemblage enhances chances of successful biological control. Complexity often equates to diversity, and there is abundant evidence that lower diversity is associated with lower complexity, so that agricultural simplification (associated also with fragmentation, disturbance and increased isolation) may decrease diversity. More diverse suites of natural enemies commonly occur in more complex landscapes, where 'complexity' includes a variety of plants and floral structure

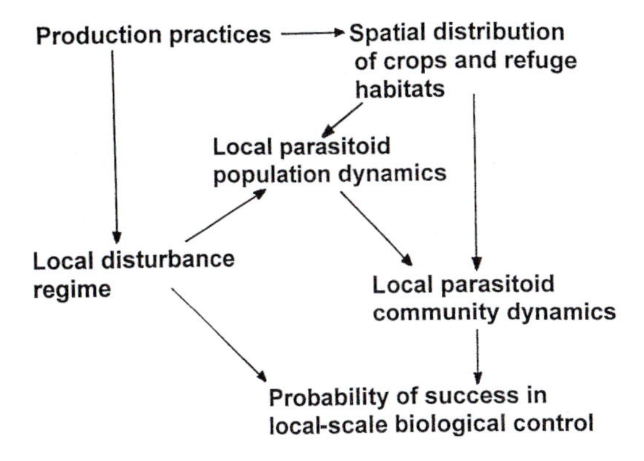

Figure 6.5. Farming practices and parasitoid populations: diagrammatic representation of influence of farming disturbances on local-scale biological control (Landis & Menallad, 1998).

(such as crop, weeds and nearby trees). The abundant literature dealing with the effects of multiple cropping systems on insect dynamics has led to progress in designing cropping systems with vegetational attributes that enhance natural enemy populations and their effects both within and around the crops (see Letourneau & Altieri, 1999; Landis *et al.*, 2000). The considerably diverse studies on these topics have focused mainly on single-species cases, so that the generality of some conclusions remains unclear.

Many features of planting arrangement, and of the physical and chemical features of particular plants, have been deployed in one or both of the twin contexts of 'resistance' to pests and enhancement of their natural enemies. Features such as foliage density, the texture of the leaf surface, general architectural complexity, and others may influence accessibility of prey or hosts to enemies or the suitability of the plants to the pests. The effectiveness of predators may depend on something as apparently simple as their ability to grasp a smooth leaf surface or to move across hairy foliage. One practical interpretative problem is that most experimental trials have examined single variables (for example, smooth versus hairy leaves) rather than the entire physical and chemical environment influencing an interaction between species. A given plant trait may also affect the pest and its consumer in different ways, with different influences at different times or on different age classes or growth stages. Even feeding by a herbivore may change the physical profile of a plant and its attractiveness to predators. Particular plant varieties or cultivars may thus facilitate natural enemy attack, but such varieties are also an important

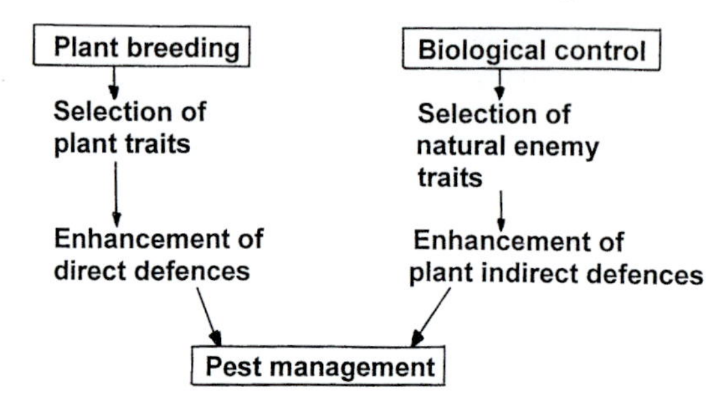

Figure 6.6. Indirect defences of plants against insects (Cortesero *et al.*, 2000). The normally separated themes of plant breeding and biological control may be combined constructively to enhance pest management.

component of pest suppression in their own right if they are developed with characters that confer resistance to pest attack.

Although a discussion of plant defences to herbivores and the factors involved in plant resistance is beyond the main scope of this book, one recent trend (Cortesero *et al.*, 2000) has been to bridge the disciplines of plant breeding and biological control more effectively by selecting plants both for their 'direct defences' to pests and for the components that enhance effectiveness of natural enemies as 'indirect defences' (Fig. 6.6), so selecting for the plant attributes that foster parasitoids and predators. The variety of such components is indicated in Table 6.1. Both morphological and chemical factors may be amenable to alteration. Natural enemies can be affected by various aspects of host-plant resistance to pests, and the view is widespread that a combination of plant resistance and natural enemies may confer higher levels of pest suppression than either of these strategies alone (Bottrell *et al.*, 1998).

Some of the wider principles of natural enemy biology, and their potential for manipulation, were discussed by Schellhorn *et al.*, (2000) (Table 6.2). The evidence that cultural practices can influence the diversity and impacts of these agents is thus substantial, and many IPM practitioners view their role as a primary component for future development. Five key issues emerge in implementing habitat management, both within and beyond crops, as major foci for future research (Landis *et al.*, 2000):

- Selecting useful and compatible plant species. This may necessitate determining the preferences/optima of particular 'key' natural enemies and providing suitable amounts of pollen or nectar, either over short seasons or over more extensive periods, perhaps involving a decision over the relative merits and values of annual or perennial plants.

Table 6.1. *Plant attributes that may influence or enhance natural enemies of insect herbivores and that might be amenable to modification to increase natural enemy impacts in integrated pest management (IPM) (after Cortesero* et al., *2000)*

Influence	Plant features
Provision of shelter	Leaf domatia; ant domatia; other morphological traits, such as foliage density
Mediation of host/prey accessibility	Pubescence; waxy surfaces; glandular trichomes; general architectural complexity
Provision of search cues	Visual cues; chemical cues from volatile chemicals
Influences on host/prey suitability	Nutritional quality; presence of toxic or repellent chemicals; transgenic plant varieties
Mediation of host/prey availability	Transgenic plants; low nutritional quality
Provision of supplementary foods	Floral and extrafloral nectar; pollen; specific food bodies

Table 6.2. *Aspects of natural enemy biology possibly amenable to manipulation to increase their impacts on pests (after Schellhorn* et al., *2000)*

Colonisation potential, varying with location, nearby numbers and habitats, suitability of area for dispersal and living on non-target hosts or prey, 'attractiveness' of location
Suitability of habitats for establishment, increasing attractiveness through high-quality and preferred foods, more abundant food, availability of food over longer periods, favourable microclimatic conditions
Increased fertility/fecundity and/or longevity, higher offspring survival, often mediated through food quality, as above
Natural enemy diversity may depend heavily on cultural practices, which also affect the characteristics of the particular species represented

- Behavioural mechanisms of the natural enemies, and how they may be influenced and manipulated. For example, the use of kairomones (such as attractants) is now recognised widely and seems to have even greater future potential.
- The spatial scale over which the habitat manipulation is relevant, with implications for distribution of refuges, food resources and other important needs. The optimal dimensions of habitats are usually not understood clearly.

- Negative aspects of habitat manipulation – for example, the removal of land from production for use as reservoir habitats may cause initial economic loss.
- Producer acceptance. Practices such as deliberate establishment of weeds in cropping areas may be viewed with suspicion as incompatible with primary, economic motives.

The first three of these topics are founded in ecological principles and knowledge, with the last two emphasising the need for that knowledge to be credible and to be seen to work. Part of the ecological understanding is involved with the dynamics of consumers and the features that may influence diversity in communities. Vegetational diversity and its effects are a key theme, particularly in appraising the ways in which monocultures can be augmented for wider benefits from IPM, and one practical paradox is that diverse vegetation can at times be associated with rather low herbivore abundance. Root (1973) formulated two rather different hypotheses to help explain this. First, his 'resource concentration hypothesis' implies that insect herbivores with a low host range (i.e. that are relatively specialist feeders) are more likely to find and exploit their hosts when these are in pure or nearly pure stands rather than 'hidden' in more diverse vegetation associations or the searching insects confused by chemical diversity. Second, Root (1973) nominated the 'enemies hypothesis', whereby more complex vegetation provides more resources for natural enemies, with the consequence that increased herbivore presence would be countered rapidly by the larger species pool and abundance of predators and parasitoids there.

Both principles have influenced development of IPM, and Russell (1989) reviewed the ramifications of the enemies hypothesis to conclude that the two hypotheses are complementary mechanisms in reducing herbivore numbers in diverse agricultural systems. Russell reviewed 18 experimental trials of the enemies hypothesis, most of which verified the theoretical prediction that enemies caused higher mortality to herbivores in polycultures (in which two or more plant species are grown simultaneously) than to those in monocultures. Russell also noted that 'mechanisms postulated to underlie the enemies hypothesis largely remain intuitively reasonable, but need more testing', and that quantitative studies are needed to move toward more general reliability of predicting natural enemy activity in diverse systems.

The response of parasitoids to plant diversification was addressed by Coll (1998), in the context of determining whether changes in plant community richness, structural complexity and plant density consistently

Table 6.3. *Roles of orchard cover crops in pest management and that might influence the choice of crop in a given context (Bugg & Waddington, 1994)*

Not harbouring important pests at times when these do not infest the crop
Diverting generalist pests away from the crop
Confusing specialist pests by providing conflicting visual or olfactory stimuli, and thus reducing their colonisation of the crop
Altering host-plant nutrition and thus reducing pest numbers and performance
Reducing dust and drought stress, so reducing likelihood of outbreaks of some pests
Changing local microclimates
Increasing natural enemy diversity, abundance or efficiency

affect the rates at which parasitoids attack herbivores. This, together with allied or parallel increases in predator abundance, is assumed inherently in most of the intrafield changes in cultural practices and vegetational management that are now commonplace – and, indeed, in some instances, traditional facets of sustainable agriculture once again coming into practical use after being largely abandoned during the insecticide era. These topics are treated individually below. Many have a variety of cultural benefits extending well beyond pest management. Cover crops in orchards, for example, may help to reduce soil erosion, add or retain soil nitrogen, facilitate availability of some minerals, increase organic matter and improve water filtration and access (Bugg & Waddington, 1994). Even within the restricted benefits for IPM, orchard cover crops may be selected for a variety of different roles (Table 6.3). Many of these have yet to be verified experimentally, but Table 6.3 indicates the considerable range of possible effects for crop protection and, in many cases, for wider environmental benefits from this practice through providing resources that help to sustain numerous native biota.

The overall theme has two main components: modifications within the crop and manipulations to the wider environment such as crop surrounds. The second of these, together with various methods whereby connectivity of the crop and periphery can be increased effectively, is treated in the next chapter. The discussion below summarises some of the more conventional components of cultural controls undertaken within cropping areas.

Modifications within the crop

The topics listed here include many traditional practices of sustainable agriculture, some of them now being developed in more sophisticated ways as a consequence of greater ecological understanding of the interactions between species in communities. The main strategies may be thought of as spatial and temporal manipulations and were introduced broadly by Teetes (1981).

Spatial arrrangements: intercropping

This broad topic includes all aspects of replacement of monocultures by polycultures (above); it thus includes themes such as trap crpping, use of cover crops, strip and other mosaic plantings, companion planting of crops and other schedules that allow increase in crop diversity in a cultivated area. The allied topic of 'weed culture' (Andow, 1991) has much of its applied ecological basis in common with intercropping, in that the intermixed vegetation may provide resources, corridors and/or refuges for natural enemies. The particular combinations of crops, the relative areas of each crop, and any special-purpose requirements may dictate the design of multiple cropping systems. Numerous studies on this topic demonstrate a great variety of cropping systems formulated for different purposes of pest suppression or natural enemy enhancement, with the dispersal powers of the various interacting invertebrates of considerable importance in the success or otherwise of each operation.

As two contrasting examples (after Fye & Carranza, 1972):

- Abundance of natural enemies was not generally increased by diversifying soybean crops by maintaining uncultivated strips as potential corridors for these to enter the crops (Kemp & Barrett, 1989). The predatory anthocorid bug *Orius insidiosus* was more abundant in control (soybean only) plots than in plots with strips of successional weedy vegetation. This vegetation seemed to constitute a 'sink' for *O. insidiosus*, with the bug moving into this, so leading to its decreased abundance in the diversified soybeans.
- In some other cases, the interplanted vegetation acts as a source for natural enemies, with increased abundance of predators or parasitoids in crop areas close to the interplantings. Sorghum interplanted in cotton can be a source of the predatory ladybird *Hippodamia convergens* because the abundant alternative food there early in the season allows build-up of ladybird populations before they are needed in the cotton crop as control agents.

Numerous other cases of both of the above trends have been documented, so that increasing crop diversity in agroecosystems can lead to very variable results in increasing or decreasing abundance of natural enemies and their impacts on crop pests (McLaughlin & Mineau, 1995).

Polyculture leads to increased ecological complexity, because the number of potential interactions between species increases. Andow (1991) described these cropping systems as 'the most ecologically complex of all the systems of vegetational diversity', and noted that they have been given considerably more research attention than other forms of vegetational diversity in agriculture, with the collective studies illuminating the ecology of arthropod interactions in agroecosystems. However, it is indeed difficult to generalise on outcomes of the practice of intercropping. Andow (1991), for example, summarised 209 studies (collectively dealing with 287 phytophagous arthropods in crops). About half of these (149 species) had lower population densities on plants in polycultures, and only 44 species had higher population densities on plants in polycultures (see also Andow, 1986).

Trap cropping

Trap cropping is the use of plants that attract pests or control agents and that are planted deliberately to protect a target crop from damage. The classic approach (Stern, 1981; Hokkanen, 1991) is that such attractive plants can be used to concentrate the pests in small areas, diverting them from the target crop, and where they may be destroyed (for example, by spraying the small areas of trap crops rather than the large areas of target crop, or by exposing them to abundant natural enemies). Hokkanen emphasised that the trap crop must be more attractive than the target crop, and this might occur in either of two main ways:

- the use of a preferred plant species or preferred cultivar grown at the same time as the target crop, so that a mobile pest arriving at a vulnerable crop is diverted from it;
- the use of the main crop plant, but in additional plantings timed to be at their most attractive at the critical times for pest attack, when the main crop is not attractive. This involves tactics such as early or late plantings in relation to predictable seasonal pest attacks.

Hokkanen (1991) suggested that a dual strategy, involving both of these options, may be useful. In general, about 10% of the cropping area is used for the trap crop, but this will depend on the intensity of pest infestation and the mobility of the key pests. However, the technique tends to fail in the trap area generally being too small rather than too large for optimal effects (Hokkanen, 1991).

Dispersion of the trap crop may also be important. For example, an extensive mosaic may be needed for highly mobile pests. For temporal

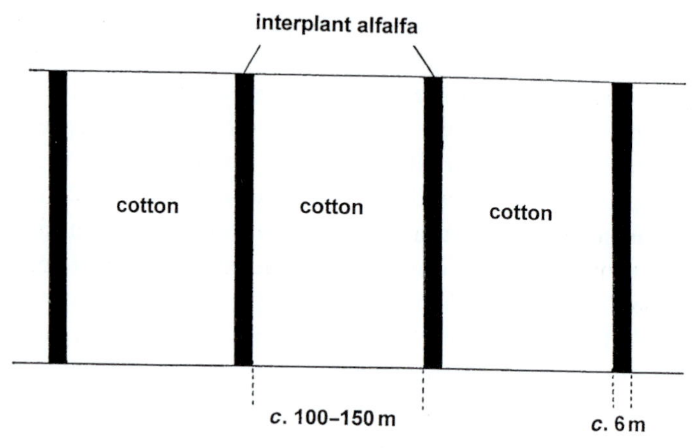

Figure 6.7. Control of lygus bugs by planting of strips of alfalfa as a trap crop in cotton in California (after Stern, 1981).

variations, the trap crop may be better placed within the main crop as strips or marginal plantings (the latter as a buffer if the crop is already vulnerable to pest damage). Roles of trap crops may be enhanced by use of chemical attractants, such as kairomones or food supplements, to increase their attractiveness further. Pheromones of the pest have also been used in some cases. However, Hokkanen (1991) noted that only about 11 pest species (in four major cropping systems: cotton, soybeans, potatoes, cauliflower) had been controlled successfully, with the most important applications being in cotton and soybeans. Despite numerous suggestions that the method is viable against pest Lepidoptera, the above successes comprise only Coleoptera and Hemiptera.

The early classic case of lygus bug control, as key pests on cotton in California, by interplanting with alfalfa was instrumental in popularising the technique. From the late 1960s onward, alfalfa strips of around 6 m width sown at intervals of 100–150 m in cotton were used to control *Lygus* spp. (Fig. 6.7). The alfalfa was maintained as attractive to the bugs by irrigation, and sprayed to kill the pests, largely eliminating any need to spray the main cotton crop. Difficulty in keeping the alfalfa sufficiently attractive thwarted wider adoption of this promising crop combination.

Trap cropping thus also has constraints, with its greatest potential against polyphagous pests or those that demonstrate clear preferences for particular plant-growth stages that can be manipulated by planting times. The major environmental benefit from trap cropping has been to reduce and localise pesticide applications, including reducing the spread of pesticides into the wider environment. However, Corbett (1998) noted the relevance of modelling the behaviour of biological control agents as a tool in helping to improve the development of intercropping in various ways, by working toward rational decisions over topics such as (1) the ratio of the crop to interplanted vegetation, (2) the spatial arrangement of interplanted vegetation within the crop, (3) the timing of planting the various plant components of the system, and (4) the proximity of sources of natural enemies.

Temporal manipulations

Various strategies for providing trap crops at particularly suitable times to affect the wellbeing of the target crop were noted above. They work by disrupting synchrony between seasonal pests and their hosts, or by favouring natural enemy build-up before pests reach damaging numbers. Cultural pest control involves a number of such approaches. These may include changes of crop-planting schedules in addition to more extreme crop rotations, in which a target crop is not planted for a season or more in a pest-prone area. In addition, harvesting may be undertaken on a 'strip basis' rather than across the entire crop area. The example noted above, of interplanting alfalfa strips in cotton for lygus bug control, is also a sound one for strip harvesting, with the alfalfa mown several times each season. If only half the strip width is mown on any occasion, then the pest simply moves to the remaining alfalfa, whereas if the whole strip is mown, the bugs move to the cotton, where damage may ensue (Reynolds *et al.*, 1982).

Differences in growth and maturation rates are important crop-varietal characteristics, which influence both pests and beneficial invertebrates (Hani *et al.*, 1998). Simply selecting optimal varieties for a mixture or for a given area in relation to pest phenology may be a strong aid in crop protection. Thus (1) late-sown maize develops faster than maize sown early and may avoid attack by frit fly (*Oscinella frit*) and (2) attack by another chloropid (*Chlorops pumilionis*) on early cereals can be avoided either by early-sown spring crops or late-sown autumn crops (Hani *et al.*, 1998). In general, early planting may be effective against late-season pests, and the converse. A third option (Teetes, 1981) is sometimes possible: that of adjusting planting times in relation to pests that undergo several discrete generations each year, so that susceptible crop stages occur during periods of low pest 'intergenerational' incidence.

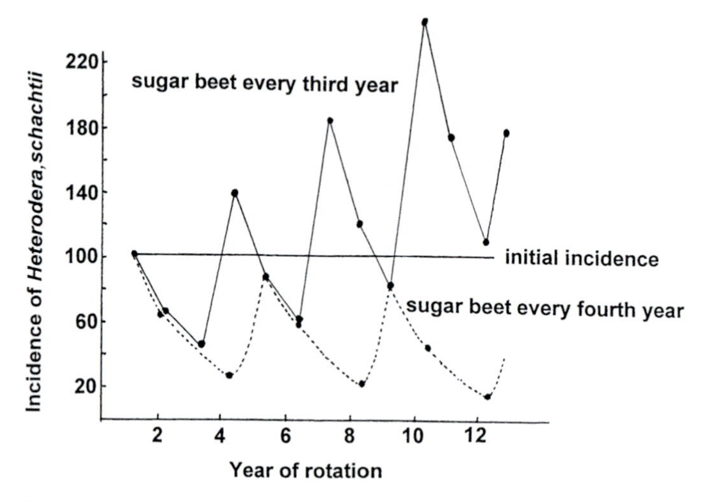

Figure 6.8. Crop-rotation effects on the plant-parasitic nematode *Heterodera schachtii* (after Hani *et al.*, 1998) (see text).

Crop rotation can be important in disrupting the continuous population build-up of pests that stay in the same general area throughout the years. It is most effective as a crop-protection strategy if based on growing the same crop on a plot only every few years, and where the crop is not grown on adjacent sites in successive years, as well as if interspersed with crops not related to the target and on which the pest species cannot be sustained. The latter (so-called 'jumping rotation') can be used to avoid thrips problems in peas, for example (Hani *et al.*, 1998). If rotation includes closely related crops (for example, different varieties of legumes or cereals), then the pest population may be sustained on both (or more) crops used. Many rotations thus include alternations of a cereal, a legume and a root crop, as categories that tend not to share most pest species. In addition to herbivorous pests, rotations can be used to help control soil pests, such as nematodes. Fig. 6.8 shows the influence of crop rotation on the plant-parasitic nematode *Heterodera schachtii*. Increasing rotation interval for sugarbeet crops from three years to four years markedly reduced the incidence of the nematode (Hani *et al.*, 1998).

Planting density

Although maximising yield is a prime concern to growers, the proximity of plants within a crop stand can influence pest movements (and attack), so that spacing of plants (even though sometimes leading to potentially decreased yield) may be a worthwhile cultural control measure in some contexts. Plants growing very close together, with contiguous foliage, provide a closed canopy through which pests can move easily – as, of course, so may their natural enemies. At the other extreme, isolated plants may be more visible as precise targets for invading pests. A'Brook (1968), for example, noted that the landing response of some aphids depends on the contrast between the plant and its background, so that isolated plants attract more pests than densely planted crops. Changes in microclimate may occur with planting density, and the overall significance of the practice for pest management and conservation remain somewhat unclear. Each case may be considered in relation to (1) health and potential yield of individual plants, (2) crop growth and development, and (3) microclimate and biology of key pest species for a particular crop and region.

Crop residues and tillage systems

Most of the impetus for management of crop residues has come from needs to control soil erosion, increase soil quality and conserve water by reducing erosion and run-off, each of which may be important in reducing crop-production costs. Crop residues also affect invertebrate populations and have attracted attention in pest management either by direct treatment or associated soil tillage. These two practices are often difficult to separate, because crop residue levels are most commonly associated with the extent of soil tillage (Forcella *et al.*, 1994). Those invertebrates that live in soil or plant residues, overwinter in soil (perhaps as diapausing stages, such as pupae of pest Lepidoptera) and are likely to be affected by tillage, and there is some correlation between reduced tillage and increased number of insect pests. Many insects increase to damaging levels in no-tillage systems, but reduced tillage also tends to lead to increases of natural enemies of those pests. Reduced tillage favours many kinds of invertebrates, including some Carabidae, spiders, predatory mites (Gamasidae), Collembola and others, as well as fungi (Entomophthorales) attacking aphids (the fungi can survive only on the soil surface) (Hani *et al.*, 1998). In contrast, as noted above, ploughing or intensive cultivation of soil is useful in reducing a considerable range of pests with soil-dwelling stages. These pests, therefore, may benefit from lower levels of disturbance as conservation tillage or no-tillage systems. Again, generalisation is difficult, with many different species-level responses occurring. Thus, different species of Carabidae in Michigan,

USA, responded in different ways to different levels of tillage (Clark *et al.*, 1997).

Tillage and crop residues can also affect invertebrates indirectly, by altering the density of weedy species that host and attract them. Thus, some important pest Lepidoptera (such as the North American corn pests *Elasmopalpus lignosellus* (stalk borer) and *Pseudaletia unipuncta* (armyworm)) lay eggs on grassy weeds, on which the caterpillars initially feed before moving to young corn plants. Corn planting into grass crops or into areas with grassy weeds has increased the risks of economic injury (Forcella *et al.*, 1994). Crop residues can also alter the effectiveness of pesticides by intercepting them.

As Stinner & House (1990) emphasised, tillage is one of the oldest traditional agricultural practices. The most commonly used techniques include 'moldboard' ploughing to turn over and invert the top 20–30 cm of soil to bury surface plant debris. From about the late 1960s, reduced tillage or no-tillage systems became more common, in part because weed suppression could be undertaken effectively by chemicals rather than needing physical intervention. In the most common sense, 'conservation tillage' entails leaving at least 30% of previous crop residues on the surface. The major effects of tillage on invertebrates are threefold:

- Mechanical disturbance – typically by ploughing up to several times each year.
- Residue placement – debris and organic matter tend to be concentrated on or near the surface in conservation tillage systems and incorporated more extensively into soil in conventional tillage.
- Effects on weed communities – as noted above, tillage may affect weed-dependent or weed-utilising invertebrates through direct removal of resources.

In these ways, tillage can have strong effects on invertebrate diversity, distribution and abundance on and near the soil surface. The various effects can reduce (through habitat destruction, increased exposure to climatic extremes or predators) or enhance (through destruction of natural enemies and their refuges) pest incidence and impacts. Conservation tillage is associated commonly with increased abundance of soil-frequenting and litter-dwelling predatory arthropods that are lost by conventional tillage. For example, a study in Georgia, USA (House & All, 1981) showed carabid beetle density in soybean crops was up to fourfold greater in conservation-tillage than in conventional-tillage treatments; another, similar study revealed the mean densities of carabids in soybeans to be 17.9/m^2 and 0.38/m^2 for the same respective tillage regimes. Higher

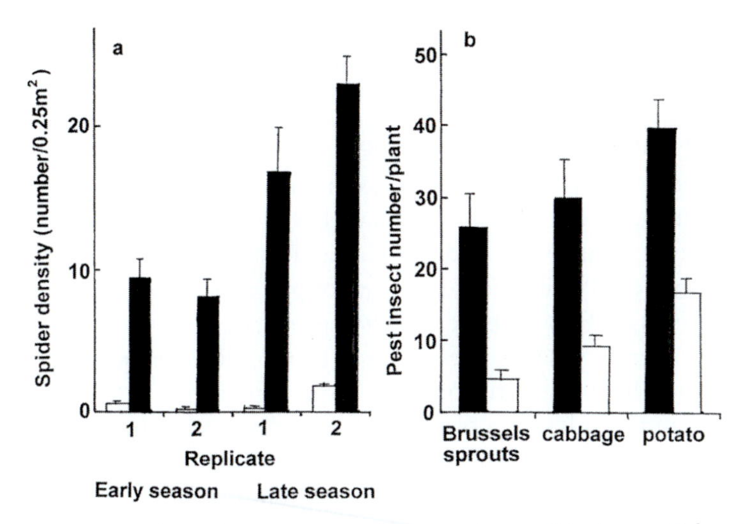

Figure 6.9. Effects of mulching on spider populations (Riechert & Bishop, 1990): (a) density of spiders sampled in mixed-vegetable crop plots, with plants grown on bare ground or with mulch and flowers added. Both early-season and mid-season replicates yielded far higher densities of these generalist predators in the augmented plots; (b) these numbers correlated with numbers of pest insects found in plots of various vegetables, implying a higher level of control associated with higher spider numbers.

diversity of soil-surface macroarthropods than in conventional-tillage systems is a frequent occurrence (Blumberg & Crossley, 1992).

Broader aspects of sanitation, including removal of weed vegetation and fallen fruit, in addition to the conventional crop residues associated with tillage systems, may also be relevant in pest management. Likewise, mulching – although associated primarily with retention of soil moisture and preventing erosion – may provide refuge habitats for pests and/or their natural enemies as organic constructs analogous to crop residues. Thus, mulching potato fields can lead to lessened damage to potatoes because of the greater opportunities for natural enemies to be present and attack Colorado potato beetle, *Leptinotarsa decemlineata* (Brust, 1994). Early-season application of mulch to annual cropping systems in North America can increase spider numbers and species richness, in turn leading to reduced pest numbers and less damage to plants (Riechert, 1998), with these applications coinciding with peaks in spider aerial dispersal and colonisation of crop areas. Similar trends in spider numbers have been reported in German vineyards (Kobel-Lamparski *et al.*, 1993). Spiders require high humidities and moderate temperatures in order to thrive in crop systems and, as Fig. 6.9 demonstrates, mulching both reduces soil-surface temperatures and increases humidity in comparison with bare, tilled backgrounds (Riechert & Bishop, 1990).

Chemical manipulations within the crop

Manufactured fertilisers have, to a large extent, replaced natural manures in cropping areas, because of the precision with which they can be applied, the relative ease of transport and costing, and the fact that many are relatively odourless. Possible consequences for pest management include (1) effects on soil quality and, hence, physiological state of the crop plants, (2) the influences of these changes on pests, and (3) the consequent impacts on natural enemies. Herbivores may respond positively to fertilisers that increase nitrogen levels in plants, and such changes in plant quality and attractiveness may thus flow to predator or parasitoid wellbeing.

Numerous options thus occur for cultural controls, and many have potential to be designed to effectively counter key pests for a particular crop and environment. Many also have important roles in promoting wider conservation of invertebrates but, in order for this to achieve greater potential, the crop must be considered as part of a wider landscape environment.

7 · *Extending beyond cropping areas*

Impacts of crop protection measures can extend well beyond the immediate cropping areas. Cultural control strategies for pests in particular, increasingly incorporate considerations of the wider environment – especially the features of field margins. This chapter summarises aspects of these wider needs for enhancing habitat quality and re-creating natural habitats for invertebrates near agroecosystems and the need to consider these components as parts of a larger system for management and conservation. Detailed appraisal of the roles of boundary features such as hedgerows and headlands in Europe has helped to clarify many aspects of their roles as refuge habitats, barriers or dispersal corridors for numerous invertebrates, including natural enemies of crop pests.

Introduction

Aspects of cultural and conservation biological control extend well beyond the physical confines of the crop and can be important in integrating agricultural productivity with the wider landscape: 'There is clear evidence that plants outside the cultivated field may provide the necessary resources to increase the impact of natural enemies' (Ferro & McNeil, 1998). In particular, modifications to the physical structure and vegetational composition of field-margins and nearby areas have received considerable attention. As Wratten *et al.* (1998) noted, field-margin refugia (as well as within-field refugia, as noted in the last chapter) have substantial values in influencing the interactions between pests and their habitats and natural enemies. Wratten *et al.* recognised five main mechanisms relevant to considering these interactions, as follows:

- the provision of overwintering or aestivating sites, or other, more temporary refuges;
- the enhancement of the amounts and kinds of pollen and/or nectar available to sustain parasitoids and predators;

- the provision of alternative prey for predators or of alternative hosts for parasitoids;
- the provision of plant food for carnivores that are occasionally polyphagous;
- The interactions involving tritrophic level interactions between the crop and/or non-crop, the pest and/or non-pest, and the predators and parasitoids.

They thus expand cultural control into the broader landscape but emphasise the importance of controlling agricultural peripheries in a variety of ways that lead toward integrating agroecosystem management with wider environmental values of importance for wellbeing of a much greater variety of native biota. They also help to understand how associated and planned biodiversity (Chapter 2) interact. Although most attention to understanding the biology of field margins has been prompted by aspects of conservation biological control, wider interactions with more natural ecosystem components is a major benefit – even though these are treated most frequently as tangential to the main focus of improved crop protection. Much cropping area in temperate zones is covered with vegetation for only a few months each year, so that seasonally active natural enemies may need to move between crop fields and adjacent parts of the landscape in order to find resources and a place to live.

Permanent plantings around the borders of fields, orchards, vineyards and the like can thus enhance cultural and biological control measures as well as provide wider environmental benefits. They can be viewed as components of 'farmscaping'. Farmscaping (Bugg *et al.*, 1998) is 'the modification of agricultural settings, including management of cover crops, field margins, hedgerows, windbreaks, and specific vegetation growing along roadsides, catchments, watercourses, and adjoining wildlands'. Whereas these modifications, particularly plantings, are not always viewed as augmenting or constituting IPM, much of their design has deliberately sought to attract, sustain and augment impacts of beneficial species and other wildlife. An allied topic has been to foster access of beneficial organisms to the crop by providing direct conduits – such as through strips of natural habitat established by planting weeds or establishment of traversing 'beetle banks' (see p. 226), corridors and refuges from which natural enemies may reach pests on crops far more readily than if left on the crop periphery. Specific designs to further this aim include a considerable variety of native and exotic trees, shrubs, annual and perennial herbs and grasses, which provide numerous opportunities for associations between invertebrates and non-crop plants in the vicinity of agroecosystems and, indeed, also within the cropping areas on occasion. Weeds have received considerable attention, not least because many are rapid-growing and fast-colonising plants that can provide foods and microhabitats not available in large weed-free crop monocultures. Some crop pests are more abundant in weed-free fields than in more diversified crops, and crop fields with dense weed cover and high diversity commonly have much higher levels of predatory arthropods than do weed-free crops, so that crop diversification by retention (or even promotion) of weeds can enhance IPM.

The themes treated in this chapter, essentially those of habitat-quality enhancement and the re-creation of natural habitats around and within agroecosystems, lead naturally to broader application of restoration ecology over larger areas, and that development is treated in Chapter 8. They fundamentally augment and increase natural habitats and resources within the agroecosystem-dominated landscape, with the broader maintenance of natural areas between agricultural units (such as by networks of separated or interconnected fields), sometimes referred to as 'ecological compensation areas' (Nentwig *et al.*, 1998). These areas are so-named because they 'compensate' (in part) for the negative effects wrought by agricultural establishment, increasing the species diversity in the landscape, enhancing resources for native biota, and providing refuges, dispersal centres or more permanent habitats for numerous invertebrate species. Their design for IPM values and benefits has many lessons for invertebrate conservation – both for management of individual species and for the general principles of enhancing local biodiversity. The areas may be dispersed in various ways, but field margins have been a prime focus (Boatman, 1994); examples of ecological compensation areas include sown weed strips and wider areas such as conservation headlands, in addition to hedgerows and other boundary features. Although viewed widely as 'natural' (or, at least, 'semi-natural') in relation to cropping areas, these are nowadays largely artificial constructs in that they have been designed and structured deliberately for IPM benefits and, in fewer cases, also for conservation. Thus, as Nentwig *et al.* (1998) noted, 'even rare species can occur in intensively managed agricultural landscape if the minimum habitat area they require is available and their specific ecological requirements can be filled'. There is no overall 'formula' for the extent of this land component, but these authors also suggested that ecological compensation areas, defined as 'totally unmanaged (and unused) or only partly managed areas', should (at a minimum level) be 5–10% of an agricultural landscape. Attaining this proportion can involve considerable sacrifice of productive land and almost always necessitates considerable augmentation of existing areas in these categories.

Two main options have been suggested for this augmentation (Nentwig *et al.*, 1998), as follows:

- separating selected areas, perhaps involving removing them from cultivation, and allowing normal plant succession to proceed; and
- sowing a given seed mixture to help direct succession in particular directions.

The first of these is regarded as 'natural', and the second as 'semi-natural'. The latter has the possible advantages of establishing beneficial plants deliberately, avoiding additional weeds and controlling existing weeds more readily than in natural succession areas. It may also be the more practical approach for use in intensively managed small-scale areas, whereas natural succession may be preferred for larger, landscape-level management. Both processes are components of both augmentation and re-creation of habitats, with the major options for the latter approach being threefold (Kaule & Krebs, 1989):

- Habitat creation through promoting natural succession, i.e. by preventing further disturbance and impacts and allowing natural processes to resume. The outcome of this 'leave-alone' approach is likely to reflect the condition of the particular habitat and its proximity to source areas of natural biota.
- Transplantation of ecosystems. If habitats have been destroyed, or in places where new habitats are desired, it is sometimes feasible to transplant complete habitats – for example, of grassland or hedgerows – to provide the foundation of natural ecosystems present earlier in the area. Several early examples were discussed by Buckley (1989).
- Planting and sowing local genetic varieties of plants. Many habitats adjacent to agricultural areas contain some 'natural habitats' but with substantially reduced species diversity. Their quality and extent may be enhanced through enrichment by introducing additional plant species.

In addition, high priority is needed to preserve existing natural remnants within the agricultural landscape. Many of these are likely to involve field margins, emphasising yet again the values of these areas for wider conservation.

The values of ecological compensation zones for pest management are enhanced by reducing field size (Hani *et al.*, 1998), whereby the controlling agents may invade crops more easily. Several workers have recommended a mosaic of such areas (Duelli, 1992), with varied components such as hedges, wide field margins and subdivision of large cultivated areas by strips of natural or planted vegetation to constitute 'artificial borders'. Hani *et al.* (1998) also noted the considerable values of permanent meadows (with low fertiliser inputs and high botanical diversity) serving as permanent reservoir habitats for biological control agents.

Field margins

Much of the pioneering research on the values of field margins for IPM was undertaken in Europe, corroborated by studies in North America (Bugg *et al.*, 1998), with the principles now recognised and applied in many places. Whereas, for example, hedgerows have long been planted to delimit property boundaries, as windbreaks (see p. 241) and to provide food and (with coppicing of incorporated trees) firewood, their values to beneficial arthropods have been considered increasingly in their design, so that enhancement of biological control by manipulating their structure and composition is now commonplace. In many instances, local native plant assemblages earlier reduced by agriculture have been re-established. In addition to marking property and field boundaries and contributing values to landscapes, one of the most important historical functions of hedges has been to constitute stock-proof barriers to enclose cattle and sheep. In Britain, hedges are still most prevalent in dairy-farming regions (Doubleday *et al.*, 1994), where they may contain a high proportion of thorny plant species, which help to enhance their containment values for stock. Additionally, they provide shelter for livestock in inclement weather. Elsewhere, in arable cropping regions, such roles are minimal, and it is there that the greatest losses of hedgerows has occurred. Hedges are human constructs, with their useful practical attributes long-recognised (see Table 7.1), but another recent trend (reflecting labour costs) is that traditional management of hedges has been replaced by mechanical slashers/trimmers/flail cutters used in ways analogous to a coppice cycle. In general, arable cropping has been associated with loss of hedges and decline in their maintenance, with hedges regarded widely as being of little practical interest and, indeed, undesirable in places as sources of crop pests and weeds. A narrow 'sterile strip' is a common modification, as a managed (e.g. sprayed, mown, or both) gap between the hedge bottom and the crop headland, and has the main functions of:

Table 7.1. *The 17 most important attributes of a hedge, as formulated by Malden 1899 (from Chapman & Sheail, 1994)*

The hedge would:

1. Develop in a reasonably short time.
2. Be long-lived.
3. Be easily repaired, if neglected.
4. Be uniform in growth.
5. Be easily kept within suitable bounds.
6. Present a compact front.
7. Prevent animals from escaping, ideally by having thorns.
8. Be easily grown from seed.
9. Be adaptable to most soils.
10. Be able to withstand severe weather.
11. Afford shelter to livestock in cold winds.
12. Produce shoots close to the ground, for containing small animals.
13. Afford little harbourage for insects.
14. Be able to withstand fungal and other diseases.
15. Have reasonably compact rooting systems.
16. Be able to withstand browsing by livestock or game.
17. Be able to regenerate, when cut down to or near its stumps.

- facilitating control of physical spread of weeds from the hedge bottom, by spraying, mowing or cultivation;
- reducing risks of applying agricultural chemicals to the hedge bottom by providing a clear gap;
- constructing a clear track for machinery passage and access (Helps, 1994).

In Britain, hedges have been regarded as the most ecologically significant kind of field margin, but it is worth noting that they are only one of several rather distinct features that constitute such boundaries (Table 7.2). Riverside and streamside vegetation contributes significant biodiversity to many landscapes and is managed in a variety of ways – including fencing off to varying extents or with bordering land grazed or cropped right up to the water's edge. Roadside verges resemble hedges and streamsides in supporting species-rich communities and may contain more species than either of those (see Bunce & Hallam (1993) for Britain). Walls have received comparatively little attention but must be considered among the array of linear features in many agricultural landscapes. Morris & Webb (1987) pointed out some additional values of dry stone walls in Britain,

Table 7.2. *Types of linear field margin that may need to be considered for management in Britain (Bunce* et al., *1994)*

Hedgerows
Streams
Roadsides
Walls
Grass strips/fences/banks
Boundary margins, such as uncultivated strips
Green lanes

particularly in areas such as moorlands, where trees may be scarce or absent. The many spaces between stones can provide retreats or refuges, and it is possible that the lichens growing on walls provide both cryptic resting places and foods for certain insects. Likewise, grassy strips and boundary margins may harbour plants (and their consumers) lost from adjacent cultivated areas. 'Green lanes' are sunken old roads, usually long disused, and can harbour woodland plants otherwise lost from lowland landscapes in Britain (Dover *et al.*, 2000). Any of these margins may be homogeneous or more varied in character, for example by supporting trees intermittently along their lengths, and each can thus have ecological significance and values.

A generalised pattern of a field margin, to illustrate the main components and terminology, is shown in Fig. 7.1. The boundary (hedge, wall, or other, as above) may be accompanied by a constructed ditch or drain. The boundary strip inside this includes any track for machinery passage and typically a mown or sterile strip, as above. The crop includes a headland perimeter (below), a marginal area in which agricultural practices may be modified substantially, and the 'crop proper'. The term 'headland' refers strictly to the area for turning of farm machinery and thus present normally on only two (opposite) sides of an area, but it is now often used more widely. For aspects of cultural control and conservation, hedgerows and headlands have received far more attention then other boundary features.

Hedges and conservation headlands

In Europe, particularly, hedges have been the most common type of field boundary (Pollard *et al.*, 1974). However, they were reduced progressively

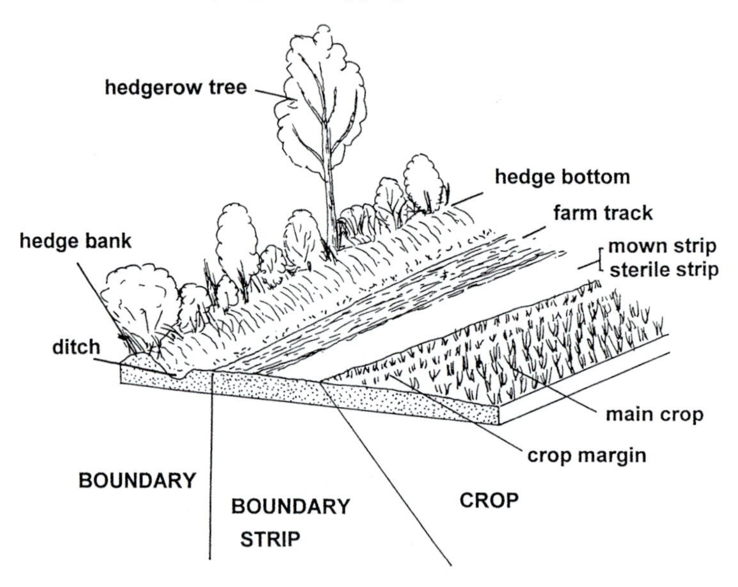

Figure 7.1. Generalised structure of a field margin in Great Britain, showing components and terminology (after Greaves & Marshall, 1987).

over much of the continent, with the increasing efficiency of agriculture in the mid twentieth century dictating the need for larger fields for cultivation and these being constructed mainly through amalgamation of smaller units through removal of their intervening boundaries. Their restoration in more recent years is an important theme. Likewise, 'conservation headlands', also a European innovation, are a more recent modification (Boatman *et al.*, 1989), but – particularly in combination with hedges – they have contributed substantially to aspects of conservation near arable systems.

The early impetus for the conservation headland concept (Boatman *et al.*, 1989) was the realisation that pesticide use on crops reduced the amount of insect food that constituted the major part of the diet of grey partridge (*Perdix perdix*) chicks during their early development. Increased rates of chick mortality were related both to the insecticidal effects directly reducing food supply and to herbicides removing the weed hosts of many prime food species of arthropods. From 1983 onwards, experiments were undertaken in which pesticide use was modified in 6-m-wide strips around the perimeters of cereal fields, these edges being the preferred habitats for partridge chicks. Such modifications led to greater numbers of insects in these areas, with subsequent dramatic increases in

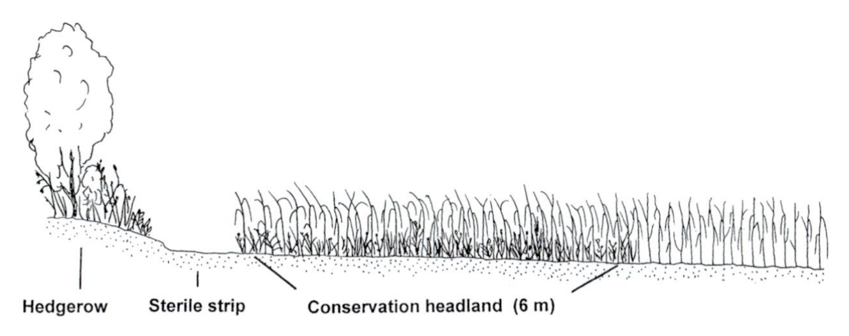

Figure 7.2. Principle of conservation headlands (after Boatman *et al.*, 1989). The outer length of a tractor-mounted spray boom is turned off, so that the width of the strip is determined by this non-sprayed area.

partridge survival rates, in addition to increased numbers of another game bird, pheasant (*Phasianus colcichus*). Implementation of the technique from 1983 to 1986 increased density of partridge breeding pairs threefold, and it has been refined progressively since that time to incorporate a wider variety of conservation values.

The basic principle of conservation headlands (Sotherton *et al.*, 1989) is shown in Fig. 7.2, exemplifying that the headland width is essentially that of a then conventional spray-boom unit length, so that the selective sprays can indeed be applied without modification of the practice elsewhere in the crop. Thus, as Boatman *et al.* (1989) pointed out, most fungicides had no insecticidal activity so that their use was continued, and this practice was partially responsible for the change in terminology from 'unsprayed headlands' to the later 'conservation headlands' now employed almost universally. Likewise, selective herbicides can be used to control particular problem weeds, but (for example) broad-spectrum weedicides formerly used against dicotyledons are largely prohibited. Guidelines for users dictate which herbicides to use, and at which times of the year, without lessening conservation benefits.

Realisation that this approach could help conserve species whose populations were previously reduced by agrochemicals has had wide impact, with early trends in increased abundance and diversity reported for groups as diverse as butterflies, mammals and rarer arable flora. The butterfly example is of particular relevance here, not least because it demonstrated the wider effects on a non-target group of a practice that could be implemented primarily for applied reasons and IPM benefits, and introduced the real importance of field margins as habitats for wildlife as a current important component of agroecosystem-level conservation practice.

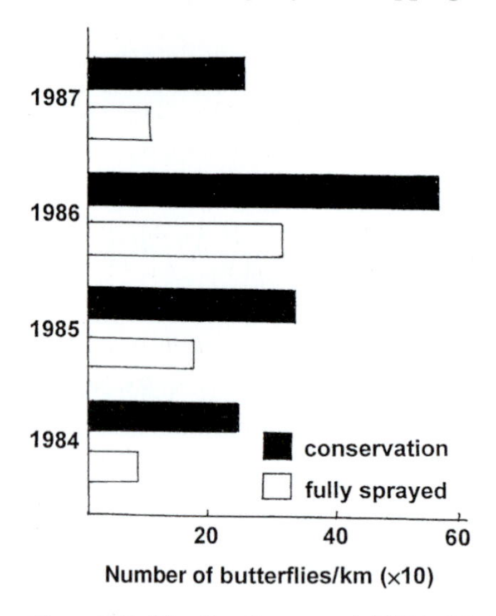

Figure 7.3. The first four years' (1984–87) information on trends in butterfly abundance from monitoring in conservation headlands around cereal fields in Britain (Boatman *et al.*, 1989). The numbers seen in conservation and fully sprayed headlands differ significantly ($P < 0.001$) for each year.

Fig. 7.3 (Rands & Sotherton, 1986; Boatman *et al.*, 1989) demonstrates the initial findings (first four years, 1984–87) of monitoring butterfly abundance on conservation headlands and fully sprayed crops as paired treatments in southern England, based on weekly survey counts during the summers. In all four years, significantly more butterflies were seen on conservation headlands than on sprayed headlands. Observations on butterfly behaviour helped to emphasise the importance of conservation headlands to the insects; green-veined white (*Pieris napi*) females, for example, were observed to feed almost entirely in conservation headlands rather than in the nectar-impoverished nearby sprayed regions. Dover (1994) noted that around half the British butterfly species have been recorded from arable field margins and emphasised, again, the conservation roles of headlands. On average, during a five-year survey (Dover, 1991), 68% of all butterflies seen on paired transect series were in conservation headlands. Dover attributed the greater range of behaviour observed in conservation headlands to the presence of additional nectar sources (with nectar regarded as a limiting factor for butterflies in croplands) and larval food plants. Augmentation of such resources (see below) has clear potential to increase habitat-carrying capacity for a considerable variety of invertebrates in field margins.

This principle has wide application in seeking to manipulate field margins to encourage natural enemies by raising their local densities adjacent to cropping areas. Provision of flowering strips along field borders can, for example, increase local densities of aphidophagous hoverflies (Syrphidae) and their species richness (Harwood *et al.*, 1994). Using yellow pan traps at different distances from a field boundary into the crop to sample adult Syrphidae suggested that their densities can be increased by planting field margins with wild flowers. However, some results were anomalous. The most abundant aphidophagous species (*Episyrphus balteatus*) showed no increase, and the most significant treatment effect (for species of Eristalinae) was for species whose larvae are not aphid-feeders. Whereas there is clear evidence that such plantings increase syrphid numbers, the overall relevance of the practice to aphid management is complex. For syrphids, in trends probably relevant also to other groups, Harwood *et al.* (1994) raised four main general considerations abut such artificial augmentations:

- Does providing floral resources really increase the local total population size of hoverflies, or does it simply influence their spatial distribution by concentrating them on superabundant resources? At the time of Harwood *et al.*'s paper, this was regarded as unclear, with work needed to determine whether fertility and fecundity were indeed affected by the quality and quantity of pollen and nectar available.
- Where are the floral resources best positioned for maximum control of aphids? Harwood *et al.* pointed out that this question had never been considered directly, and knowledge of the dispersal of syrphids (including any impediments, such as barriers to movement) is critical. One experiment cited (Fig. 7.4) showed that linear features could indeed act as barriers to syrphids (see also Chapter 8), in that higher proportions of flies that had ingested *Phacelia* pollen were caught in pan traps 10 m from a *Phacelia* strip than when a road or creek plus hedge were present in this short interval.
- Is provision of flowering strips for aphidophagous hoverflies likely to make economic sense? Strips, for example of *Phacelia*, are cheap to provide, need little maintenance and need not reduce crop area. Any reduction of pesticide use resulting from the practice may be economic gain.
- Are there potentially deleterious effects of providing additional floral resources along field margins? Two possible deleterious effects were noted: (1) possibly reduced levels of pollination in native wild plants because of concentration of pollinators in the planted strips, and

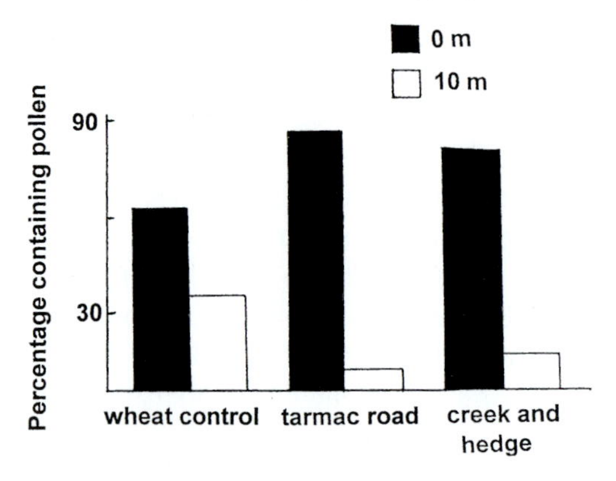

Figure 7.4. Effects of linear features on abundance of syrphid flies (Harwood *et al.*, 1994). A higher proportion of syrphids containing *Phacelia* pollen were trapped (yellow pan traps) when no barrier was present (a) than when either a tarmac road (b) or a creek and hedge (c) were present between the pollen source and the traps. Traps all 10 m from *Phacelia* strip.

(2) whilst concentration of syrphids could provide local (i.e. field-scale) reductions in aphid density, this might also result in increased regional density because of decreased predation in other fields resulting from displacement of the hoverflies by attracting them through resource enhancement. A possible parallel (Holland *et al.*, 1994) is that parasitoid Ichneumonidae were also considerably more abundant on planted *Phacelia* strips than on adjacent wheat crops.

Regenerated field margins (either natural regeneration or enhanced by seeding or planting) are important also for pollinators of crops, with, for example, bumblebees and honeybees preferring different flowering species in such areas (Kells *et al.*, 2001). In a strictly defined sense, uncropped strips differ from conventional conservation headlands in that they are outside the crop, but the two are often confounded. They can provide substantial nectar resources for foraging pollinators on areas where nectar is otherwise scarce. Initially, such areas in Europe were designed to be 'buffers', serving to conserve arable flora with their associated consumers in areas freed from pesticide use and with reduced competition from other species (Boatman, 1994). The broad values of field margins for pollinating insects have been confirmed on many occasions, with strong evidence of the preferences of different insect taxa for different vegetation types or flowering species (Lagerlof *et al.*, 1992) emphasising the needs for protecting margins from herbicides and using other means to cultivate a wide variety of wild plant species there. The key management issues bearing on the values of small areas with flowering plants between arable fields are how to avoid weed problems and how to retain and encourage the most attractive floral mix with minimum effort and costs. For bees, Banaszak (1992) also recommended increased diversity in agricultural landscapes, with different types of shelter belts needed and total cultivated land not to exceed three-quarters of the total land area, with the remainder serving as refuge.

More broadly, field margin management has two main target roles in conservation, according to de Snoo & Chaney (1999):

- To create unsprayed margins to help conserve rare or 'red list' species. This approach implies focus on species of defined conservation significance or interest, so that management can be targeted directly toward the needs of such species. The measures needed, however, may be regarded as drastic and incompatible with the highest levels of crop protection – for example, if crop protection is compromised by non-use of pesticides or anticipated yields are reduced by lack of fertiliser use. Such target species may utilise margins as permanent habitats or as more transient refugia.
- To create unsprayed margins not for particular focal species but to compensate for the indirect effects of pesticides by providing refuges and resources. As de Snoo & Chaney emphasised, this tends to have the aim of maintaining common species as common and is largely epitomised in the conservation headlands approach (see p. 00).

However, in order to focus field-margin management more effectively, rather than rely solely on the basic general principles (often, in practice, the only way to proceed), more research is needed on spatial aspects and roles that should be accorded priority. To answer such questions, 'we not only need knowledge of single species and populations, but also scientific knowledge about metapopulations, agricultural foodwebs, agrocommunities, agroecosystems, both at the farm level and the wider landscape' (de Snoo & Chaney, 1999, p. 9).

Numerous studies have documented the diversity of selected arthropod groups on a variety of different field margins, with the results having important implications for invertebrate communities, particularly those associated with European cereal fields and their boundaries. This is historically perhaps the most intensively studied context (Hassall *et al.*, 1992). Details of the margins, the approaches to sampling and evaluation, and the levels of analysis differ widely (Dennis & Fry, 1992), but the questions addressed commonly include:

- Does habitat quality in field margins determine the density and/or diversity of natural enemies overwintering there?
- Are the features that render field margins good for supporting natural enemies also beneficial in fostering/sustaining wider arthropod or other invertebrate communities?
- Does the overwintering density of natural enemies determine their distribution and abundance in crops the following season?

Whilst the last of these points is of greatest practical interest to farmers, the others have clear and wide conservation relevance. Many studies have thus compared the abundance and diversity of focal invertebrate groups in margins and at various distances into the crops, and the ways in which these change during a growing season. Strong seasonal dependence of invertebrates on field margins is relatively common, with the immediate economic values of field margins in arable land related to chances of greater predation on crop pests.

Within-field habitats for natural enemies

Various kinds of within-field habitats have been demonstrated to be important for natural enemies of crop pests, and their design and construction is being adopted increasingly to facilitate contacts between potential biological control agents and their target pests and to provide habitats in which agents may breed and take refuge.

Thomas *et al.* (1991, 1992a, 1992b) drew on knowledge of the important features of field margins of cereal fields in Britain in particular, that a raised boundary with rough (tussock) grass cover can augment densities of spiders, predatory beetles and earwigs) to re-create 'beetle banks' (Sotherton, 1995) within the crops. These are grass-covered raised earth banks. Initial designs were ridges 290 m long, 0.4 m high and 1.5 m wide, leaving areas of 20 or 50 m of cultivated crop at each end to allow for passage of machinery. Banks were hand-sown with various grass species (taking care to avoid aggressive weedy species likely to contaminate the crop). Beetle banks provided effective overwintering refuges for predators and in the first year of trials were found to harbour many species of Araneae, Carabidae and Staphylinidae, with predator densities up to 150/m^2. These densities increased greatly in the second year, to exceed 1500/m^2, thus considerably exceeding those from previous records in field margins. As the banks were treated as permanent features of the fields, successional changes by the third winter led to some turnover of the species present, with some replacement of open-field Carabidae by more boundary-frequenting taxa, possibly reflecting greater stability within the assemblage.

Another, related approach has been the use of sown weed strips in crops. Sown weed strips were advocated by Nentwig *et al.* (1998) as a form of ecological compensation area (see p. 215) and, as well as being incorporated in field margins, can also be used to divide large fields with distance between strips around 50–100 m. The sown strips are commonly around 3–8 m wide, with a selection of regionally appropriate plants included. The prepared seed mixture recommended for Switzerland (Nentwig *et al.*, 1998), for example, comprises 29 species of wild flowers and is available commercially. Selection principles for the plants included using different-sized plants (low, middle-sized, tall), early to late flowerers, and annuals, biennials and perennials. They also needed to be able to survive in agricultural soils and in complex mixtures of species, and be able to support and foster beneficial arthropods. Species that may act as reservoirs for crop-plant pathogens were avoided, but those that might harbour potential pest insects were not avoided. Suggested subsequent management included alternating mowing of half of each strip every second year to slow succession, and early cutting effectively reduced undesirable emerging weedy species. Sown weed strips are important refugia for natural enemies of crop pests but also support numerous additional plant and animal species.

Both these examples reflect the general trend, associated with promoting lower-impact farming, of attempting to enhance biodiversity by reducing the size of large fields, sometimes using smaller cropping areas in an integrated rotation – an approach that Hart *et al.* (1994) stated 'lends itself ideally to the use of flower or grass margins' to divide large fields and to provide resources suitable for natural enemies. Hart *et al.* regarded raised margins (0.3 m high) as preferable to flat margins because they allowed for better drainage and provided more distinctive habitat; they trialled a variety of different plant combinations ranging from two to seven species. These ranged from a grass mixture recommended as standard for beetle banks and others representing a 'headland mixture' and an 'enhanced fallow' regime to control margins treated with the herbicide glyphosphate to eliminate plants. Preliminary results comparing numbers of selected arthropod groups in these trials indicated that spiders were significantly more abundant in sown margins, and Staphylinidae were significantly less abundant in sterile and enhanced fallow margins than in grass and wildflower margins. The latter were thereby regarded as likely to be more attractive to beneficial arthropods.

Interactions between in-crop refuges and cropping areas can be complex, and refuges can have strong impacts on assemblages of taxa such as carabids (Lee *et al.*, 2001) in being considerably more attractive than bare fields as overwintering sites. Such refuges may therefore protect populations from disturbance and serve as sources for natural enemies for disturbed areas. As with field boundaries, they may be far more permanent landscape features than the adjacent ephemeral crops and participate in 'cyclic colonisation' of natural enemies from more permanent to more temporary habitats. Such areas may buffer the consequences of insecticide applications affecting carabids in the crop: assemblages in crops may be reduced rapidly by toxic chemicals, but presence of beetle banks as refuges facilitates exporting carabids to crops, with post-insecticide treatment beetle densities being much higher than in the crops (Lee *et al.*, 2001).

8 · *Field margins and landscape ecology*

This chapter expands on the themes introduced in Chapter 7 to treat the wider aspects of landscape ecology involving cropping areas and field margins together as parts of a wider landscape, and their relevance in invertebrate conservation. Aspects of invertebrate distribution, movement patterns and population structure are integral to considerations of wider conservation in and around agroecosystems.

Introduction

Conservationists are concerned with properties of landscapes, because the interactions between their various components and features affect strongly the ecological systems present and the sustainability of ecological processes, and influence the wellbeing of various species occurring there. So-called 'patch dynamics' entails defining not only the properties of 'patches' (usually equated to 'suitable habitats') but also the ways in which boundaries and intermediate areas of the overall mosaic influence connections and communication between them through facilitating or impeding movements of organisms. As Chapman & Sheail (1994) noted, 'a landscape of hedgerows of varied density, size and composition provides an obvious test-bed for such explorations in landscape ecology'. Broadly, the properties of linear features have substantial implications in broader conservation, and some exploration of those properties in relation to invertebrate wellbeing is highly relevant to wider conservation management of agroecosystems. Fry (1994) illustrated a variety of ecological features of field margins relevant to conservation management, emphasising the dual roles of margins as conservation assets and having roles in crop production and sustainability in farming systems (Fig. 8.1). He thus pointed the way to more holistic landscape management, in which agroecosystems are important and inevitable components. Many of the main management systems have resulted directly from greater ecological understanding and have been formulated from studies on particular species (or

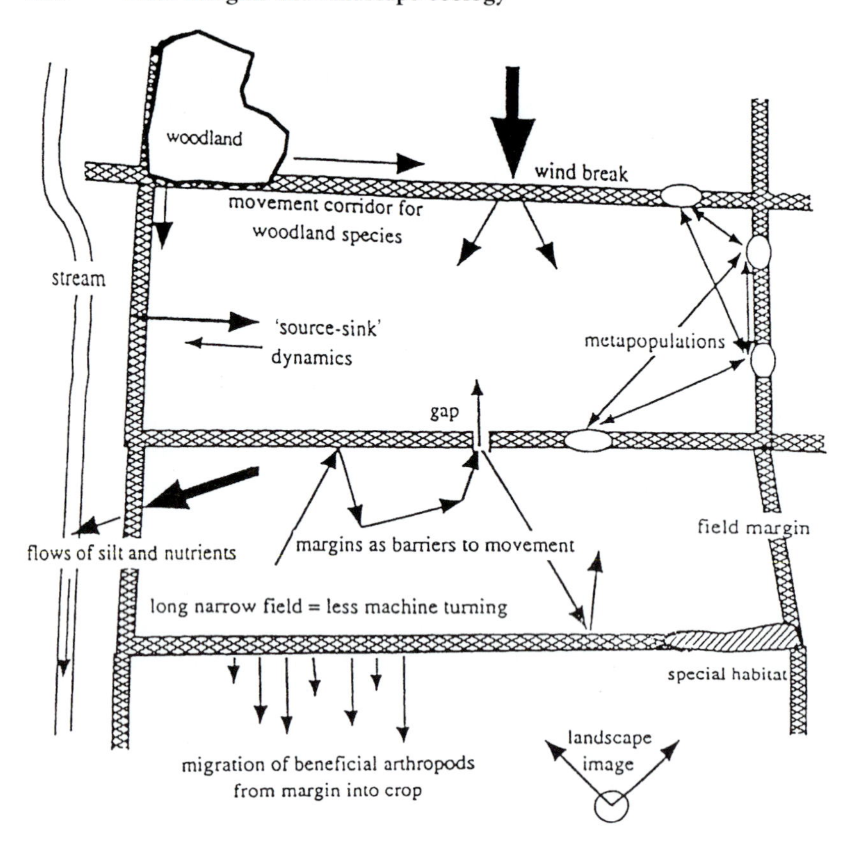

Figure 8.1. Landscape ecological features of field margins and their functional attributes in relation to management for invertebrates (Fry, 1994).

groups of species), so that ecological variety and species characteristics are also a prime concern in determining how they interact with landscape components. The underlying principle, discussed by Benton *et al.* (2003), is that landscape heterogeneity is important in fostering and supporting biodiversity and that management should seek to re-create and conserve heterogeneity as a key to sustaining native biota in agricultural ecosystems.

Lubbe's (1988) categorisation of the more natural habitats in agricultural landscapes into 'linear features' (including margins, tracks, windbreaks, watercourses and others) and 'insular features' (such as remnant patches, groups of shade trees, ponds and fallow fields) is a useful broad basis for stimulating discussion on their functions and ecological roles. Within each of these broad categories, a variety of different microhabitats greatly influence invertebrate incidence and diversity, and many of these microhabitats reflect the structure and composition of vegetation, augmenting the mosaic nature of such broad habitat categories. The dispersion of species in the landscape, their capacity for dispersal, and their population structure are all pertinent and are alluded to in Fry's diagram. One helpful categorisation of animal species in agricultural landscapes (Macdonald & Smith, 1990) recognised three patterns of dispersion:

- Species restricted largely to cropped habitats: crop-dwelling species.
- Species restricted to the semi-natural interstices between cultivated land (including field margins): interstitial species.
- Species that cover both the major components of the landscape: mosaic species.

Macdonald & Smith (1990) used these categories to help focus on the different ways in which species may partition their activities and lives within the landscape, whilst recognising that the actual pattern of land use in an area might cause a species to fit into different categories in different places. Mosaic species may be particularly difficult to allocate, because of their variety. As examples, (1) they may consist of populations within which some individuals occupy interstices and others occupy crops, or all individuals may use both habitats; (2) they may depend on the crop and interstices for different activities (such as feeding and reproduction), or at different times of the year; or (3) they may be facultative or obligate users of the various components. As Macdonald & Smith (1990) observed, obligate mosaic species are of particular interest because they functionally require, rather than simply tolerate, the agricultural/natural habitat combination. They cited the carabid beetle *Agonum dorsale*, which requires the prey supplies provided in the crop in summer (for both larvae and adults) but also needs the protective microclimate of the field margins in order to overwinter successfully. Loss of either habitat type could affect its wellbeing.

Many studies on invertebrates in farmlands are dominated by common species, be they pests, beneficials or more innocuous taxa such as butterflies. This is unsurprising, as Dover (1999) noted, simply because most rare species are ecologically specialised, requiring specialised habitats, and many such habitats have been almost entirely eliminated from many arable areas. Dover also observed that the more uncommon species reported in farmland surveys tend to be those breeding in adjacent natural habitats and utilising field margins only transiently, perhaps as corridors. For some butterfly species in field margins, numbers may be limited more by lack of nectar resources for adults than by lack of larval food plants. Changes to landscapes resulting from deintensification can benefit numerous invertebrate species, but the mechanisms whereby such benefits are conferred are not always clear. As one example, Dover *et al.* (1997) emphasised the importance of simply providing shelter for many British butterflies, with hedges a prime component of this in open agroecosystems, and particular hedge characteristics such as shorter lengths and thick base flora with some flowering herbs being particularly valuable (see also Feber & Smith,1995). Of the economically valued groups, none has received more attention than Carabidae. Ground beetles are diverse and occur within fields, in adjacent areas, and interact between the two regimes. They are affected by many changes to agricultural systems. In addition, and in functional contrast to butterflies, they move largely on the ground (Holland, 2002); many species, indeed, are flightless. Together, butterflies and ground beetles have helped to illustrate many of the principles of much wider relevance to invertebrates in agroecosystems.

Key issues thus centre on the spatial arrangement of landscape features and habitats within agroecosystems and the encompassing mosaic, particularly on the hospitality of the various components to invertebrates and the facility with which these animals may move around to colonise and exploit them (Duelli, 1990; Mader, 1984). Habitat management in landscapes dominated (or otherwise influenced) by agroecosystems thus devolves largely on spatial arrangements and their influences on population dynamics of the species present. These influences are mediated largely through their structure – the shape, size, diversity and density of the various landscape components. As Fry (1994) noted, field margins vary in type, width, height, length, composition, diversity of habitats (and microhabitats) that they provide and their connectedness, and they range from being very simple (such as fence lines) to very complex (as in older hedgerows containing many plant species). Flows of materials and exchanges of species between landscape components reflect both the structure of landscapes and the properties of species, and how these may vary over time – both naturally and in response to management. Field margins may change through vegetation succession, neglect, modification or removal, or management may maintain them in more constant form over relatively long periods. Their interactions with species of animals (and plants) reflect the extent to which they are 'connected' (within the dispersal capabilities of the species concerned) or separate fragments in the wider environment, together with the population structure of the particular species and their ecological specialisations. These themes are discussed in this chapter.

The major functions noted in Fig. 8.1 are mainly self-evident and are discussed further below. As summary comments, (1) linear landscape elements such as field margins may constitute corridors for species able to move along them and for which such habitats are relatively natural; (2) conversely, they may constitute barriers for many species seeking to move across them, and thus restrict their dispersal unless gaps are present to allow passage; in such instances, the barriers may help to funnel the organisms toward those gaps; (3) they may provide permanent or refuge habitats for predatory or parasitoid arthropods of value in IPM, and facilitate their movement into adjacent crops, or for species displaced from elsewhere in the modified landscape; (4) they may constitute or provide habitats for populations or metapopulation units (see p. 257) of resident species that typically occupy only small areas of habitat (which may be fragments remaining after agricultural conversion); (5) they may constitute special habitats for ecologically specialised species of even more limited distribution; and (6) they may participate in source-sink dynamics, supporting species that are declining in farmland. Field margins may thus affect the host-seeking behaviour of pests that have to find their host plants in time and space – as for intercropping (see p. 204), with the analogy that vegetated field margins are more permanent strips of non-crop vegetation that may have values in reducing the apparency of the target crop.

One pertinent implication of this is that IPM can increasingly be planned at the landscape level and combined meaningfully with wider conservation considerations. Fry (1994) envisaged the context where the distance that natural enemies are able to penetrate a crop to control pests at requisite times from margins might become a basis for optimising spacing of field margins in the landscape. Because the spatial pattern affects within-field dispersal of such agents, it also affects the costs and benefits balance of field margins in agriculture.

Influences of field margins on arable crop fauna sometimes appear to be most pronounced when the borders are marked by a hedge or similar substantial physical feature such as a row of trees, which can be considered as an 'ecotone'. Thus, whereas the edge is the physical boundary between the different communities or landscape elements, with its wider influences termed 'edge effects', an ecotone is the narrow overlap zone between those communities (both vegetation and associated fauna) often linked with that edge. Ecotones may occur on various scales and are important considerations in landscape structure because they may provide specific conditions linked with corridors, and opportunities for dispersal not otherwise present, in reflecting aspects of edge quality.

Ecotones can thus harbour greater numbers of species and individuals than surrounding habitats and may support specialist species occurring only there. Their properties can thereby overlap strongly with edge effects, and definitions can become blurred. For Carabidae, Hance *et al.* (1990) noted that hedgerows and the like differ from 'true ecotones' in constituting abrupt changes between environments; nevertheless, they are also stable features of the landscape in which crops to either or both sides may change (such as by rotation), so that hedgerows are sometimes the most permanent features of agricultural landscapes and also increase the heterogeneity of the local environments. Considered as permanent refuge habitats, for example, their importance is demonstrated by presence of resident species that could not (or do not) tolerate changes elsewhere. Thus, Hance *et al.* (1990) found that even a small hedge (200 m long by only 50 cm wide) harboured ten species of Carabidae not found in adjacent arable habitats. Habitat-edge management is an important facet of invertebrate conservation in agricultural landscapes for this fact alone. Its importance is enhanced by its roles in facilitating inter-habitat movement, as a refuge for numerous non habitat-specific taxa, and influences on the wellbeing of species of direct interest to farmers. The influences of such boundaries on invertebrate ecology are thus of considerable relevance in understanding their practical roles in landscapes. Whereas the diversity, abundance and guild structure of natural enemies in arable landscapes may be moulded in large part by the crops present (Booij & Noorlander, 1992), the assemblage may also depend on the nature of the boundaries to those crops (see also Kromp, 1999).

Contour analyses (Thomas *et al.*, 2001), based on frequency of individuals of particular taxa from grids of (in their example) pitfall traps, help to clarify distributions and detect levels of aggregations and changes over time. Fig. 8.2, as examples from the much wider data illustrated by Thomas *et al.*, illustrates a carabid species (*Harpalus rufipes*) found mainly in the boundary hedgerow (Fig. 8.2a) and two time intervals for *Nebria brevicollis* to demonstrate changes from within-hedge (Fig. 8.2b) to much wider (Fig. 8.2c) distribution within the adjacent cereal crop. In both these species, and others, local densities are patchy and change over the season.

Movement patterns of natural enemies

The patterns and roles of dispersal of invertebrates are both immensely complex themes, but understanding them has important implications for

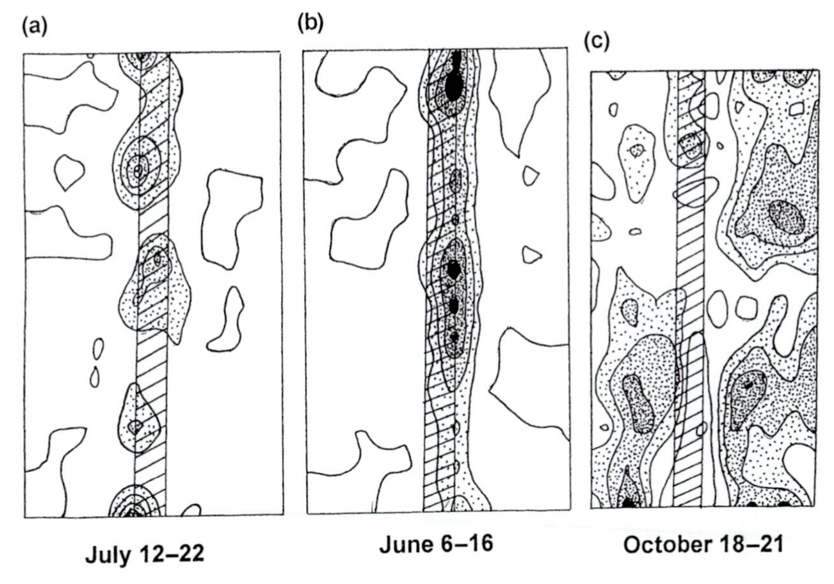

Figure 8.2. Contour analysis for plotting densities of carabid beetles in fields. Density measurements, based on a grid of pitfall traps, are shown as contours in relation to a hedgerow (hatched) separaing two arable fields: (a) *Harpalus rufipes*, indicating higher density in hedgerow than in fields; (b, c) *Nebria brevipennis* at two different times, indicating changing densities from June (b, concentrated in hedgerow) to October (c, concentrated in fields) (after Thomas *et al.*, 2001).

pest management and for conservation. At one level, two major categories of dispersal of predatory beetles from field margins into crops have been inferred (Coombes & Sotherton, 1986), namely:

- species that walk and progress on a 'wave' of dispersal outwards from overwintering habitats in field margins into crops, gradually progressing into the fields;
- species that disperse by flight may exhibit rapid dispersal across the whole field.

Other parameters may determine the particular form of dispersal exhibited by any given species. Thus, Wratten & Thomas (1990) distinguished six classes of movements of beneficial arthropods in agricultural landscapes (Fig. 8.3), as follows:

- Colonisation of crops from non-crop habitats in spring, and autumn emigration from cultivated land to overwintering or refuge habitats.

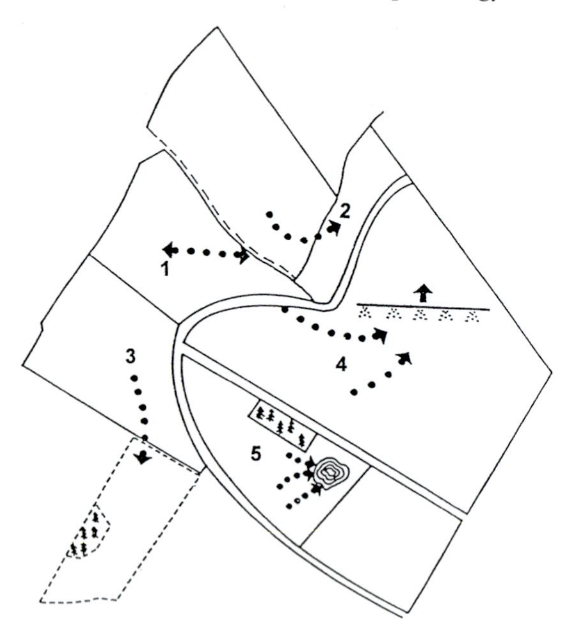

Figure 8.3. Classes of movements in landscapes (Wratten & Thomas 1990): the five major categories of beneficial arthropod movement recognised are shown schematically as (1) colonisation of crops from non-crop habitats in spring and reverse movements in autumn; (2) movement between crop types during the growing season; (3) colonisation of new habitats as area under intensive cultivation changes; (4) recolonisation of land as effects of broad-spectrum pesticides on beneficial insects decline with time; and (5) reproductive and aggregative numerical responses to local areas of high prey density in crops.

- Movement between crop types during a growing season, driven by the crops being out of phase in their developmental stages.
- Colonisation of new habitats as the area under intensive cultivation expands and contracts.
- Recolonisation of land as the effects of broad-spectrum pesticides on the arthropods declines over time.
- Aggregative and reproductive movement responses to local areas of high prey density in the crop.
- Large-scale migration into cropping areas.

The first of these categories has received far more attention than most of the others, with numerous studies on the movements of pest arthropods and of their natural enemies (particularly of predatory beetles, such as Carabidae and Staphylinidae) showing progressive movement on to crops from boundaries and later-season cohorts returning to the margins. Different species may differ considerably in vagility and dispersal behaviour, so that rates of movement between and within habitats are very variable. However, all of the above themes emphasise the importance of dispersal and how this may be affected by dispersion of landscape elements – a central paradigm in both cultural IPM and practical conservation. In addition, hedgerows and other vegetated margins can harbour woodland or forest species, with studies such as that by Burel (1989) emphasising the variety of dispersal patterns that can occur along such putative corridors. Thus (see also Burel & Baudry, 1990), non-flying forest Carabidae in Brittany fall into three categories according to the distance they move along adjacent hedgerows:

- Forest core species were found only within 100 m of the forest edge.
- Forest peninsula species used hedgerows for up to 500 m but were not found further away from the forest.
- Forest corridor species used hedgerows as corridors and could be found at any distance from the forests; Burel & Baudry reported samples up to 15 km away.

Carabidae have been especially well studied in cropping environments in Europe, with influences of field margins on their wellbeing investigated in considerable detail (Thiele (1977) reviewed early studies)(Lys & Nertwig, 1991). The interactions of crop-frequenting carabids with hedgerow environments are very varied and complex. Consider, for example, some of those noted by Jepson (1994) and how these may bear on more general difficulty of seeking generalisations applicable to arthropods above the individual species level. The examples cited by Jepson include:

- Some commonly occurring crop Carabidae are associated with hedgerows but not restricted to them.
- Other species have only very limited interactions with hedgerows.
- Wooded strips around fields contain important forest assemblages that rarely enter agricultural crops.
- Carabidae from adjacent agricultural crops have much lower densities and activities than those in wooded strips.
- Field boundaries are important overwintering sites for some species that invade crops in spring.
- Because many species have very narrow habitat preferences, the structure and vegetation of field boundaries may influence the assemblage composition and beetle densities substantially.

In a somewhat more complex example, Pollard (1968) suggested that the lifecycles of nocturnal Carabidae (active when the crops are absent in autumn) were influenced more by hedgerows than those of diurnal species active at the same season or of nocturnal species active earlier in the year whilst the crop was present. In such ways, the behaviour and phenology of particular species may influence interactions with local habitats. This variety implies strongly that movements of predominantly non-volant arthropods may be affected substantially by linear landscape features – in addition to movements of flying insects being influenced. At one extreme, waterways (irrigation channels, canals, rivers) or walls may be total barriers to ground-dwelling species. However, hedgerows and the like reveal a range of different levels of 'permeability' to different taxa. Permeability is the proportion of individuals of a species that cross a putative barrier when they encounter it, and it is thus an index of the effectiveness of landscape features as barriers to dispersal. Table 8.1 indicates some aspects of permeability for some European Carabidae and indicates the wide variety of responses that may occur or may be supposed to occur. The effects of the field boundary on carabid movements depend on a wide spectrum of factors, and Jepson (1994) noted that short-term studies on levels of interchange may not be sufficient on which to base firm conclusions. Rather, it should be measured over a whole generation, because phenological factors and behavioural responses to particular field conditions may be important – in addition to the physical permeability of the boundary itself. For example, there may be important differences between carabids that overwinter in the boundary and those that overwinter in the crop area. Determining the role of field boundaries on distribution and abundance of Carabidae is thereby difficult.

In yet another categoristion of responses, Fournier & Loreau (1999) recognised four relevant groups of agriculture-associated Carabidae:

- species restricted to hedges;
- species preferring the hedge to the crop;
- species preferring the crop to the hedge;
- species unaffected by the hedge.

Clearly, many species have particular habitat requirements. Holland & Luff (2000) suggested that many species may move relatively little, but they also endorsed the above expressions of low permeability through many linear features in landscapes. The broad grassy field margin studied by Duelli *et al.* (1990) had similar effects, leading to a wider discussion of the features of habitat edges (see p. 219), but the overall findings imply

Table 8.1. *Habitat boundary permeabilities for certain Carabidae (Jepson (1994), based on literature records and 'guesses')*

Reference	Boundary type	Permeability (%)	Carabid species
Thiele (1977)	10-m woodland strip	20	*Pterostichus melanarius*
Duelli *et al.* (1990)	3-m dirt road	51	Smaller species
	6-m tarred road	40	Smaller species
	Grass strip	41	Smaller species
Mader *et al.* (1990)	1.2-m grass track	55	*P. melanarius* and others
	1-m gravel track	15	*P. melanarius* and others
	0.5-m paved road	23	*P. melanarius* and others
	5.7-m rail embankment	9.8	*P. melanarius* and others
	5.7-m rail embankment	17.4	*P. melanarius*
	5.7-m rail embankment	0	*Nebria brevicollis*
Guesses	Canal or irrigation ditch	0	? all species
	Crop/same crop interface (strip cropping)	100	? all species
	Overwintering boundary	50	Boundary-overwintering species

that vegetated field margins can have various effects on beetle species, and these effects may be difficult to interpret as well as being confused easily. Thus, it may be extremely difficult to decide whether the habitat is inhibiting movement or whether it is simply more attractive than its surrounds so that the beetles prefer it. In this way, the habitat boundary can become a 'sink' (if it is also suboptimal for breeding), so that linear margins can then impede movements between population units and foster overall reduction in reproduction.

Similarly, grassy banks have lower permeability than cereal crops to several common crop carabids (Mauremooto *et al.*, 1995), but the reasons for this are unclear. Studies comparing the permeability of grassy banks and barley crops in Norway showed that the banks had lower permeability to three large carabid species (*Harpalus rufipes, Pterostichus melanarius, P. niger*) (Frampton *et al.*, 1995). Even within a cereal crop, rows of wheat plants can affect movement patterns of these three species (Lys & Nentwig, 1992). Again, the causes of this are unclear, but several factors were suggested as possible influences on permeability, because they are likely to affect beetle movement:

- changes in speed and/or direction in response to microclimatic factors – due to both features of the boundary and its vegetation;

- relatively high prey availability in the boundary;
- differences in beetles' burrowing behaviour between the crop and boundary;
- physical structure of the grassy bank – for example, a steeply sloping face may lead to avoidance by some species;
- presence of a bare soil strip between the crop and the bank.

Frampton *et al.*'s trials showed that the grassy bank was less permeable than an equal width of barley crop to carabids and could thus influence the extent of field-to-field movement by *H. rufipes* and *P. melanarius*. Although there was no evidence for direct relationship between the width of the grass bank and its permeability, some trials indicated clearly that permeability decreased as bank width increased (Frampton *et al.*, 1995). The species investigated in this study are among the largest and most mobile European carabids that commonly do not fly, and it is likely that the numerous smaller species, many of them lacking functional wings, would be influenced even more strongly by field boundaries. The clear implication is that landscape-feature manipulation could have significant effects on carabid populations – and, by analogy, on those of many other groups of natural enemies by affecting their dispersal and population linkages.

The permeability of a field margin thereby has considerable effects on movement patterns and dispersion of many normally vagile invertebrates of direct economic relevance. A further category of influence devolves on how margins affect local microclimates. Pioneering work in Britain (much of it summarised by Pollard *et al.* (1974) and Helps (1984) elucidated the various shelter effects of hedges. Thus, Fig. 8.4 (Helps, 1984) demonstrates the differing shelter effects of solid and permeable hedges. A solid margin is associated with a relatively small sheltered zone, but one in which turbulence may also occur, whereas a permeable margin provides a much greater shelter zone. The primary climatic effect of such boundaries is to alter wind speeds in adjacent areas, with substantial effects on local microclimates.

Other invertebrates

Studies on a variety of other invertebrate groups have led to broadly similar inferences as for Carabidae, with suggestions that linear boundaries can indeed impose unexpectedly subtle effects on population continuity and connectivity. Linear features, such as roads, may deter movement of

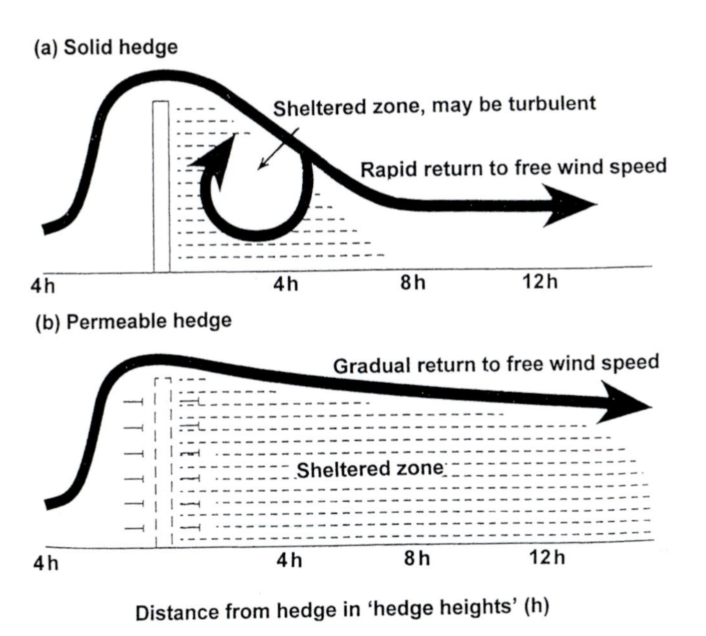

Figure 8.4. Wind effects and shelter capabilities of (a) a 'solid' and (b) a permeable field margin (Helps, 1994).

many kinds of invertebrates between field margins. For example, Dennis's (1986) study of the orangetip butterfly (*Anthocaris cardamines*) in England showed that a motorway reduced overflights by about 92%, but such impediments are by no means general, and Munguira & Thomas (1992) considered that major roads posed little impediment to dispersal of some other species.

Likewise, some woodland specialist butterflies in Britain, as well as open grassland species, exploit some aspects of hedgerows, where this ecotone habitat may be particularly important to such habitat specialists during dispersal or after failure of critical resources in their primary habitats. However, such species remain linked primarily with their main habitats (Dover, 1994). Within field margins (in this case, mainly hedgerows or grassy strips), the factors affecting distribution of butterflies (three related British Satyrinae: *Aphantopus hyperantus*, *Pyronia tithonus*, *Maniola jurtina*) included shelter, presence of nectar sources, insolation, incidence of farm tracks, and variable effects of 'habitat quality'. Factors affecting the partitioning between field margins included shelter, a small uncropped area between a copse and hedgerow boundary traversed by a farm track, and nectar sources (Dover, 1990a; Dover *et al.*, 1992). Each

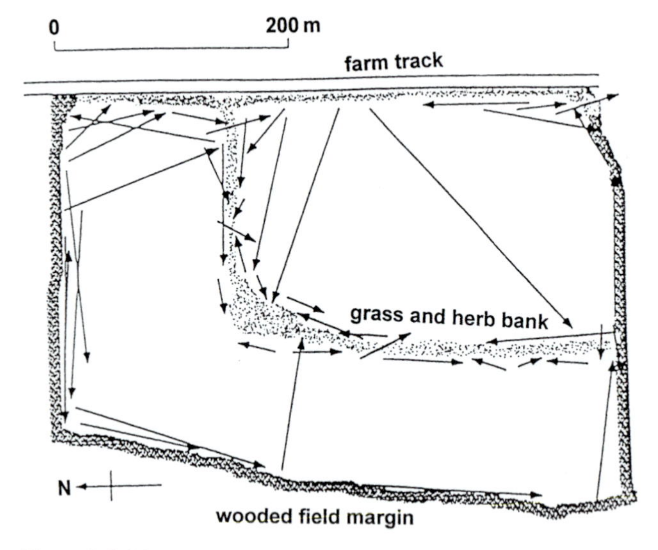

Figure 8.5. Movements of a lycaenid butterfly *Heodes virgaureae* in arable fields (Fry *et al.*, 1992). Each arrow shows the track (15 minutes observation) for an individual insect. The butterflies are associated closely with edge habitats, and very few crossed boundaries – either the bank through the centre of the field or the wooded/track perimeter.

of the three satyrine species could move readily over distances of more than 1 km in farmland, but other butterflies may differ in behaviour, with Dover (1990b) showing that some pierids used the field margins as flight corridors, apparently preferring these to flying over crops – and probably reflecting the greater shelter they confer. Males of territorial nymphalids (such as *Aglais urticae* and *Inachis io* in Britain) used the field margins as territorial posts, thereby maximising encounters with females seeking oviposition sites along the margins.

Mark–release–recapture studies on the above satyrines demonstrated movement of individuals between adjacent colonies of butterflies (Dover *et al.*, 1992). Many butterflies are associated clearly with edge habitats on farms. Fig. 8.5 (Fry & Main, 1993) shows movements of individuals of the scarce copper (*Heodes virgaureae*) in southern Norway, indicating the close correlation of butterfly movement with such edges and showing that very few butterflies crossed such boundaries (which included a farm track with roadside verge, a grassy bank, and a coniferous field margin). Further studies on permeability of a range of different edges to *H. virgaureae* (Robson 1992), cited in Fry & Main (1993) revealed considerable variations in the roles of different boundary features (Fig. 8.6)

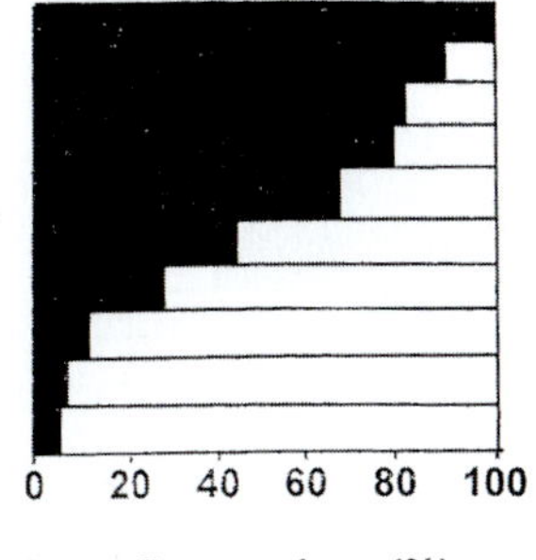

Figure 8.6. Movements of *Heodes virgaureae* (Fry & Main, 1993, after Robson, 1992). The permeability of different field margins is represented by the proportion of butterflies crossing measured for 100 encounters with each edge category; control 1 comprises random lines across meadow, and control 2 represents edges between two crops with no defined boundary zone.

(Fry *et al.*, 1992), some restrict severely the movements of butterflies between fields, and interspecific differences confirm that a given feature can act as a corridor for some species and a barrier to others. Forest-frequenting butterflies, such as the satyrine *Erebia ligia* in Robson's (1992) study, were not as severely restricted as *H. virgaureae* by taller tree vegetation and crossed woodland boundaries at about twice the frequency of the latter species. However, the isolation of some butterfly populations imposed or enhanced by linear margins may be important in accelerating their decline.

Clear parallels to Frampton *et al.*'s (1995) study on carabids are thus evident for butterflies, in that the structure of a field boundary has considerable influences on movement. Whilst butterflies have benefited considerably from the resources provided in conservation headlands (Chapter 7), the roles of some other boundaries, such as hedgerows and shelterbelts, are less well defined, despite evidence of ecotone uses by a variety of different butterfly species, as above. In southern Norway, Fry & Robson (1994) recognised eight structural categories of field boundaries and studied the responses of butterflies encountering them, whether they were leaving the meadow, entering the meadow, or 'rebounding' and not crossing the boundary. Their results (Fig. 8.7) showed considerable variation with boundary structure, and the proportion of butterflies crossing each natural boundary was correlated with the percentage of boundary length for which vegetation was less than 1.5 m high. Most

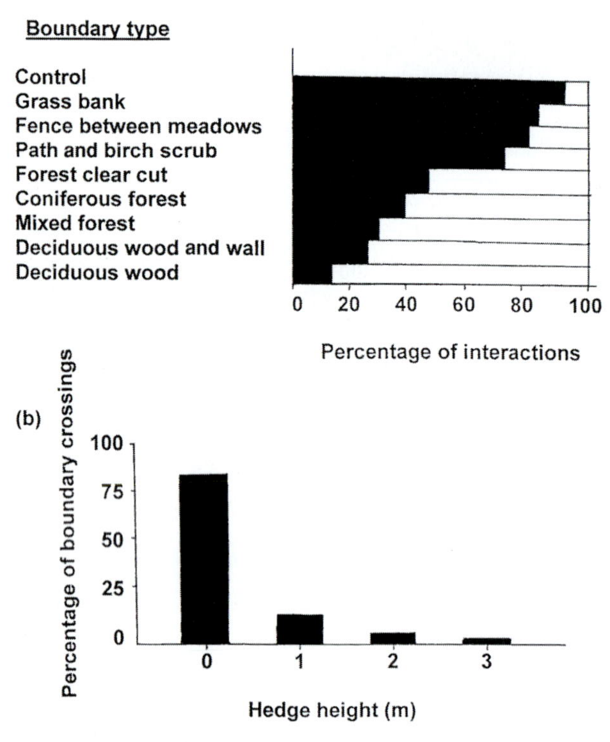

Figure 8.7. Boundary structure – effects of tall vegetation (Fry & Robson, 1994): (a) percentage of butterflies crossing (black) and rebounding from (white) different field margins in Norway (n = 1047 butterflies) (cf. Figure 8.6); (b) percentage of butterflies crossing the different experimental height treatments (n = 1808 butterflies).

crossings occurred through gaps in the boundary vegetation, although the size of those gaps was apparently not important. Fry and colleagues also undertook trials using an 'artificial hedge' of green tarpaulin stretched between wooden stakes, to heights of 1, 2 or 3 m, with controls in which the hedge was absent. This experiment confirmed clearly the barrier effect of tall vegetation (Fig. 8.7b) (n crossings = 1808), with even a 1-m artificial hedge substantially reducing permeability of the landscape to butterflies. Responses varied somewhat between species and, although it is not always clear whether the impediment effect is due to physical or behavioural influences, the outcome of reduced dispersal seems clear. Hedges may thereby increase isolation of butterfly populations.

More generally, a patch of habitat (such as a field, or the field margin itself) may be bounded by either a hard edge or a soft edge. A hard edge is a boundary that individuals of a given species encountering it are unlikely to cross to enter adjacent habitats. A soft edge is a boundary that is permeable and does not necessarily or predictably constitute a barrier to such dispersal.

Despite the attractions of doing so, it is difficult to generalise on edge effects because different invertebrate species, even when related closely to each other, may variously (1) be affected positively or negatively by edges, so that assemblages differ in composition and richness at different distances from a defined edge; (2) show no response to edges; or (3) manifest changes that are not detectable because other variables lead to a 'cancelling-out' effect (Murcia, 1995). Nevertheless, it is common for assemblages to be more diverse at habitat edges, largely reflecting increased abundance of many of the species characteristic of the disturbed habitats and an increased influx of species from elsewhere. In contrast, many of the invertebrates characteristic of interior habitats (of forests, for example) are 'edge-avoiders'.

A hard-edge field margin implies that there may be no measurable population exchange between natural areas and cultivated fields, as partitioned habitats. However, Duelli *et al.* (1990) recognised a greater variety of soft-edge categories between natural areas and cultivated fields, as indicated in Fig. 8.8, which summarises the number of carabid species in each of several categories (see below) in each, based on pitfall-trap captures over 300-m transects through fields and their surrounds. In this figure, the various soft-edge categories (2–5) represent: (2) species that usually do not leave the preferred habitat (either natural or agricultural) and even avoid border areas; (3) the contrary case, of species that thrive in the preferred habitat and invade elsewhere to varying extents; (4) both these effects may occur together to produce mutual influence (e.g. in *P. melanarius*); (5) in a few cases, the ecotone (edge between the habitats) seems to be the preferred habitat; and finally (6) 'no edge' implies more or less even distribution in both habitats, reflecting either high-mobility or ecologically generalised species. The species categories in Fig. 8.8 are based on an index of naturalness (calculated for the 113 species for which more than 20 individuals were captured), based on catches from equal numbers of traps in annual-crop fields and perennial semi-natural habitats (wetland, pasture, grassy slope, dry meadow). Species with more than two-thirds of individuals in such perennial habitats were deemed 'natural' because of their abundance; species with more than two-thirds of

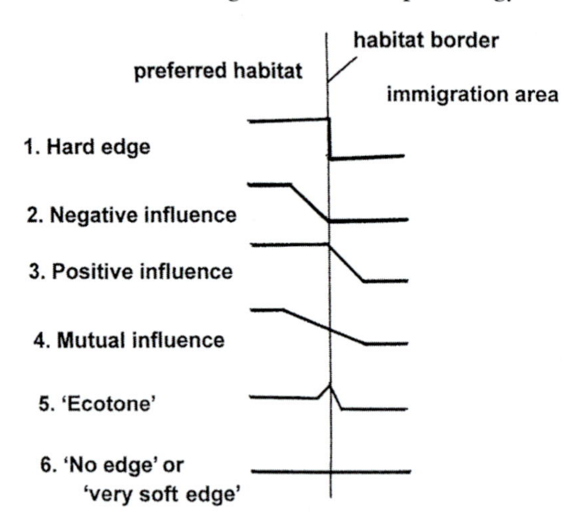

Figure 8.8. Edge categories in relation to dispersal of Carabidae in fields (Duelli *et al.*, 1990). Six categories of species are recognised according to their distribution around habitat borders.

individuals in crops (here, wheat, rape, maize) were considered 'agricultural'; others were 'intermediate'.

The proposal that hedges act as corridors is inherently appealing (Jones, 1991) but often very difficult to validate, with many purported cases notwithstanding critical appraisal. In some instances, hedgerows may serve as 'stepping stones', for example because of their roles as windbreaks. Reduced wind speeds on the leeward side of hedges (see p. 241) may lead to many insects 'dropping out' of the air there, resulting in increased concentration and enhancement of hedgerow insect diversity. The various physical and behavioural factors involved in species' reactions to habitat edges have been discussed extensively and modelled (see Thomas & Hanski, 1997). However, there have been relatively few detailed experimental studies on individual invertebrate species to determine how and why they may opt to cross or not cross when confronted with a boundary.

One such study involved the European Roesel's bush-cricket (*Metrioptera roeselii*) in Sweden (Berggren *et al.*, 2002). Artificial test arenas (Fig. 8.9), each of 5.75 m diameter, were created in grassland with clover. The mown matrix was approximately 5 cm high and the corridor was unmown (*c.* 60 cm high). Some test arenas had a soft edge, an intermediate corridor boundary of 30-cm-high grass, whereas others had a hard edge as an abrupt transition from 60-cm to 5-cm grass. The centre

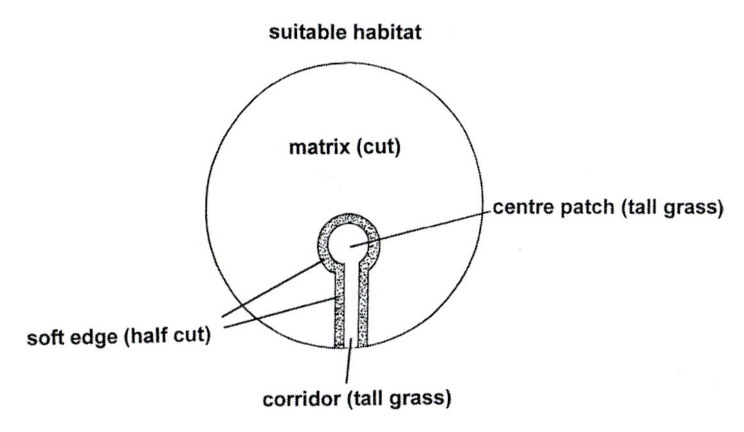

Figure 8.9. Design of test arenas for appraising movements of the bush cricket, *Metrioptera roeselii* (after Berggren *et al.* 2002). Arena radius, 5.75 m; central circle radius, 75 cm; matrix cut to approx. 5 cm high; corridor of tall grass approx. 60 cm high; buffer zone (soft edge) cut to approx. 30 cm high.

patch at the head of the corridor approximated the size of a normal male cricket territory, and all the test animals were flightless, brachypterous individuals. The natural habitat of *M. roeselii* is tall grass, equivalent to the corridor vegetation and central patch of the arenas, and the cut matrix was regarded as an unsuitable habitat (Berggren *et al.*, 2002). Individual insects were released in the centre patch, and their movements were monitored. The corridor was the preferred option for dispersal, with approximately 30% more individuals using it (to move to its end) than would be expected if dispersal was random. The edge quality (hard or soft) had no influence on whether crickets moved through corridors or the matrix, and those that moved across the matrix did so faster than through corridors – possibly reflecting likelihood of greater exposure to predators or desiccation. Many individuals showed avoidance behaviour when encountering the edge and turned back. The features of a habitat patch and of an adjoining habitat (the 'context' of the patch) are both important in influencing permeability.

Two further studies, by Duelli *et al.* (1990) and Mader *et al.* (1990), both demonstrated that linear landscape features in agricultural systems affect the movements of ground-dwelling animals, such as beetles and lycosid spiders (Maelfait & De Keer, 1990). Again, the population consequences of these movements are not always clear. Thus, adults and subadults of the lycosid *Pardosa amentata* leaving the road-verge habitat return without crossing a field track (Fig. 8.10), which thus constitutes an effective

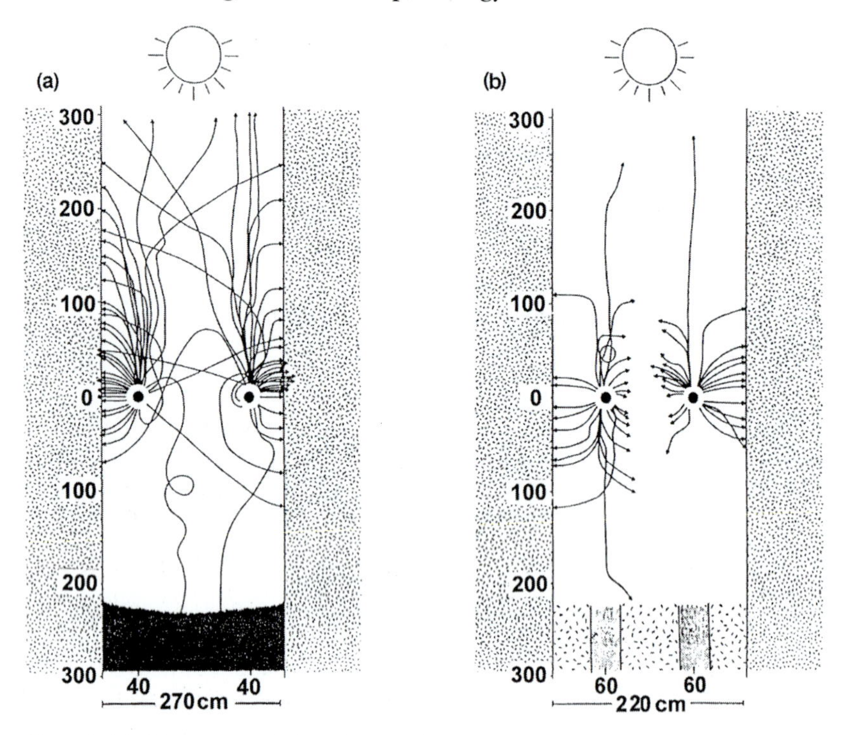

Figure 8.10. Movements of the spider *Pardosa amentata* in relation to road boundaries (Mader *et al.*, 1990): (a) paved field track – released spiders (at point indicated) all sought edge rapidly and entered vegetation; (b) grassy field track – spiders entered side vegetation or central grassy strip.

barrier to dispersal. However, immatures may disperse aerially, so that a diversity of dispersal strategies at different life stages may counter environmental restrictions on any one of these. Nevertheless, a major point of practical interest for conservation arises from Dempster *et al.*'s (1976) claim that some species may have lost their ability to disperse in modern, highly altered landscapes, as with the swallowtail butterfly, *Papilio machaon*, in Britain. Although the interpretation is tentative (and intriguing), Dempster *et al.* found significant morphometric differences in size and thoracic shape of swallowtails from the Norfolk Broads and from the extinct but formerly isolated population at Wicken Fen; tentatively, differences in thoracic shape may have been associated with changes in mobility, thus 'helping' to stabilise a normally highly mobile species in a relatively small habitat area. Dempster *et al.*'s study was undertaken in part to help evaluate the likelihood of successfully re-establishing *P. machaon* at Wicken from Norfolk stock, and led to a suggestion that these might, indeed, be too mobile to stay in the fen. Dempster (1991) speculated that if mobility is determined genetically, then isolation of populations (as with the Wicken Fen swallowtails) might increase selection against mobility, because emigrants would not easily find other suitable habitats and immigration would be rare.

Wider values of unsprayed field margins

In addition to their influences on disruption or continuity of populations, as corridors or barriers in the landscape and as refuges for biota, field margins and crop edges can have additional beneficial effects, each of which may influence needs for management. One such benefit, discussed by De Snoo & Chaney (1999), is as buffers reducing pesticide drift to adjacent areas such as hedges, ditches and other water courses. Thus, creation of a 6-m unsprayed strip in wheat can give up to 70–75% reduction in the amount of pesticide reaching adjacent hedges (at wind speeds of 4–6.4 m/s). Other trials in the Netherlands also produced similar reductions, of up to 99% reduction in spray drift to adjacent ditches separated from a 6-m unsprayed crop edge (De Snoo & De Wit, 1998). Unsprayed crop edges often have increased floristic diversity, sometimes associated with greatly increased abundance of many flower-visiting insects. Numerous other authors have commented on particular buffer widths and their influences. Thus, a 2-m-wide buffer around a crop is sometimes sufficient to prevent drift to most plants from fields sprayed with herbicides, although wider buffers (of 6–10 m) are sometimes recommended (Marrs *et al.*, 1993). For some caterpillars, buffer zones of only around 1 m may reduce mortality substantially but, again, much wider buffers (8–14 m, sometimes up to 23 m) have proved necessary to avoid insecticide drift (De Jong & van der Nagel, 1994; Davis *et al.*, 1991). In general, whereas relatively narrow buffer zones markedly reduce the impacts of crop pesticides, use of highly toxic compounds can not be countered completely unless much wider buffers are used.

The values of hedgerows and other margins reflect their varied structures and the range of plant species they can support. Jones (1991) divided structural diversity of hedgerows into four main components, as in Fig. 8.11, namely:

- the ‘hedge body’, a more or less continuous row of shrubs and bushes, which may be sufficiently wide to include interior habitat for some

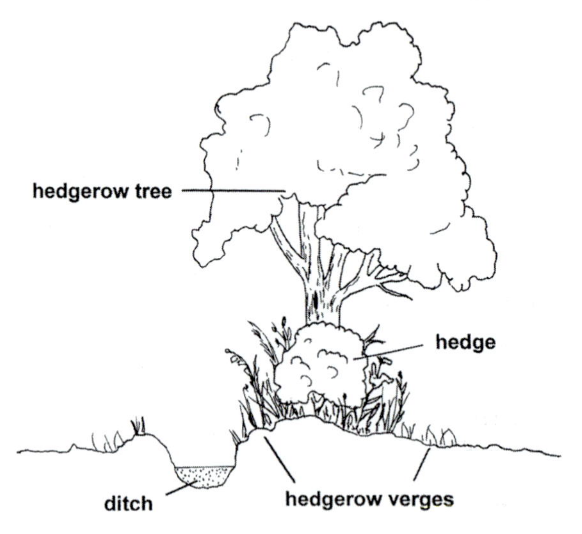

Figure 8.11. Structural components of hedgerows (after Jones, 1991).

invertebrates, as well as edge habitats. This, and associated banks, can also provide important nesting sites for Hymenoptera;

- hedgerow trees;
- hedgerow verges, the strips of land either side of the hedge body and that are relatively undisturbed by many agricultural activities;
- ditches. In low-lying areas, hedges may be accompanied by parallel ditches, which provide habitats (sometimes ephemeral) for a variety of aquatic and semiaquatic invertebrates.

In Britain, the relatively advanced knowledge of the fauna enables nomination of particular tree and other species valuable for conservation of many species of individual conservation interest in hedges, with management specifiable in some detail. Jones (1991) noted that approximately 1000 plant species have been recorded from British hedgerow habitats, with about 250 of these being associated closely with hedges. Management may then target particular species or be more general to foster diversity or other ecological attributes. For the former, the needs may extend beyond aspects of floral composition, to include topics such as the method, frequency, timing and scale of cutting, extent of management of trees, whether and how pesticides are used in and near the hedge, the width of the verges, and the removal or disposal of trimmed debris. For example, flail-cutting is now used widely and has largely replaced traditional hedge-management practices such as layering, which were commonly undertaken on a smaller, rotational scale. This relatively insensitive technique is not always beneficial to hedge-dwelling invertebrates.

Refining the design and management of field margins can therefore have significant influences on many invertebrate species. For butterflies, management affects the species, distribution and structure of plants used for larval food and adult nectar sources. Either of these can be valuable as predictors of butterfly abundance and, by inference, that of other less conspicuous invertebrate groups. Thus, Feber *et al.* (1999) (see also Feber & Smith, 1995) emphasised the importance of spatially variable management to create patchy habitats for two nettle-feeding British nymphalids, *Aglais urticae* and *Inachis io*. As with the grassland butterflies discussed elsewhere (see p. 271), these two species differ in the height of vegetation (nettles, *Urtica dioica*) that they prefer for oviposition; this preference serves to segregate the species, with *A. urticae* preferring shorter nettles. Aspect of the site may also be important, with south-facing (insolated, in Britain) sites particularly important to the butterflies in spring. Feber *et al.* (1999) noted that spatially variable management of field margins provides greater opportunities for the butterflies, providing variety in both nettle height and the nutritive quality of foliage for the caterpillars. Optimising the conservation values of field margins can thus depend on the aims of specific target species approaches, but the broader view of maintaining diversity may require considerable flexibility in management. Haughton *et al.* (1999) suggested the following broad management considerations in order to increase abundance and diversity in arable field margins:

- Natural regeneration may be a better option if a good local flora and few weed problems exist, particularly on lighter soils.
- A less frequent cutting regime, avoiding a summer cut and leaving the hay in situ, is preferred, where the integration of patchy rotational management would optimise uncut areas for invertebrates.
- Where weed control becomes necessary, spot treatment can be used to target problem plants.

Dense low herb cover on field margins, as well as benefiting invertebrates, has agronomic advantages because the cover helps to suppress or exclude certain aggressive annual weeds of arable land that thrive in disturbed field margins.

However, the major conclusion from the many and varied studies involving arthropods in field margins in Europe is that the margins are indeed important in increasing general arthropod diversity on farmland (Dennis & Fry, 1992). This importance extends throughout the year. In summer, margins can provide a complex and relatively stable habitat for species that could not survive in the open farm landscapes with crop habitats alone. In winter, they provide refuges for many species, including economically important natural enemies, whose wellbeing provides economic justification for the presence and maintenance of the margins.

Remnant habitat patches

Frequent reference has been made above to the 'linking features' of field margins between patches of remnant or other natural habitat. Before considering the values and desirable properties of these linkages, it is pertinent to recapitulate briefly (see also p. 60) on the features of such remnants, considered as the outcomes of fragmentation and habitat loss and now serving to focus wider landscape management for enhancement and conservation of invertebrates. As Didham (1997) remarked, in a comment of much wider relevance, 'exactly what impact forest fragmentation has on animal assemblages, particularly invertebrates, is poorly documented empirically, and theoretically not well understood'. Local responses to fragmentation will clearly differ with site and taxon differences, but the general perception is that about five parameters determine fragmentation/patch suitability, with the implication that increasing the extent of fragmentation will lead to (1) fewer species, (2) smaller populations and (3) less genetic interchange between populations. The major processes that affect communities and the physical environment within fragments have all been discussed at length elsewhere, and are:

- area effects
- edge effects, in part linked with
- the shape of the fragment or remnant
- the extent of its spatial isolation, associated with
- the degree of connectivity in the landscape.

Remnants or fragments, by definition, have reduced habitat area over the more continuous or extensive habitats that preceded them, although the level of occupancy of that habitat by particular invertebrates is often restricted and itself patchy rather than extensive. Discussions over the effects of area on species diversity in habitats draw heavily on principles derived from studies on island biogeography. However, as Didham (1997) emphasised, area *per se* is by no means the whole determinant of invertebrate assemblage structure, although many studies do indeed link species richness positively with area. Studies on forest fragmentation in Australia (such as that by Margules, 1992) demonstrate that particular invertebrate taxa may respond very differently to the fragmentation process. Even small fragments may continue to provide sufficient high-quality (often core; see p. 255) habitat, with many invertebrates occurring in low densities or in small, highly localised populations rather than being distributed across the entire apparently suitable habitat area.

Nevertheless, the inevitable consequence of increased fragmentation of natural habitat resulting, *inter alia*, from agricultural intensification is that many invertebrates must now depend on resources provided within those fragments. It is therefore necessary to understand the spatial scale of their natural movements if effective, specifically designed conservation is projected. This process is sometimes of considerable applied value in agriculture. As well as for movement of natural enemies, pollinators are a major consideration. For example, bumblebees (*Bombus* spp.) are important pollinators of arable crops and of wildflowers, and their effectiveness can depend heavily on their ability to traverse no-crop areas (and to reach the centres of crop fields) from their nesting sites in more natural habitat fragments (Saville *et al.* 1997). Bumblebees tend to have high patch 'fidelity', but some marked bees were recaptured up to 350 m from their initial point of capture, although most were found again within about 25 m of that point. They clearly do not always forage close to their nest and may cross inhospitable areas to reach suitable nectar sources.

Landscape linkages

Field margins constitute an important category of what Bennett (1999) termed 'linkages in the landscape', facilitating possible connectivity between fragments of natural habitat in a highly altered agricultural landscape. Nevertheless, invertebrates do not recognise a link as such but opt to frequent or avoid such features largely on their qualities as habitat or, often more serendipitously, in the extent to which opportunistic occupation affords protection from hazards more predominant elsewhere. As will be noted later (Chapter 10), promoting connectivity is one of two major strategies for promoting landscape-level conservation in agricultural systems, with the other being to promote greater hospitality within the wider intervening matrix. As exemplified and inferred in this chapter, much of our understanding of how margins act as connectants has been derived from few insect groups – paramount among them being butterflies and carabid beetles – and the roles of margins as barriers or corridors for many other groups are largely inferential. However, such roles are assumed widely to be important, but the enormous variety of different responses by individual species to a given landscape feature renders generalisations difficult without such direct knowledge. The principles of countering the impacts of habitat fragmentation (essentially, landscape patchiness and widespread low hospitality of intervening areas) by providing structure that could function as linkages is fundamental, largely reflecting the widespread acceptance that gaps between suitable habitat patches are associated with progressive isolation of populations and gene pools and reduced capability for them to be dispersed to (and interchange with) other habitat patches. Structural connectivity in landscapes, and the role of linkages, therefore reflect the size and number of gaps (patches of unsuitable habitat), the presence of connectant pathways or networks between suitable habitat patches, and the presence of 'nodes' in the system (Forman & Godron, 1986; Bennett, 1999), in addition to the structure of the purported linkages themselves. Gaps may be permitted to persist if the habitat fragments are still sufficiently close to permit inter-patch movement, or if intervening patches can function as 'stepping stones' by being within the usual dispersal range of the otherwise isolated organisms. The influence of any gaps will depend also on the extent of contrast between the suitable habitat and the conditions present in those gaps – to the extent that a narrow gap of highly unsuitable or contrasted habitat (such as, in examples noted earlier, a paved road or a watercourse) may have much more severe effects than greater expanses of moderately changed conditions. Attempts to reduce such contrasts may include construction of buffer habitats (see p. 15). 'Nodes' of favourable habitat within linkages can also increase the values of those linkages through helping to provide more permanent or refuge habitats, including opportunity for resident populations to establish. In the present contexts, nodes may be formed by vegetation at T-junctions and cross-junctions of field margins (Fig. 8.12). It is obvious that they should provide resources needed by a variety of species if they are to be deliberately established or designed in restoration projects, but many nodes in agroecosystems are more fortuitous in occurrence, reflecting areas too tight for turning large machinery and other restrictions. Equally obviously, small or remnant habitat patches may already possess the qualities desirable for nodes, so that it is preferable for nodes (and, indeed, other linkage attributes) to incorporate existing natural vegetation or habitat wherever feasible rather than rely entirely on reconstruction. Many resources for invertebrates cannot be created simply by planting – for many taxa, resources such as deep leaf litter or rotting or fallen wood are the products of long-lived communities.

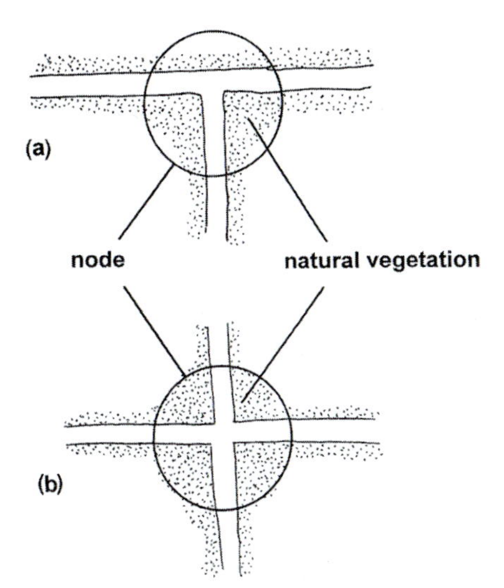

Figure 8.12. Facilitating linkages in landscapes: examples of nodes of natural vegetation associated with junctions of field margins.

The dimensions of linear features, and the sizes of isolated habitat patches such as putative stepping stones, tend (almost by definition) to be subject to edge effects through having very high edge-to-area ratios, so that the vulnerability of linkages to such influences may be considerable. Simply, there may be little or no undisturbed 'core' or 'interior' habitat present unless those areas are wide in relation to the particular edge disturbances of the local environment. The principle is shown in Fig. 8.13 (Moen & Jonsson, 2003), which indicates how the proportion of core habitat is influenced by the different shapes of an area-constant site. In this case, in which an edge effect of 50 m is exemplified, the amount of core habitat (i.e. that further than 50 m from an edge) is considerably less for each of the three rectangular-shaped plots (with length/width ratios of 2 : 1, 4 : 1 and 10 : 1) than for a circular plot. Edge effects may include many kinds of localised microclimate changes, but a major significance is that lack of interior habitat in many field margins precludes such linear features functioning as refuges for many invertebrates through not providing resources needed by many species living naturally in the area.

Their roles as linkages or barriers may not be compromised by this in relation to many transient species. However, the most effective means of reducing edge effects is to increase width of such linear features, with the expectation that their carrying capacity for species pools will thereby increase and that those species that 'enter' a linkage in order to disperse

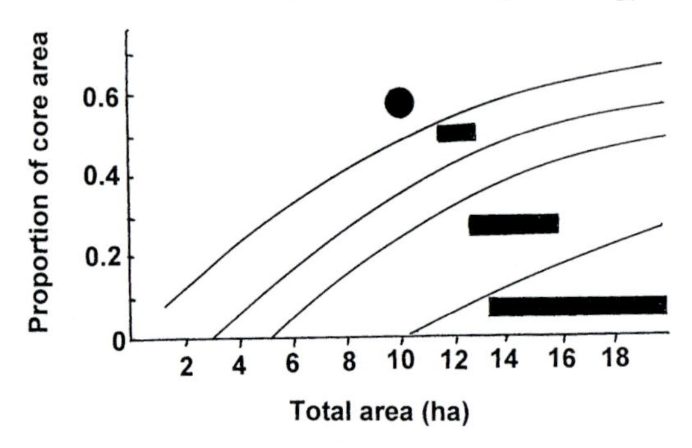

Figure 8.13. Influence of habitat patch shape on amount of core habitat present (Moen & Jonsson, 2003). The proportion of core area is related directly to shape; in this example, (with an edge effect of 50 m), the round fragment has the lowest perimeter: area ratio (shape index 1) and the three rectangular examples have length: width ratios of 2 : 1, 4 : 1 and 10 : 1, with shape indices (relationship between sides) of 1.2, 1.41 and 1.96, respectively. Thus, a 2 : 1 patch has to be more than 3 ha in area to have any interior habitat, and a round patch of 3 ha contains 25% (0.75 ha) of interior habitat.

along it away from the exposed edges may also be fostered. Little information on this aspect is available for invertebrates, but a number of studies on vertebrates – for example, mammals – have shown that wider linkages function adequately as corridors whereas narrow ones do not. In general, three major categories of invertebrate inhabitants of habitat patches can be recognised (Laurance & Yensen, 1991):

- Interior species, whose presence is correlated with occurrence of core areas, and are generally those of direct conservation concern in relation to corridor widths.
- Edge species, whose presence and abundance is often related strongly to the length of the habitat edge.
- Other taxa, which depend mainly on the core habitat but are not influenced strongly by edge effects.

It is commonly unclear without direct investigation which invertebrates belong to the first and third of these groupings, but many edge species are simply absent from apparent true core habitat – or, in contrast, their presence signifies absence of this as true interior core. Distribution of invertebrate populations, and their variations in space and time, are key considerations for conservation management.

It follows that the composition of the corridor area may be of critical importance in determining whether 'habitat' is indeed provided, and it may also influence the degree of management needed. Bennett (1999) summarised a categorisation of corridors based on their origins. Thus, natural habitat corridors (such as riparian vegetation) are natural features following some defined environmental feature or gradient; remnant corridors are also natural in composition, but their retention reflects wider changes to the local environment. More artificial corridors include planted ones, with a gradient from predominantly natural to highly unnatural vegetation composition. 'Disturbed habitat corridors' include features such as roadsides, transmission lines and railway cuttings, some of which may constitute valuable conservation areas for invertebrates, albeit sometimes being very small in area. This variety dictates that a framework for studying ecological boundaries, such as field margins, must incorporate relevant features of those boundaries, considered broadly as both the limits to patches (fields) and the ecological zones between those patches (Caldenasso *et al.*, 2003). Six very general characteristics are then pertinent:

- Boundaries may be entirely distinct or may share features to varying extents with the patches they separate.
- The ecological gradients within the boundary separating two patches are steeper than within either patch.
- Boundaries may be narrow or wide, depending on the gradient of changes between the patches.
- The boundary for one characteristic may differ in magnitude and location from the boundary defined by another characteristic.
- The function of a boundary is determined by an organism, or by material, energy or ecological process affected by the boundary gradient.
- Boundaries are best considered as three-dimensional entities.

Metapopulations

As indicated above, the importance of field margins to invertebrate dispersal relates to the fragmentation of the agricultural environment and its effects on the 'connectivity' between populations or population segregates, together with the ways in which they counter the effects of population isolation. Some level of dispersal is vital to the persistence of many (perhaps, most) invertebrate species and is an intrinsic component of the population dynamics of many varied taxa. Many taxa form wide-ranging populations. The traditional view of many species of butterflies, for example, is that their populations are 'open', so that individuals can disperse widely over the countryside to embrace a variety of habitats, as relative ecological generalists. Others, following a survey by Thomas (1984), have been shown to form highly localised populations or colonies in very small habitat patches. Some butterfly populations, indeed, can persist on areas of a hectare or less, so that small remnant habitat patches in altered landscapes (as long as they remain suitable) may be of critical importance. In the past, many such species have been presumed to have 'closed' populations, with little or no immigration or emigration taking place – so that the protection of those particular, often isolated but occupied remnants has been an important conservation focus. However, with the more recent realisation that many butterfly species manifest as 'metapopulations', the entire landscape has assumed greater importance in practical conservation, particularly in seeking to increase connectivity between habitat patches in the wider matrix. The critical emphases for conservation then include also the effects of habitat patch size and quality, together with the effects of isolation by prevention of migration. As Thomas & Hanski (1997) commented, the two key metapopulation processes (namely, colonisation and extinction) have perhaps been studied better in butterflies than in any other comparable taxon. The lessons learned are likely to have much wider relevance to invertebrate conservation in and around agroecosystems. In essence, many species occur across a landscape in a mosaic of inhabited and uninhabited habitat patches, often with those patches appearing to differ little or consistently in quality. Over time, local (patch) extinctions occur as natural events and so, in contrast to much conventional wisdom in conservation, such extirpations need not necessarily cause alarm.

The most basic principle of a metapopulation is that a species persists in a variable number of population units, each occupying a habitat patch in the wider landscape as a predominantly independent demographic unit. Not all suitable habitat patches are usually occupied. At any one time, a particular patch may or may not be occupied, and the species persists in the landscape by rolling cycles of colonisation and extinction involving the multiple habitat patches that are present. The dynamics of each population segregate (local population) are density-dependent within a patch, but asynchronous among patches, so that migration (broadly, dispersal) among patches links them together. Inter-patch movements are thereby critical to facilitate long-term persistence of the metapopulation. If migrations between patches are frequent, then the dynamics of each population unit will be less distinctive, and they will tend to resemble those of a single, large population. If, conversely, migrations are infrequent, then they may not be sufficient to ensure colonisation of vacant habitat patches in which extinction has occurred or to augment the gene pools of extant colonies. In both cases, these conditions influence the condition of the metapopulation, and in the first case help to drive the entire metapopulation to extinction (Wiens, 1997). Potential corridors, with facility to link habitat patches and thus keep these within the accessible range of rare species, may be of critical importance in conservation of such species. As Thomas *et al.* (2002) commented: 'Although a metapopulation is an easy image to conjure up, it is more difficult to pin time-scales and spatial scales on the dynamics.'

Consider the implications of the conditions summarised in Fig. 8.14. Fig. 8.14a indicates occupied habitat patches in an undifferentiated matrix. The persistence of each unit reflects migration between the patches, with the success of this depending, *inter alia*, on movement rates and capabilities and the distance between patches. In practice, the landscape is usually not as featureless as implied here but comprises a mosaic of different land-use units, incorporating the preferred habitat patches (perhaps as remnants) in a multiple-featured landscape (Fig. 8.14b) in which barriers, corridors and areas of intermediate permeability co-occur. Movement success and distance are thus additionally influenced in various ways, as indicated by the arrows in the diagram. Patch edges and general configuration may both be important influences on connectivity.

Studies on butterflies, particularly in Europe and North America, have played pivotal roles in elucidating how such metapopulations operate (Thomas & Hanski, 1997). Amongst other important outcomes has been increased appreciation of the critical habitat features they need, revealed when the butterflies occupy only very small areas in a much greater and apparently uniformly suitable environment. Thus, (1) the skipper *Hesperia comma* in Britain is restricted largely to south-facing dry calcareous grasslands in southern England, where oviposition occurs only on the grass *Festuca ovina* and only on plants of a particular size growing in particular microclimatic regimes – so that only a small subset of the existing calcareous grasslands are suitable habitats (Thomas *et al.*, 1986); (2) *Euphydryas editha bayensis* in California occupies only small patches of serpentine grassland (Harrison *et al.*, 1988), with most of the area lacking critical, desirable features of topography and/or vegetation for the butterfly. In these species, and many other butterflies, mark–release–recapture studies have shown that most individuals remain within their natal colony or site (97% of individuals of *E. e. bayensis*) (Harrison *et al.*, 1988), and the small proportion that disperse may not lead to effective gene flow.

(a)

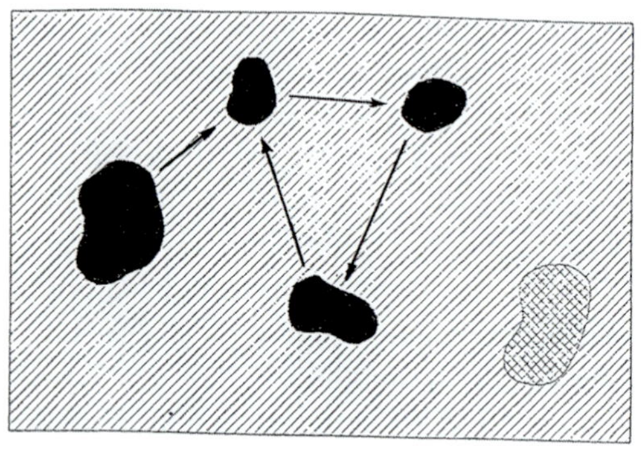

(b)

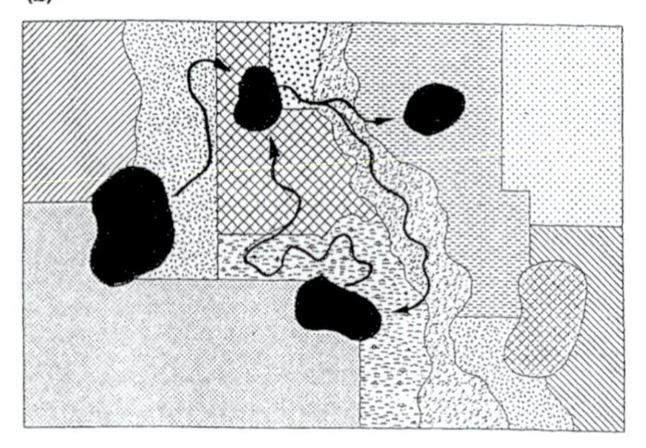

Figure 8.14. Structure of metapopulations: (a) occupancy of habitat patches in apparently uniform suitable environment; (b) occupation of suitable habitat patches in variegated, largely unsuitable environment (Wiens, 1997). In (a), the solid patches (occupied) are linked by sporadic migration, and the shaded patch is at present unoccupied, and it is assumed that the background matrix has no effect on movements between patches. In (b), the matrix is varied, so that interpatch movements may be variously facilitated or impeded by the different landscape features present.

For this checkerspot, Ehrlich & Murphy (1987) noted that fewer than 5% of individuals marked in each of three metapopulation segregates were recaptured in another habitat patch, and opportunities for mating after dispersal are low. *E. e. bayensis* immigrant males have little chance of mating because most female butterflies will have mated already with non-migrant males. Any such matings are likely to occur mainly late in the reproductive season, under conditions in which larval development will be less favourable. Ehrlich & Murphy (1987) suggested that it is unlikely that any but the earliest female migrants may contribute genetically to the recipient population or be effective as founder colonists in new habitat patches.

Three aspects of metapopulation studies are important considerations in agricultural landscapes (Bunce & Jongman, 1993):

- The dynamics (extinction and immigration rates) of the subpopulations: if a habitat patch is small and highly isolated, the extinction rate might exceed recolonisation, so that a subpopulation becomes extinct.
- The connectivity between patches, with important landscape features involving barriers and corridors.
- The spatial and temporal variation in habitat quality, this being influenced by the absences or presences of disturbances due to land-use practices.

Relationships between habitat patches in a fragmented landscape can be complex. Three patterns, defined early in the study of metapopulations, exemplify the dynamics that may occur (Fig. 8.15). Fig. 8.15a represents an 'archipelago' or non-equlibrium metapopulation, with interchange occurring between all the individual 'islands' of habitat present. Fig. 8.15b is the classic (Levins, 1970) model, in which some islands are occupied and others are vacant, so that examination at different times may reveal different patterns of patch occupancy over the array as dispersal/extinction at different times occurs. Fig. 8.15c represents the 'continent and island' analogy, where a large habitat patch is occupied permanently and furnishes dispersers that colonise small, isolated habitats elsewhere. In this case, chances of extinction in the large habitat patch are assumed to be negligible, and the small patches undergo repeated colonisation and extinction events.

Each small habitat patch will vary in quality (suitability to particular colonising species) over time, and vary from being optimal to marginal in both space and time. In the more marginal conditions, perhaps reflecting ecological succession or loss of key resource species (such as the host plants for specific herbivores), chances of extinction may be increased.

Hanski & Simberloff (1997) noted that emphasis on understanding metapopulations has, to a considerable extent, replaced earlier emphasis on island biogeography in conservation, an emphasis evident simply by the terminology used above in metapopulation dynamics. Perhaps the most important change has been replacement of the emphasis on species–area relationships, so that very small patches of habitat sometimes previously regarded as unimportant for reservation or care because

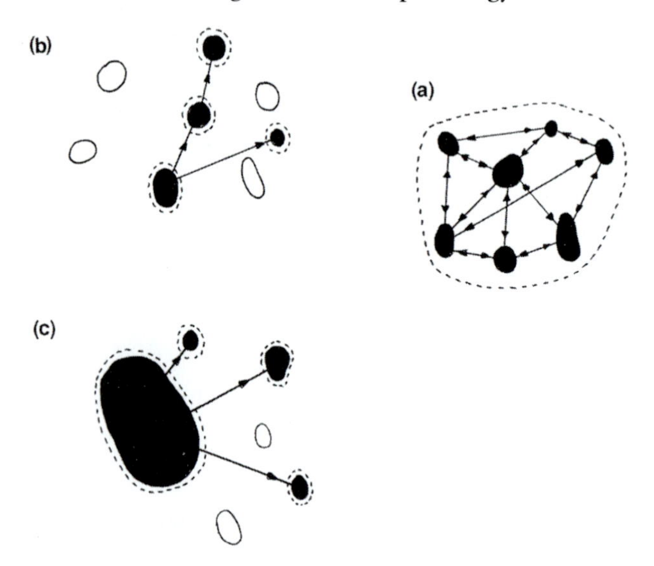

Figure 8.15. Some structures of metapopulations in the landscape: (a) patchy population in which the boundary (dotted) surrounds all subunits, but these are localised within the larger habitat; (b) classic Levins model, in which movement between patches results in patterns of local colonisation and extinction, so that only some patches may be occupied at any time; (c) continent–island model, in which the large source area/population (continent) provides a continual flow of individuals to small island habitat patches to constitute a source-sink model, whereby large and closer patches may be colonised more rapidly or frequently than those that are smaller or further away (shaded patches, occupied; unshaded patches, vacant; arrows indicate direction of movement).

they were deemed 'too small' to sustain populations are now considered much more positively. The metapopulation concept recognised the considerable importance of such small sites, even if they are not currently occupied, within the population range of a mobile species, so that it is not now necessary to confirm residence in order to postulate their importance. The main effect has been to focus attention in practical conservation to the roles of 'networks' in the landscape, as opposed to small isolated sites. The major desirable features of designing reserves, mostly postulated or stimulated by island biogeography, indeed incorporate a range of aspects of size, shape (reflecting the influences of edges), isolation and connectivity, and dispersion, so that dispersal of organisms in relation to colonisation is a critical consideration. Nevertheless, as Hanski & Simberloff (1997) remarked, increase in citations of metapopulation studies in conservation biology parallels decline in those on island biogeography and 'these trends represent a paradigm shift'. In relation to area, relatively high rates of butterfly extinctions have been observed in small populations on small habitat patches, with Thomas & Hanski (1997) citing studies on a variety of species in which this has occurred.

'Connectivity' refers to the extent to which a landscape facilitates or hinders movement between such isolated habitat patches, and is thus often related to coverage and dispersion of those patches. It reflects also the presence of possible corridors (including field margins) between them. If a continuous habitat (of the nature implied by Fig. 8.14a) is broken up, then the initial effects may simply be of decreased carrying capacity with connectivity not affected; continuing loss of habitat, however, progressively increases the effects of isolation to the extent where all connectivity may be lost.

Extinctions in metapopulations

Many causes of local extinctions have been postulated, with the most commonly cited cause (of universal concern for conservation) being loss or change of natural habitats. Extinctions of local populations (extirpations) are often linked with 'chance' (stochastic) events, where some unexpected or extreme happening may cause loss of the population (either by inflicting direct mortality or by a massive change in the local environment), even when wider conditions remain constant and suitable for it to persist. Small populations, and those on small habitat areas, may be particularly susceptible, with a variety of studies on butterflies indicating the widespread nature of this occurrence.

Environmental stochasticity, represented by unpredicted events such as drought or aseasonal extreme weather, is also known to have led to butterfly extirpations. A third factor, listed by Thomas & Hanski (1997), is the interaction between environmental stochasticity and habitat heterogeneity, reflecting that stochastic events may act very unevenly across a variable habitat in which particular topographic, vegetational or microclimatic features may each help to buffer populations against extirpation. Extending from this principle, a metapopulation may occupy a variety of more distinctive habitats with varying likelihood of extinctions from unusual environmental events. Extinctions in these circumstances may occur most often in suboptimal habitat patches. Over time, vegetation dynamics and management (such as grazing (see p. 270) or cropping) may render particular patches increasingly unsuitable, consequently increasing the likelihood of extinction on these. Thomas & Hanski (1997) considered that, for butterflies, this was the most common apparent cause of decline for medium-sized and large local populations and thus is of particular importance for the persistence of metapopulations. Such changes may occur at any of a variety of scales and, in general, loss of all suitable habitat is more likely for a small habitat patch than for a large one. Finally, isolation has a strong correlation with extinction, reflecting lack of immigration and any deleterious genetic effects arising from population restriction. Lack of immigration (1) fails to replace individuals lost to emigration and (2) fails to 'rescue' the population from genetic deterioration and erosion. Reproductive failure may result simply from inability to discover a mate.

Once a metapopulation unit has become extinct, patches within an existing metapopulation range may be colonised relatively easily. In these instances, the 'rescue effect' is common, and isolation may not constrain recolonisation from normal individual movements among patches. At least in some meta-populations, patch isolation seems generally not to be large. However, in these, only a small fraction of patches are vacant unless patches are very small (Thomas & Harrison, 1992). Different circumstances prevail beyond the normal bounds of the metapopulation, where isolation may be more effective, and larger distances between habitat patches (1 to >20 km may provide effective barriers for different butterfly species) (Thomas & Hanski, 1997) may ensure that any recolonisation may be very slow or simply not occur.

In another case, a habitat network may be sufficiently extensive to sustain a metapopulation, but the whole network may be too isolated to have been colonised. Simulation studies on the silver-spotted skipper butterfly, *Hesperia comma*, by Thomas & Jones (1993) demonstrated this scenario and implied that the butterfly could take more than 100 years to recolonise one particular suitable area naturally. The life stages not usually thought of as strong dispersers may also contribute: for cinnabar moth (*Tyria jacobaeae*) feeding on ragwort (*Senecio*), van der Meijden and van der Ven-van Wijk (1997) noted that the fourth and fifth instar caterpillars can cover hundreds of metres to reach undefoliated sites. Although the female moths appear not to be strong fliers, they are indeed effective colonisers. Studies on *Tyria* have confirmed the importance of large spatial-scale studies in order to understand population dynamics and local extinctions.

Determining such influences, and postulating the extent to which they may be extrapolated to other taxa, is an important component of management of threatened species, particularly in fragmented habitats, and most of all in small patches in which edge effects may become pervasive. Such effects (see p. 00) may extend for up to 0.5 km in forests (Ranney *et al.*, 1981), so that small patches of woodland and forest in agricultural landscapes may not contain any habitat suitable for true 'interior' species.

9 · *Pasture management and conservation*

Management of grazing areas, in addition to the cropping areas emphasised until now, is important in invertebrate conservation. Differing grazing regimes reflect different intensity of impacts and management of more or less altered communities. Grazing affects floristic composition, and features such as sward height may also be critical for particular invertebrates, so that management of pastures and rangelands can influence a wide variety of taxa and assemblages.

Introduction

Several of the previous chapters have emphasised the influences of pest-management practices and the impetus for their refinement for greater efficiency and reducing environmental side effects. Most examples refer to cropping systems, but the principles apply also to pasture pests (which may, for example, take more than 20% of gross economic value (Landy, (1993) for Australian sheep pasture)). However, other agricultural practices associated with pastures, such as grazing, mowing (haymaking) and pasture improvement, also have important implications for conservation, with these practices being used widely for more general management of grasslands in many parts of the world. The effects of such practices, for example in the Western Australian wheatbelt (see Hobbs & Saunders, 1993) are often sufficiently severe to cause dramatic loss of vertebrate and native plant species, as taxa appraised much more frequently than invertebrates, from vast areas. In general, major features affecting invertebrates in managed grasslands include floristic composition and diversity and sward structure.

Pasture landscapes, of course, encompass far more than just grassy fields, with many low-intensity pastoral systems based on rangelands or open woodland or forest ecosystems. Thus, in Europe (Finck *et al.*, 2002), changes in land use related to agricultural intensification have had

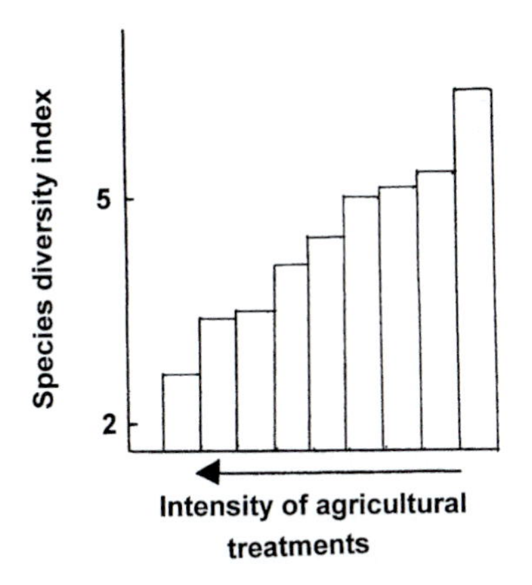

Figure 9.1. Management trends of pasture, and effects on abundance of herbivorous insects: diversity of leafhoppers in Poland (Andrzejevska, 1979).

substantial impacts on 'semi-open landscapes' linked to grazing of large domestic herbivores in open forests. The systems have been referred to as the 'New Wilderness', with its success reflected in the values of grazing as preferred land management over mechanically based management in conserving ecological variety and heterogeneity through regulation of stocking rates and seasonality of grazing pressures (Kampf, 2002). This practice differs considerably from more conventional and intensive pasture use and incorporates habitats that support massive additional invertebrate diversity. Areas not conventionally regarded as 'agricultural' are thus incorporated into the agricultural estate, so that places such as the New Forest of southern England are subject to management through grazing (Spencer, 2002), together with the numerous local or endemic invertebrates that occur there.

Intensively managed grasslands tend to be dominated by disturbance-tolerant arthropod species, whilst the invertebrate communities of more extensively managed grasslands tend to be more complex, mirroring the complex and varied habitat present (Curry, 1994). Such trends have long been demonstrated in both herbivorous insects (Andrezejevska, 1979) (Fig. 9.1) and in microarthropod decomposer communities. Curry (1994) cited his earlier studies reporting a lightly grazed grassland having 220 000 microarthropods/m^2 (representing 200 species) compared with an intensively managed sward on a similar soil type with only 16 000 microarthropods/m^2, with the latter representing 100 species. In contrast, larger species of earthworms may benefit from more intensive management. Extensively managed meadows can furnish valuable habitats for specialised insect species, such as some Heteroptera (Di Giulio *et al.*, 2001), and support high diversity. In contrast, and apparently typifying a much more general case, intensively managed meadows are dominated by more widespread, relatively generalised species. For these plant bugs, frequent and early cutting reduces abundance of many species, whereas some generalist species do not respond consistently to such treatments, and may even be enhanced by them. Extensive management of grasslands, as exemplified in Switzerland by Di Giulio *et al.* (2001), can effectively maintain insect diversity in agricultural landscapes, with both local and regional diversity enhanced. As in numerous studies of species within assemblages, individual species vary considerably in their response to 'management' (O'Neill *et al.*, 2003).

Intensity of grazing management may be focused to ensure particular forage species composition, such as the balance between annual and perennial species in relation to whether seasonal or year-round grazing is needed. In basic ecological terms, other processes are also important – simplistically, too much grazing may lead to loss of native biodiversity and to general land degradation, and too little grazing may not prevent successional changes in vegetation. Particular grazing regimes may become important tools in invertebrate conservation management. Both management regime and landscape structure can enhance or deplete invertebrate diversity and particular species wellbeing in grasslands subject to agricultural modification. Grazing can lead to floristic changes, through selective feeding on more palatable plants, but it also has immense effects on the structure and height of the sward, so affecting the size, microclimate and complexity of the habitat available to above-ground invertebrates. Effects are perceived widely to be most marked for vegetation- and litter-frequenting biota, but soil populations may be affected additionally by trampling and microclimatic and resource changes resulting from removal of vegetation and litter. In addition to grazing, pasture-management practices such as re-seeding, mowing and fertiliser applications also influence grassland invertebrates, with such changes commonly employed to 'improve' pasture quality. On more natural pastoral grasslands, as Morris (1971b) (drawing on a series of his studies, e.g. Morris (1969, 1971a)) emphasised, 'grazing is the most usual form of grassland management in Britain', with the particular grazing species and intensity sometimes having substantial effects on invertebrates and also being critical in managing succession.

Grazing is the historical form of management for grasslands over much of Europe and has, perhaps, received most attention in relation to its roles in habitat maintenance for butterflies inhabiting chalk grassland in southern England (BUTT, 1986). Unsuitable grazing management may rapidly eliminate the butterfly species from sites, and any pervasive grazing regime also has implications for the maintenance of botanical richness on sites. The major variables are (1) the method of grazing (continuous stocking, rotational stocking, seasonal grazing or opportunistic/sporadic grazing) and (2) the kind of grazing animal(s) used (in Britain, rabbits, sheep, cattle, horses or others) used in combination to provide regimes suitable for particular species and sites. In some areas, the form of grazing can be controlled closely and its effects monitored, but where this is not possible, BUTT (1986) noted the practical values for butterflies of providing permanent 'exclosures' within a large grazed area to act as refuges for invertebrates, which can benefit from the juxtaposition of grazed and non-grazed areas. BUTT (1986) also noted that changes can occur under very heavy grazing regimes through provision of supplementary foods (such as hay) on high-quality grasslands through trends such as weed introduction, heavy treading and increased nutrient inputs from dung. As with other forms of pasture management, unexpected subtleties over effects of grazing appeared in this British study, and its inferences have much wider values and relevance in optimising pasture management for invertebrate conservation. The various grazing options are summarised in Table 9.1. These regimes, and the different grazing methods of the various herbivores, lead to different outcomes and impacts by each in relation to pasture and butterfly wellbeing. Thus, BUTT (1986) noted the very considerable importance of rabbit grazing on chalk grassland because of their selectivity. Rabbits can create patchy mosaics of small areas of different heights, whilst their scraped areas of bare ground provide favourable microclimatic conditions for basking and oviposition. Sheep affect grassland by trampling and close nibbling, and careful control of grazing intensity (stocking rates and timing) may be needed. Cattle may be preferable for summer grazing on some rich butterfly sites, because they create a diverse sward height if low-density grazing is undertaken, and their trampling can variously have beneficial or detrimental effects. Horses and ponies graze patchily. There is thus a need to select grazing species as an integral part of grassland management, especially for the management of habitat for particular rare or vulnerable invertebrates living in the area.

Table 9.1. *Grazing practices as options for management of chalk grassland for butterflies in southern England (BUTT, 1986)*

Mode of grazing	Comments
Continuous stocking	Stock density critical; a low stocking density may maintain a mosaic of shorter and longer turf. May be suitable (1) on tiny isolated sites with low stock density; (2) on large sites where labour and division fencing limited; (3) on other sites with high sward density and habitats for many butterflies; and (4) where heavily grazed sites are important only for colonies of one or two species adapted to uniform habitats
Rotational stocking	Temporary or permanent splitting of sites into compartments, with stock moved around these. Allows sward regeneration by 'resting'; need to divide resources (larval food plants, nectar plants) spread among several/all compartments to avoid critical reductions. Suitable for most sites where fencing and labour are made available
Seasonal grazing	Time of year may be critical: (1) winter grazing generally considered least damaging; (2) heavy summer grazing to be discouraged, except in cases of specific management need; (3) may be other critical periods to avoid for particular species
Sporadic/opportunistic grazing	Use of grazing on an *ad hoc* basis as becomes available: (1) uncritical use may be unwise; (2) if whole site 'blitzed', may be considerable harm to butterflies; (3) should be integrated with wider management

The BUTT (1986) survey, and the studies on British butterflies that underpin it, demonstrate clearly the subtleties of grassland-grazing management needed to sustain ecologically specialised rare insect species. Whilst the topography of a site may be critically important, maintaining a mosaic matrix of food plants, bare ground and turf of particular heights may also be vital. Thus, *Lysandra bellargus* (the Adonis blue) is confined almost wholly to areas where the sole larval food plant (the vetch *Hippocrepis comosa*) grows in turf less than 3 cm in height and especially only 0.5–1 cm tall, sometimes when the plant grows in sunny, sheltered depresssions (Thomas, 1983a). Many extinctions have occurred when sites became overgrown, particularly after loss of rabbit grazing resulting from the onset of myxomatosis. Grazing is viewed usually as essential to maintain suitable sites, with continuous grazing commonly advocated. Rotational grazing can be used to help maintain *L. bellargus* when it coexists with other species that need taller growth. Thus, the small blue (*Cupido minimus*) survives mainly on ungrazed or irregularly grazed areas and is usually absent from more heavily grazed areas, including those subject to more traditional grazing regimes. Sheep and other grazers are hostile to this species because they eat off the flower heads (on which caterpillars feed) of the vetch *Anthyllis vulneraria*.

The variety of management regimes that may be required for the British chalk grassland butterflies is implied by Figure 9.2, in which the turf-height preferences for each species are summarised. The taxa range from those that depend on very short turf (as with *L. bellargus*, above) to those needing turf 30 cm or more in height, essentially completely ungrazed. For the Lulworth skipper (*Thymelicus actaeon*), for example, absence of grazing to foster the tall (and only) food plant (tor grass, *Brachypodium pinnatum*) is preferable. However, it is possible for *T. actaeon* to coexist with *L. bellargus* by imposing light grazing, either continuous or patchy, because stock tend to avoid mature *B. pinnatum* when other food is available. Exclosure may be needed if any heavier grazing takes place. Recent studies (summarised by Asher *et al.*, 2001) suggest that this skipper can tolerate many kinds of grazing, but parts of its core areas are sometimes simply too steep for grazing to occur (Bourn *et al.*, 2000; Thomas, 1983b).

Many other studies have confirmed that management for one species of arthropod in grazing systems may not benefit other species. For example, Volkl *et al.* (1993) compared mown, sheep-grazed and abandoned calcareous grasslands, particularly in relation to endophytic insects developing in flower heads of Cynareae (Compositae). Responses of different insects differed according to features of their lifecycles and specific habitat. Most species were more abundant in abandoned sites, but two were more abundant on mown or sheep-grazed sites. The tephritid fly *Urophora cuspidata* reached high density on mown sites cut early in the season, i.e. in time for a new supply of flower buds to be grown as oviposition sites in the same season. Several other species also seemed to benefit from increased bud production by *Centaurea scabiosa* after mowing. However, mowing also reduced the abundance of some univoltine insects on *C. scabiosa*, with these effects reflecting removal of all available flower heads at cutting

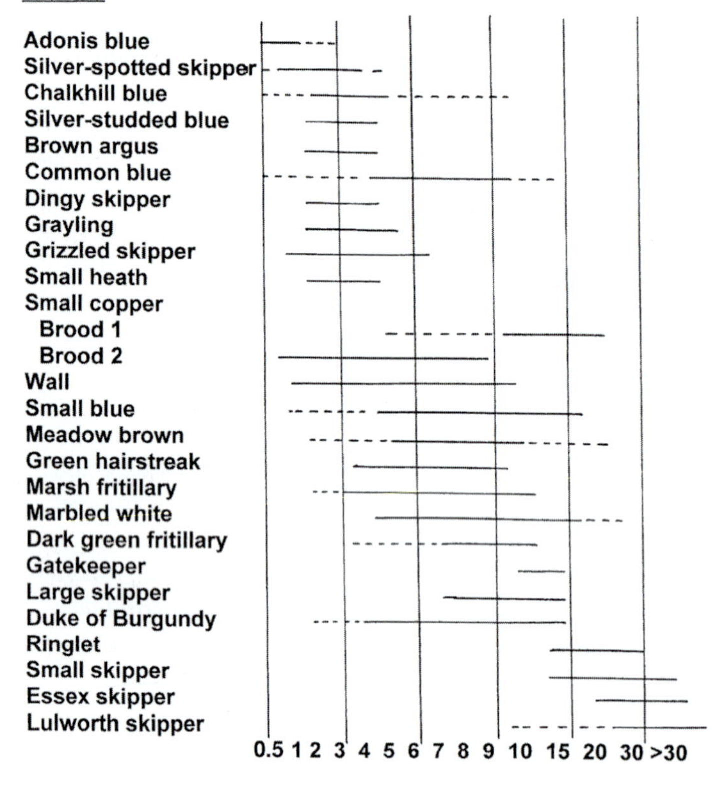

Figure 9.2. Management of chalk grassland for butterflies in Great Britain: the importance of sward height for individual species characteristic of the habitat (BUTT, 1986). Turf-height preferences (cm) on typical sites are shown. Species names, from top of diagram, are as follows: *Lysandra bellargus, Hesperia comma, Lysandra coridon, Plebejus argus, Aricia agestis, Polyommatus icarus, Erynnis tages, Hipparchia semele, Pyrgus malvae, Coenonympha pamphilus, Lycaena phlaeas, Lasiommata megera, Cupido minimus, Maniola jurtina, Callophrys rubi, Euphydryas aurinia, Melanargia galathea, Argynnis aglaja, Pyronia tithonus, Ochlodes venata, Hamearis lucina, Aphantopus hyperantus, Thymelicus sylvestris, T. lineola* and *T. acteon.*

and subsequent microclimatic changes. In contrast, species developing in the stemless *Cirsium acaule* were not affected directly by mowing or grazing but were affected by consequent changes in microclimate. Both management practices maintain areas as short turf, characterised by low humidity and substantial day/night fluctuations.

Wider grassland management

Pasture management of this nature intergrades with wider aspects of decline and changes of native grasslands in many parts of the world and wider rangeland management. In general, research on wider management of grasslands needs to be sufficiently broad to assume that the results are not simply site-specific (however useful that may be in any particular case), sufficiently comprehensive taxonomically to understand impacts on a variety of different invertebrate groups, and integrated so that the links between land use, land cover and invertebrates are understood adequately (Usher, 1995).

The changes in grassland constitution brought about by establishment of pastoral industry are sometimes severe. In south-eastern Australia, native lowland grasslands are regarded as the region's most threatened ecosystem (Kirkpatrick *et al.*, 1995), with various commentators recording its reduction to estimates of 0.5–5% of the former pre-European settlement extent of some two million hectares. Remaining natural patches are mostly highly fragmented, sometimes present only in small areas protected from grazing stock by being fenced, with grants for fencing sometimes recognised as important practical conservation facilitation, particularly for flora. These reserves include pioneer cemeteries, rail reserves and other private areas.

Kirpatrick *et al.* outlined the process whereby this drastic erosion of grassland occurred:

1. Introduction of stock, particularly cattle and sheep, preceded invasion by numerous European plants, many of them transported accidentally on or in stock animals. These plants were preadapted to trampling, grazing and dunging patterns of stock, in marked contrast to many of the native species.
2. The hard-hooved stock compacted the original friable soils, rendering them unsuitable for many of the native grasses and herbs. Long-lived, deep-rooted native species were gradually displaced by more ephemeral and shallow-rooted exotic plants.
3. Continuous, intensive stocking so led to vegetational and soil change and loss of native species. During the second half of the twentieth century, in particular, widespread application of fertilisers and sowing of pasture plants (including 'nutritionally desirable' exotic grass species) led to further demise of native grasslands.

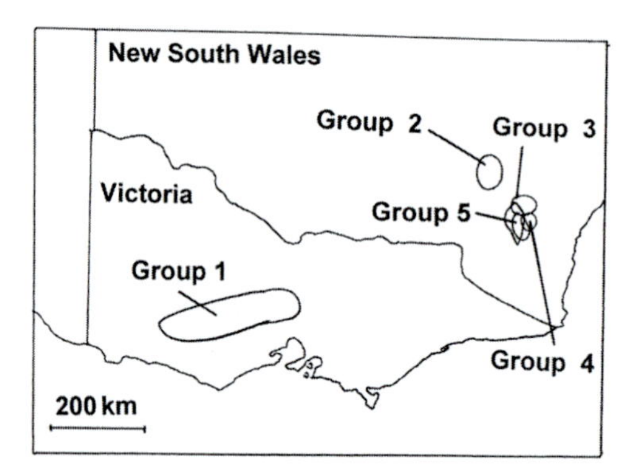

Figure 9.3. The five main distributional and genetic segregates of the golden sun moth, *Synemon plana*, in south-eastern Australia (after Clarke & O'Dwyer, 1999).

Several of the 29 lowland grassland communities of south-eastern Australia recognised by Kirkpatrick *et al.* (1995) are likely to become extinct without further protection; most of them are not represented adequately in secure managed reserves, and none was unthreatened at the time of their publication.

Their relevance here is related directly to the existence of numerous endemic Australian invertebrates found only in native grasslands, and several such taxa are of individual conservation significance in Australia because of their declining distribution and state of being under continuing threats from further habitat change. In addition, the communities present have not been evaluated fully. Key (1978) noted that some flightless endemic morabine grasshoppers had become threatened because of loss of grassland habitat by sheep grazing, and by then occurred predominantly in small, ungrazed reserves such as pioneer cemeteries, as above. Their populations were thus both highly fragmented and subject to stochastic effects. A partial parallel occurs with the flightless grasshopper *Prionotropis hystrix rhodanica* in southern France (Fourcat & Lecoq, 1998). Its habitat is *coussou*, a graminaceous steppe vegetation with abundant pebbles. Modification of the habitat for cropping and more intensive agriculture has led to once continuous habitat becoming very fragmented and populations of the grasshopper becoming ever more isolated. Fourcat & Lecoq noted that large patches (>1000 ha) are necessary for the grasshoppers and believed that they could become extinct without these persisting.

Perhaps the most intensively studied lowland grassland invertebrate in Australia is the golden sun moth (*Synemon plana*, Castniidae), one of a series of species in this unusual diurnal moth family regarded as threatened and that has disappeared from many localities where it occurred in the first half of the twentieth century, or more recently. *S. plana* is listed for protection in Victoria, New South Wales and the Australian Capital Territory and has recently been the subject of one of the most wide-ranging ecological genetics surveys of any Australian insect (Clarke, 2002), in addition to extensive field surveys and formulation of management plans (Edwards, 1994; O'Dwyer & Attiwill, 1999). Decline is linked clearly with habitat loss; in general, greater than 40% cover by native *Austrodanthonia* grasses is needed to sustain *S. plana* populations (O'Dwyer & Attiwill, 1999), with this figure providing a clear direction for restoration of degraded habitat to a suitable quality. Populations of *S. plana* are isolated, largely because the female moths fly little and almost always for distances well under 100 m (O'Dwyer, 1999). However, there are clear genetic differences across the species' range. Clarke & O'Dwyer (1999) recommended that each of the five main genetic segregates they found should be treated as a separate entity for management, with one of them probably constituting a distinct taxon. The related pale sun moth (*Synemon selene*) appears to have died out in South Australia as a direct consequence of habitat loss caused by widespread ploughing and cultivation of native grasslands (Douglas, 2000), and its distribution in Victoria declined and become fragmented for similar reasons. However, some grazing is needed to maintain castniid habitat quality, because the food plant grasses are easily overgrown by more agressive competitors.

Moths have received considerable attention also as indicators of grassland management in Europe (Erhardt & Thomas, 1991), New Zealand (White, 2002) and Tasmania (McQuillan, 1999), where changes in floristic composition and grazing regimes can render numerous species vulnerable. Thus, Tasmanian native grasslands support numerous native Geometridae, many of them endemic to the state, and McQuillan was able to allocate the 53 species of Xanthorhoini to a gradient of susceptibility to local extinction in relation to grassland change and management, with their conservation sometimes depending on maintaining native interstitial herbs in grasslands. McQuillan pointed out that in contrast to the monocotyledon-feeding Castniidae, Geometridae depend more on such herbaceous components, but (like *S. plana*, above) many female xanthorhoines are not capable of sufficiently sustained flight to colonise habitat fragments beyond the one where they were bred.

So-called 'grasslands' utilised for pasture are often botanically diverse, with numerous non-grasses present and susceptible to loss in various ways. Most arthropods regarded as pests of pastures occur mainly on 'improved' pasture used by stock. Improved pastures may still be diverse (with 'normal' Australian sheep pasture comprising mixtures of legumes, grasses and broad-leaved species (Hutchings, 1993)), with the particular mixture influencing stocking rates and periods.

10 · *Towards more holistic management for invertebrates*

Approaches to harmonising and integrating the primary aims of agricultural producers and invertebrate conservation advocates involve exploring more effectively the common ground that already exists between these disparate priorities. Increasingly, both agricultural production and conservation are considered at the landscape level, and some aspects of more holistic area-wide management are discussed in this chapter, together with the complementary roles of practices such as agroforestry. Approaches towards more holistic management for invertebrates, and progressive refinement of pest management to facilitate this, necessitate wider recognition and development of the mutual interests that are largely already evident.

Introduction

The foregoing chapters have demonstrated that agriculture is one of the primary vectors of habitat modification, and that much of such modification for agriculture has been at the expense of more natural environments, including native vegetation, and of the biota they support. Within the wider landscape, studies on invertebrates have been of considerable importance in clarifying the effects of agricultural practices (Collins & Qualset, 1999), with recent emphasis on developing cultural control measures for pests and appreciation of features such as field margins conveying the practical lessons of conservation need (albeit focused primarily on taxa seen as beneficial in pest management) and establishing it firmly on the practical agendas of agricultural producers. Thus, the appreciation of values of generalist, often native predatory arthropods has led to calls for an increased 'predator buffer' in cropping systems (Hokkanen, 1997), in essence the maintenance of a diverse and relatively stable predator guild that can respond – or be manipulated – and aggregate to pest 'hot spots'. These predators are normally sustained on non-pest and non-conservation priority prey in relatively natural environments. Field

Table 10.1. *Integrated control techniques for pests and diseases, in order of preference (Vereijken, 1989)*

Biological
Proper crop rotation to avoid soil-borne pests, especially nematodes
Resistant cultivars
Nitrogen fertilisation under the current advice to enhance crop resistance against aphids and moulds
Enhancement of beneficial organisms (parasitic and predatory insects, antagonistic fungi) by organic manure, minimum tillage and selective chemical controls
Introduction of (mass-reared) biological control agents
Chemical
Monitoring harmful species to develop control thresholds
Choice of pesticide based on efficacy, selectivity, toxicity to people and animals, persistence, mobility and price
Seed dressing and row application instead of area-wide treatment

margins are a critical focus for such aims, with their management helping to determine numbers of natural enemies in fields and also influencing the abundance and diversity of numerous other invertebrates.

The impetus for seeking the least damaging (and most economically advantageous) methods of pest management has resulted in large part from the realisation that a suite of different control options may be available, and that these can in some way be ranked for effectiveness and also for wider benefit or lower non-target damage (Table 10.1). Lockwood's (1998) ranking of risks to conservation of Orthoptera from various control practices for pest grasshoppers (Table 10.2) illustrates this point well.

Ecosystem and landscape management

Hochberg (2000) ventured that 'anything but community or ecosystem conservation is unlikely to make much headway in the global conservation of insect parasitoids', as a group that can make up high proportions of insect-species richness in agroecosystems and that has massive diversity and economic importance. This sentiment is held widely, and for many taxa, by workers on invertebrate conservation, simply because the prospects of developing sufficient resources to treat every deserving species (or even a minute representative selection of them) individually as conservation targets are utopian, and gaining sufficient knowledge to select the most deserving taxa by rational triage (other than in a few

Table 10.2. *Conservation risk factors to Orthoptera in relation to primary means of orthopteran pest management (Lockwood, 1998)*

	Risk factor*			
Approach	Time	Space	Spectrum	Lethality
Chemical				
Blanket	Low–moderate	Moderate–high	High–extreme	Moderate–extreme
Barrier	Nominal–low	Low–moderate	High–extreme	Moderate–high
Biological				
Insecticide	Nominal–low	Low–moderate	Low–moderate	Moderate–high
Organismal	High–extreme	High–extreme	Moderate–high	Low–moderate
Cultural				
Crop	Nominal–low	Nominal–low	Moderate–high	Low–moderate
Rangeland	Low–extreme	Moderate–high	Low–moderate	Low–moderate
Mechanical	Nominal–low	Nominal–low	Nominal–low	Moderate–extreme
Social**	Low–high	Low–high	Low–moderate	Low–moderate

*Risk factors are generalised relative estimates, here limited to the context of orthopteran conservation. In Lockwood's explanatory example, some chemical insecticides may be persistent hazards to human health because of bioaccumulation, but their efficacy to non-target orthopterans declines rapidly.
**Very varied approaches, including education, selling livestock, purchase of substitute forage, providing food to replace crop losses, and insurance programmes for compensation.

exceptional flagship groups, such as butterflies) is equally impracticable. According to Hochberg (2000), '. . . focused conservation of, say, 1000 individual species over this century may be numerically infinitesimal compared to the actual number of endangered species'. Emphases on developing conservation biological control provide a pragmatic reason based on the applied importance of parasitoids for seeking to conserve them in the wider landscape. Within agricultural landscapes, Tscharntke (2000) drew on many of the themes discussed earlier in this book to suggest how this may be approached. In summary, he advocated 'neutralising the island status' of annual field crops by improving habitat quality within crop fields and promoting connectivity between fallow patches. Parasitoid survival within agroecosystems could thus be promoted by aspects of low-intensity and organic farming practices, including conservation tillage, reduced pesticide use, organic (rather than mineral) fertilisers, conservation of residues, modifying the physical and temporal features of crops, and provision of reservoir habitats – in short, the gamut of IPM

Table 10.3. *A comparison of management goals for agricultural and natural ecosystems*

Focus	Agricultural	Natural
Short-term	Profit	No irreplaceable losses
Medium-term	Profit margin for maintenance of system	Minimum external inputs
Long-term overall	Persistence of system Economic viability	Persistence of system Self-perpetuation

tactics that are also of much wider importance in promoting conservation of invertebrates other than those for which they are popularly credited.

Invertebrate conservation necessitates management of ecosystems, particularly in order to ensure habitat and resource availability and to facilitate dispersal, both in relation to anthropogenic changes such as agriculture and the more intangible future changes posed by likely impending climatic changes. Such management trends gradually align more closely the primary aims and needs of both agriculturists and conservation biologists so that, for example, minimising and mitigating the effects of habitat fragmentation, of non-focused chemical usage and of invasive exotic species are now, to a large extent, common goals (Gillespie & New, 1998). Many of the divisive conflicts of the past are progressively giving way to the greater collective awareness of the needs to conserve native invertebrates in agricultural environments, not least for their beneficial values in crop protection and ecological sustainability. In this chapter, the potentials for more holistic management for invertebrate conservation are explored as a practical means to capitalise on this recent augmented awareness and mutual goodwill. However, one fundamental difference does, and will, persist between the major parties involved in land management. Simply, few conservationists are driven primarily by the need to make a profit from their recommendations; farmers must do so, and economic viability is a primary driver of their activities. Issues of commodity protection and production efficiency are understandably a greater priority than some aspects of landscape conservation. This point of difference has been emphasised, *inter alia*, by Main (1993) (Table 10.3), and any realistic compromises must respect this difference and place it in perspective.

Nevertheless, the basic needs to increase invertebrate conservation, in common with those for numerous other taxa, depend on landscape-level considerations, in addition to more particular management of selected sites and individual ecosystems within landscapes. Additionally, and largely implicit in this, the integrity of remnant communities must be protected as effectively as possible to conserve the attributes of 'typicalness' and 'representativeness' (Usher, 1986), with the realisation that even very small habitat patches may harbour numerous localised invertebrate taxa, many of which may normally disperse little but remain susceptible to disturbances of many kinds. The dual approach to management of agricultural landscapes suggested by Lambeck (1999) appears to be embracing. His first management strategy is to pursue 'general enhancement', using ecological principles (as below) as general guidelines for action. This approach to some extent obviates the need for detailed knowledge of the particular landscape being managed – and thus is realistic for almost all invertebrate assemblages in any part of the world. It has the general broad objective of maximising the number of species (basically, 'the amount of biodiversity') retained within the constraints of other, previously more conflicting, land uses. This approach may be thought of as a form of 'coarse-filter' landscape management.

The second approach noted by Lambeck (1999) was termed 'strategic enhancement', with the aims of specifying conservation targets and determining the actions needed to assure conservation – a 'fine-filter', often taxon-focused approach requiring far greater knowledge of the needs of the target biota in the landscape. Two broad kinds of objectives may be involved:

- to retain the biota, or particular target biota, currently present in the landscape: species retention;
- to reintroduce species that occurred previously in the landscape but no longer do so: restoration or reintroduction/translocation, again needing detailed knowledge of the resources and conditions needed by focal taxa, and – in some cases – to have the practical option of translocation without increasing the vulnerability of donor populations.

Management of any landscape with varied land uses is likely to be based on a combination of these sentiments and strategies.

However, and as Lambeck emphasised, the distinction between the two strategies is important because the expected outcomes and the information needed to pursue them differ considerably. The first strategy has obvious limitations, but the general principles of landscape ecology can indeed indicate the directions in which management might proceed. But, because of the lack of information on any particular case, quantifying responses is difficult and – probably – only suboptimal outcomes can be guaranteed from either an agricultural or a conservation perspective. The strategy alone may have very limited value if any emphasis is to be placed on the conservation of particular focal species rather than on 'diversity' or some other more embracing category. The focal taxon strategy thus has potential to generate specific conservation programmes based on detailed and specific information. For invertebrates (particularly of groups other than butterflies and beneficial arthropods), such attempts have been somewhat sporadic in agriculture-dominated landscapes. Both strategies, however, emphasise that protection of biodiversity is an explicit objective of sustainable land management.

One inherently attractive approach – that of using 'umbrella species' – was also discussed by Lambeck. The idea is attractive because if it is indeed possible to manage entire ecosystems or communities by focusing on the needs of one or few such species alone, then less focused needs may become redundant. However, it seems intuitively highly unlikely that any such single (or a small number of) species in any landscape could be an effective umbrella for all critical functional attributes of the ecosystem in which it (they) occurs, even though selection and detection of such taxa may be instructive and important. They may be of more practical relevance, as Launer & Murphy (1994) indicated, in conservation of threatened fragmented habitats. Other than for specific attention to rare invertebrates as flagships, most invertebrate conservation in agricultural mosaic landscapes seems destined to depend largely on the less precise 'general enhancement' strategy, as formulated by Lambeck (1999). It is therefore important to appraise these measures in relation to possible modification of agricultural practices. The Heteroptera of Swiss grasslands (Di Giulio *et al.*, 2001) (see p. 268) may exemplify this more general trend.

This dichotomy in some ways parallels Forman's (1995) 'paradox of management' of landscapes, whereby the more intensive management applied at small scales (in Forman's example, the scale of an individual caring for their own garden within a larger landscape, but here broadened to the analogy of focal species within complex communities) is less likely to ensure sustainability than larger-scale but less intensive management. Forman based this premise on the wide acceptance that larger-scale systems intrinsically self-regulate to help provide stability (and, hence, sustainability), whereas small-scale systems may lack equivalent capability and necessitate continuing management. He suggested that management and planning at an intermediate scale appeared optimal – albeit as a compromise. Forman (1995) went on to note the functional properties of environmental heterogeneity afforded by mosaics of different landscape components in promoting sustainability. Four major foci for planning emerged and, although directed mainly at human communities and activities in Forman's discussion, are of much wider significance

- Resources are separated spatially, so that specialisation and differentiation are fostered.
- Juxtaposition of different resources in adjacent patches provides complementarity, and may enhance stability by increasing possible options for communities near boundaries.
- Heterogeneity is a main cause of movements, with amounts and rates of species and material flows varying greatly across a mosaic.
- Presence of corridors additionally affects flows within a mosaic.

Characters of the landscape mosaic, the spatial arrangement of patches, corridors and the matrix of land uses and natural habitats determine much ecological integrity and are key considerations for invertebrate wellbeing. In many agricultural systems, the concept of 'matrix' as 'the background ecosystem or land-use type in a mosaic' (Forman, 1995) commonly refers to highly altered and 'unnatural' environments far larger in extent than more natural elements in the region. Thus, the matrix in a given area may itself be agricultural, as the most extensive and influential land-use component present.

Kalkhoven (1993) formulated three general 'rules' for landscape management in order to increase the chances for survival of species in fragmented landscapes. In slightly modified form, these are:

- Increase the size and quality of habitat patches in order to increase local population/metapopulation size and diminish risks of extinction.
- Increase the number of habitat patches in order to improve chances for exchange and recolonisation by individuals and lower the chances of stochastic extinction of metapopulation units.
- Decrease the 'resistance' of the landscape by including corridors and reducing the effects of barriers in order to enhance the possibilities of dispersal and of lessening the effects of habitat fragmentation and isolation.

All these broad strategies involve opposing habitat fragmentation and countering its effects, and the two main complementary approaches are (1) to increase connectivity and (2) to increase matrix hospitality, with the latter relatively unusual as a deliberate facet of conservation planning. Both goals recognise that the existing system of nature reserves, and the realistic extrapolations from these, will remain inadequate and – despite the ongoing importance of any and all such reserves and the biodiversity preservation they may help to foster – that a wider landscape approach to conservation may be the optimal, even if the most difficult, way to proceed. 'Strict segregation of ecosystems does not occur in nature' (Stary & Pike, 1999), so that crop boundaries or other ecotones where agroecosystems and more natural ecosystems meet are important in conservation and management of indigenous invertebrates. Conservation biological control has the underlying principle of designing such areas to enhance crop protection by providing resources for potential natural enemies of crop pests; and such measures have much wider values.

It has, perhaps, been inevitable that management agencies have evaluated structural aspects of landscapes (corridors, patches) rather than their less tangible functional attributes. As With (2002) and others have remarked, if organisms can move unimpeded through the intervening matrix, then functional connectivity exists. More broadly, then, considerations of landscape connectivity must embrace the characteristics of the matrix as well as the more obvious focal features. The relatively recent aspects of 'percolation theory' examining connectivity within spatially structured landscapes facilitate analysis by dispersal opportunity to dispersal capability of organisms. Particularly relevant is their ability to cross gaps, and the threshold values that functionally diagnose a 'gap' in such contexts. Thus, many vagile insects may be able to cross vegetated areas between patches of favoured vegetation (habitat) but cannot move over areas of bare ground. On a finer scale, some woodland insects may be able to cross thinned areas with sporadic trees but may be thwarted by low vegetation alone. The extent of contrast between residential habitat and the matrix may thus be important in both structural and biological (functional) ways. Landscape structure and function are related intricately and, usually, understood very poorly without direct study. In one specific example, percolation theory was applied to evaluate effects of habitat destruction/fragmentation on the Neotropical army ant *Eciton burchelli* (Boswell *et al.*, 1998). Colonies of this highly mobile insect, viewed as a keystone species in rainforests of Central and South America, were predicted to go extinct when 45% of the forest was cleared randomly in small blocks (180 × 180 m). Destruction of larger blocks (720 × 720 m), though, increased persistence of colonies because continuous habitat remained to constitute the large areas needed by the species. Habitat corridors were not regarded as able to restore connectivity, because ant colonies might become 'trapped' in local rainforest

Table 10.4. *Some measures of landscape structure, expressed as attributes of 'patches' and the entire landscape mosaic, with the distinction between these not always clear (after Wiens, 2002)*

Patch measures	Size, shape, orientation, perimeter: area ratio, context (degree of isolation by contrast), distance (degree of isolation in space), corridor parameters and incidence
Mosaic measures	Patch number, patch size and frequency of various patch sizes, patch diversity, percentage of landscapes in a given patch category, patch dispersion, edge density, fractal dimensions (area, edge), heterogeneity, gaps, spatial correlation, connectedness (network properties)

pockets and deplete resources there. Habitat corridors were believed to exacerbate fragmentation effects by providing habitat 'culs-de-sac' (With 2002) that enhanced probability of colony extinction should they fail to locate corridors. Similar principles might apply elsewhere but – as with some other recent developments in landscape ecology – the generality of extrapolating such data even to additional closely related species is often tenuous and may be misleading.

The practical management emphasis to support invertebrate diversity conservation is likely to remain largely structural in emphasis, but with underlying functional considerations gradually increasing in relevance and guiding management progressively. Connectivity is a central concern, and the importance of the matrix in fragmented landscapes relates directly to 'patch context' – with 'the recognition of the importance of patch context . . . perhaps the essence of landscape ecology' (Wiens, 2002). An earlier expression of the same belief ('no park is an island': Janzen, 1983) also emphasised the importance of the surrounding areas, either as buffer or matrix. This perspective applies to numerous aspects of practical conservation, of course, but helps to highlight the practical needs for such general measures for the largely unheralded functional biodiversity constituted by invertebrates in agricultural systems. However, the large variety of relevant parameters for appraising patch and matrix/mosaic quality (Table 10.4) helps to demonstrate the options for formulating landscape conservation by emphasising management of one or (preferably) both of these structural components.

The concept and scale of 'landscape' thus implicitly includes the mosaic of different land patches and ownerships present in an area.

Hobbs & Saunders (1993) used the Western Australian wheatbelt as an example, and this includes farm paddocks and native remnant patches as blocks and corridors along road verges and field margins, with biotic interactions occurring between these various elements. No single landscape element functions independently of others. These authors pointed out that conventional management usually deals with landscape elements in isolation – so that farmers manage their individual paddocks and properties, conservation managers manage 'their' reserves, shire engineers manage roadside verges, and so on. The common consequence is that each individual manager is likely to function independently of (and unaccountable to) other management sectors in the same region. As Hobbs & Saunders (1993, p. 304) put it, 'the fragmented landscape is subject to fragmented management', with considerable disparity between the scale of the problem and the scale of the treatment.

Similar points have been made by other commentators, so that an integrated landscape plan for conservation is seen widely as valuable and, in many instances, preferable to more small-scale operations – with the main emphasis in many parts of the world being to integrate conservation of biodiversity with sustainable primary production industries. The elements listed in Table 10.4 are therefore of much wider relevance than to invertebrates alone, although these are among the major beneficiaries. Returning to Hobbs & Saunders' (1993) wheatbelt example as one in which desirable changes have been appraised in some detail as a model for possible application elsewhere, these authors advocated moving towards changes to reflect the differences between Fig. 10.1a and Fig. 10.1b. The traditional 'rectangular' agricultural pattern could be replaced by cultivation in relation to topography and soil type, and the various elements of natural landscape vegetation linked by revegetation areas, these following natural features such as drainage lines to a large extent.

One practical problem in promoting habitat linkages is that the commonly made distinction between 'habitat' and 'non-habitat' for many invertebrates in a landscape is not always clear. In terms of the polarised landscapes (see p. 58) so typical as consequences of much modern agriculture, the distinction is often reasonably unambiguous, but studies on invertebrate metapopulations have revealed the subtleties and temporal changes now regarded as widespread in affecting habitat suitability. Thomas & Hanski (1997) noted four situations arising from studies on butterflies that reflect this continuing difficulty of meaningfully categorising landscape elements as 'habitat' or 'non-habitat':

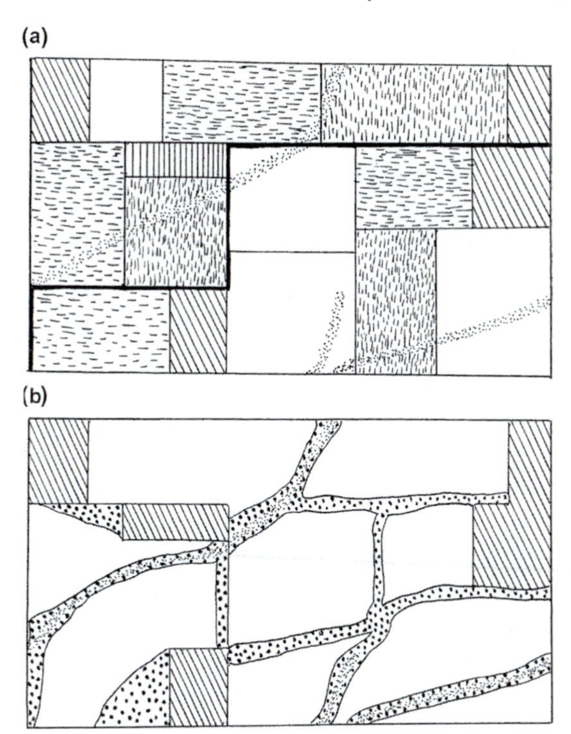

Figure 10.1. Possible changes in farming practices for greater conservation benefit: (a) traditional pattern based on rectangular modules and fencing, with paddocks (including different crops), remnant vegetation and other, mainly in discrete areas or along roadsides; (b) preferred option or potential for future landscape, whereby more natural landscape features are incorporated in definition of farming units, with increased connectivity in the landscape by linking existing remnant vegetation by revegetated areas (hypothetical example, based on the West Australian wheatbelt) (from Hobbs & Saunders, 1993). Fine dots, drainage channels; coarse dots, revegetation; diagonal hatching, remnant vegetation; vertical hatching, nature reserve; dashed lines, unshaded areas, various crop/pasture options.

- Typically, not all habitats available to a butterfly will be of equivalent quality and suitability, so that the series of habitats comprising the species' environment are sufficiently variable to blur the distinction between those suitable for occupation and those that are not adequate.
- Changes wrought by people and succession change the suitability of any given habitat patch over time. Some of these changes may be unpredictable and reversible.
- Temporal variability in the environment may affect a whole environmental gradient. For example, a drought may lead to normally

hospitable habitats being unsuitable. A variety of different microclimatic changes may influence the distribution of many ecologically specialised or 'range-edge' taxa/populations.

- Some habitat patches, although occupied, may not be suitable for populations to persist. They may be 'sinks', with populations present there only because of immigration from other areas.

It is usually impossible to contemplate reconstructing entire natural communities, but the restoration of ecological functions on degraded farmlands is an important partial move toward this utopian aim. As Fry & Main (1993) noted, the removal from agricultural production of millions of hectares of previously productive land (in Europe, for example) provides a probably never-to-be-repeated opportunity for revolutionising agricultural practices to meet the newer priorities of agroecology. The major aim of such restoration can be formulated in terms of sustaining desirable levels of agricultural production whilst simultaneously maintaining and enhancing biodiversity by creating conditions that support an ever-wider variety of natural communities and ecological functions. Setting priorities within this may help practitioners to decide, on the one hand, what is feasible and achievable and, on the other hand, which problems arising from degradation are simply too great to overcome on any wide basis – with the aim of identifying the species and communities with potential for restoration or other effective conservation. These objectives are likely to include various combinations of enhancing rare species, enhancing and sustaining diversity and richness of species, and maintaining representative communities. These options may then need to be appraised in relation to feasibility and also for their acceptability to the farming community (Dennis & Fry, 1992). Most such options are feasible only in temperate regions, but despite this restriction, succession in tropical agroecosystem areas (for example, the changes that occur during the fallow phases of slash-and-burn farming) represent 'applied restoration' (Janzen, 1988), and some earlier farmlands in the neotropics are now unintentionally restored forests.

'The most common changes that affect invertebrate assemblages are those induced by slight changes in land use' (Usher, 1995). Usher went on to exemplify this statement by noting that those changes may result from increased stocking rates, more intensive production, changes in use of pesticides, changes in the crop species, and so on. Increased concern for biodiversity in agricultural environments has led to establishment of a number of 'agri-environment schemes'. In Britain, these include

Environmentally Sensitive Areas (ESAs) and the Countryside Stewardship Scheme (CSS). The primary aim of the former is to 'maintain the landscape, wildlife and historical values of designated areas by encouraging beneficial agricultural practices', with the conservation outcomes depending on farming practices that are amenable to changes to improve that interest (Ovenden *et al.*, 1998). Management of such areas thus seeks, *inter alia*, to maintain and enhance biodiversity values for each ESA. Objectives of CSS, initiated in 1991 as a voluntary cooperative process, include provisions for sustaining and improving (including restoring and creating habitats) wildlife habitats. Such themes generally lead to shifting of agricultural methods from more intensive to less intensive, lower-input systems, with major effects listed by Ovenden *et al.* (1998) including:

- restricting use of fertilisers and pesticides;
- preventing agricultural intensification of grassland;
- requiring maintenance of field boundaries, including hedgerows, walls and ditches;
- creating uncultivated or reduced-impact arable field margins;
- restoring semi-natural agricultural habitats such as heathland and wet grassland.

Such schemes can help to realise 'generic benefits' for biodiversity by (1) maintaining existing wildlife habitats on farmland, (2) arresting habitat declines due to intensification or management and (3) helping to enhance, restore and re-create habitats. In many instances, measures can be based on needs of particular species or larger groups, with birds having become a primary focus in the UK. The broad objectives serve to emphasise the central roles of ecological understanding in land management. Most recently, the proposed British 'entry-level agri-environmrntal scheme' (at present being trialled and likely to become available officially in 2005) is intended to help reverse the loss of biodiversity in the wider countryside and also to complement existing conservation efforts by recognising that farmland is a key area for invertebrate conservation. Open to all farmers and landowners (who will receive payments based on area and amount of environmental management options undertaken), options include hedgerow management, woodland rides, buffer strips around fields, maintenance of stone walls, and low-input grassland maintenance with retention of flower-rich grasslands and other semi-natural habitats, with voluntary management agreements extending for five years.

In essence, these measures entail moving towards a more holistic scheme for land management, incorporating the major needs and

Table 10.5. *Some factors relevant in approaches to holistic land management involving agricultural land and the wider environment*

1. Reduce conflict between stakeholders
2. Seek cooperation between stakeholders
3. Reduce damaging inputs
4. Increase emphases on integrated management (for integrated pest management, on cultural controls)
5. Promote conservation biological control
6. Restore degraded habitats
7. Facilitate matrix hospitality within landscapes
8. Reduce edge contrasts and increase connectivity within landscapes
9. Clarify land-tenure issues and mutual responsibilities
10. Adopt long-term approach

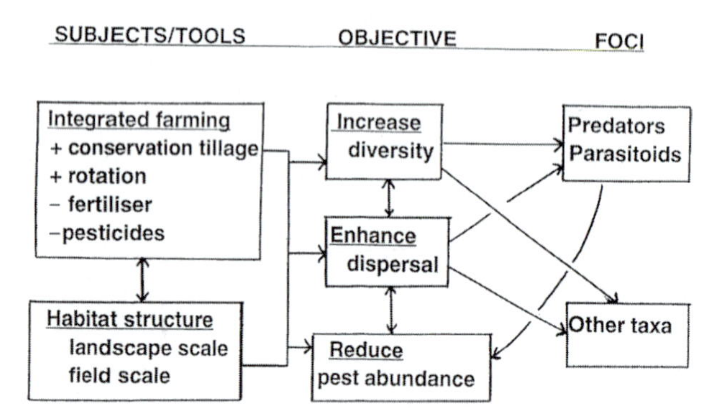

Figure 10.2. Harmonising the priorities of farming and conservation by more holistic approaches to management (after Poehling, 1996).

priorities of both agriculture and conservation for greater collective benefits. Many of the major political and ecological elements of such an effort are noted in Table 10.5. Criteria for evaluation of modified farming systems importantly include nature conservation but, as Poehling (1996) remarked, the twin aims of increasing species richness and abundance and enhancing effectiveness of natural enemies are easily confused. Fig. 10.2 indicates a basic rationale whereby beneficial arthropods are enhanced in farmlands. Habitat structure and farming practices are both manipulable to increase diversity and abundance of natural enemies, enhance and facilitate their dispersal, and lead to reduced pest abundance and impacts. However, the same or similar measures may benefit numerous other invertebrate taxa. Poehling noted these as often underrated, 'indifferent

Table 10.6. *Common elements in an integrated landscape approach to conservation (after Bennett, 1999)*

Plan at broad spatial scales
Protect key areas of natural habitat; expand protected habitats where possible
Maximise conservation values across a variety of land tenures and existing habitats
Maintain and restore connectivity
Integrate conservation with surrounding land uses, and minimise impacts from more intensive land uses

species' and advocated that these should be taken into account in attempts to enhance diversity in agroecosystems. Thus, enhancement of conservation biological control deals implicitly with issues of primary concern in invertebrate conservation – the structure of food webs in environments beyond the agricultural arena, and the communities that sustain these – a priority tallying closely with aims of ecosystem-level conservation. The major objective implicated in Table 10.6 is to increase diversity of agroecosystems through the twin strands of reducing further harmful impacts and attempting to ameliorate the conditions of lowered carrying capacity of the disturbed or degraded systems. The framework shown in Fig. 10.2 can be used to differentiate the priorities of major parties and to delimit the 'common ground' of interest. Thus, slightly modified, Fig. 10.3a emphasises the pathways to enhance pest management, with the major focal group for enhancement being natural enemies. For conservation priority, the emphasis changes somewhat (Fig. 10.3b), with other taxa given greater prominence, pest reduction not commonly heralded, and habitat modification treated as separate from farming attributes. Nevertheless, the same elements and – by and large – similar processes and aims are present, with the main differences reflecting farmers' focus on economic sustainability and on groups of direct value to support this.

Ecologically sustainable land-management practices, of course, must also be acceptable to people using the land, so that ecological practices must be socially and economically harmonious in the context of current and likely future uses of any given area. Traditionally, this has meant the separation of production areas from conservation areas to give the 'twin estates' noted in Chapter 1. However, because of the close proximity of many examples of the two systems, they are rarely completely discrete in functional terms – and perhaps nowhere is this more apparent than when considering invertebrate roles in the harmonising of crop-protection and conservation measures over the landscape. At such scales, many of

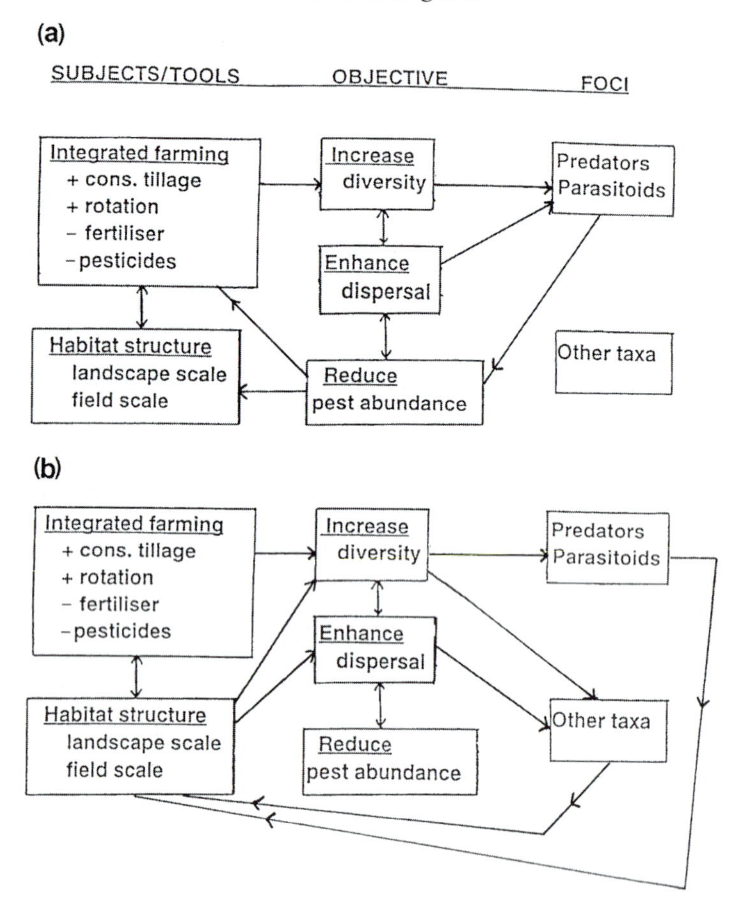

Figure 10.3. Exploring the common interests by modifications of the scheme outlined in Figure 10.2, changing the balance of processes to facilitate benefits to (a) farmers and (b) conservationists. The major change of emphasis is expansion from considering natural enemies for crop protection and consequent pest reduction to the wider community in which those organisms occur, with the wider benefits that accrue from habitat modification.

the measures needed to enhance sustainability overlap considerably as important tools in what Main (1993) termed 'landscape integration'. Hobbs & Saunders (1993) see the reintegration of fragmented landscapes as vital to restoring production and conservation systems, a sentiment with which many other writers implicitly agree.

One important trend with considerable potential to enhance such wider conservation is so-called 'area-wide pest management'. Simply, 'area-wide management implies that there is potential for influencing both biological and ecosystem processes at the landscape level, and thus also impacting directly or indirectly on endemic non-target species in the region' (Rothschild, 1998).

Area-wide pest management

There is increasing realisation that, particularly for management of highly mobile or polyphagous pests, the management efforts of any individual grower may need to be replaced or augmented by a wider consideration of the dynamics of those pests. Area-wide pest management may then become much more effective than the uncoordinated efforts of individuals with differing economic and ecological priorities and with different crops and areas of cultivation. Area-wide management must depend on effective local cooperation. However, it may also lead to the more difficult step of imposing mandatory control measures (Carlson & Rodriguez, 1984), with formal obligations on all growers to implement defined management measures on their crops. Although, idealistically, an area-wide approach may be preferable to unorchestrated and varied individual management efforts, it may be very difficult to implement properly because of the differing priorities and needs of individual stakeholders. Greater collective benefit may be accompanied by overall lower costs, due to economies of scale, but such may again impinge on the priorities of individuals, and any form of mandatory pest control is intrinsically likely to be unpopular, at least initially. Despite a general willingness to contribute to the 'common good', alternatives such as changing cropping pattern may be preferred.

Issues of scale become particularly important for highly mobile and rapidly moving pests. In addition, conflict may occur with the need to manage pests well beyond cropping areas, as pre-emptive measures, and as noted earlier for locusts. Another Australian example was the use of attractants (such as methyl eugenol) in conjunction with insecticide to eradicate the recently adventive papaya fruitfly (*Bactrocera papayae*) in northern Queensland forests in the mid 1990s. Because many native fruitflies, some of them rare, occur in tropical forests (including national parks and World Heritage Areas), they were considered vulnerable to such treatment (Abbot & Seymour, 1998), but lack of knowledge of the native fauna precluded critical evaluation of the likely impacts. This scenario is only too familiar to many invertebrate biologists. Part of the issue relates to some such pests originating far from cropping areas and demanding control in remote, non-agricultural parts of a country – in Australia, locusts (mainly *Chortoicetes terminifera*) (see p. 113) and *Helicoverpa* moths are prime examples of such highly mobile pests, whose sporadic outbreaks or more predictable incidences may necessitate control in such non-crop areas, without which more intensive control would be needed in crops. Hamilton (1998) reviewed the scope of forecasting outbreaks for *C. terminifera* and noted that it is difficult to extend to other pests, even *Helicoverpa*. A key reason for this perception is that individual landowners have long realised that they can do little to avoid locust damage when they act alone, so that management for the pest has to be determined on a larger regional scale if it is to be effective. For *Helicoverpa* in Australia, Rochester & Zalucki (1998) noted the need to measure further the impacts of migration of *H. armigera* during summer and autumn. In contrast to the impacts of *H. punctigera* migration in spring, these aspects remain poorly understood as components to evaluate for more effective regional management. More forcefully, Zalucki *et al.* (1998) stated that attempts to manage *Helicoverpa* spp. as a set of independent local problems are 'flawed from the onset when dealing with a multivoltine, polyphagous, highly mobile species' and that 'these species need to be managed on a season long area wide basis'.

A major practical and philosophical divergence between conventional IPM and area-wide IPM is simply the increased relevance of the non-agricultural matrix in considerations of area-wide management. This typically comprises all gradations from natural to highly altered environments and thus largely coincides with the target arenas of conservation interests. In essence, conservation management and area-wide IPM, although pursued largely by separate groups of practitioners, may target the same systems and for largely overlapping needs. This perspective incorporates the whole regional environment in which the pest may have effects or simply occur. It increases consideration of areas traditionally ignored in considerations of agricultural management.

Prescriptive land management

The research demands recommended in order to understand landscape features such as field margins (De Snoo & Chaney, 1999) (see p. 217), however sensible, are patently utopian. However, some aspects of field-margin management can be prescribed practically from the realisation that concerted remedial effort for intensive agricultural practices is a cost-effective way to help restore biodiversity in agricultural habitats (Smallshire & Cooke, 1999). On this premise, field margins can contribute substantially

Table 10.7. *Management options for field margins available through UK agri-environment schemes (from Smallshire & Cooke, 1999)*

Option category	Typical prescribed width	Description
Unsprayed cereal crop headlands (conservation headlands)	Usually 6 m, but ranges from 4 to 12 m	Aimed at restricting summer insecticide; prescriptions vary, ranging from basic to restrictions on fertilisers and pesticides
Uncropped wildlife strips	Typically 6 or 12 m	Field margin left uncropped but cultivated most years to encourage arable plants
Grass margins in arable areas	Typically 5–10 m, but ranges from 1 to 30 m	Margin sown with grass and managed by infrequent cutting to give a tussocky sward
Wildlife seed mixtures	Headland blocks, size unspecified	Sown with cereal-based cover, or seed- or nectar-producers
Buffer strips in grassland	Usually 5–10 m, but range from 1 to 30 m	Unfenced grass margins to buffer adjacent habitats from drift/run-off; inputs prohibited, except for specific weed control
Wildlife strips in grassland	Ranges from 5 to 30 m	Usually fenced, but managed by restricted grazing and/or cutting; inputs prohibited, except for specific weed control

to wider conservation efforts such as the targets set through the UK Biodiversity Action Plan, through which a number of 'agri-environment incentive schemes' have been instituted. Several formal prescriptions for management of field margins have thus been made, with details varying in different parts of the UK and under different authorities (Table 10.7). Thus, in Wales, the ESA scheme includes a mandatory provision for a 2-m buffer strip along all field boundaries under the agreement, and wider margins are often specified. Farmers in much of Europe have had unique opportunities to formulate and implement strategies for conservation management on farmlands – both for margins and for whole fields taken out of production as a facet of 'set-aside' and similar schemes brought about by food surpluses. Financial incentives for habitat creation may be offered. Thus, in Britain in the early 1990s, the Ministry of Agriculture offered a 30% grant (rising to 60% in designated ESAs) for farmers to establish new hedges (whereas grants for hedge removal were still being paid in the early 1980s; Raymond & Waltham, 1996). Amongst other benefits, such incentives can help to establish 'best-practice' methods for such restoration, in terms of siting, dimensions, species composition, establishment techniques and subsequent protection and management.

In 1992, set-aside became a compulsory part of the European Commission's Common Agricultural Policy for farmers wanting to claim arable area payments, and 15% of eligible arable land had to be set aside, either to be rotated annually or remaining out of production for up to five years. This step facilitated studies on wide field margins, with studies such as that at Boxworth (see p. 130) stemming directly from it. Diverse seed mixtures in wide set-aside margins provided evidence for strong correlations between numbers and diversity of invertebrates and plant species richness and the number of sown species. Data for butterflies and carabid beetles agreed in this trend (Kirkham *et al.*, 1999).

Farm woodlands and agroforestry

One important strategy to enhance conservation without reducing agricultural overproduction in the UK has been to encourage conversion of land from agricultural to timber production, with an implied assumption that the woodlands created would enhance wildlife (Insley, 1988). During the first three years of the scheme, more than 12 000 ha were approved for planting, with the average block size for individual woodlands being less than 2 ha (Usher & Keiller, 1998), so enhancing landscapes considerably. Although the long-term values of the practice can not yet be evaluated fully, Usher & Keiller (1998) sampled the larger moths ('macrolepidoptera') in 18 farm woodlands in Yorkshire and evaluated their richness and diversity from light trap catches. The data led to preliminary formulation of some 'design parameters' for farm woodlands (which may be compared with those recommended commonly for nature reserves) (Fig. 10.4), which are also important principles to consider in more general landscape enhancement and restoration for invertebrate values.

Enlarging on Fig. 10.4, point by point:

- Very small woodlands, of less than a hectare, do not support characteristic woodland moth communities, whilst woodlands of 5 ha or more generally support more stable moth communities, with inference that extinctions through stochasticity are then less likely to occur.

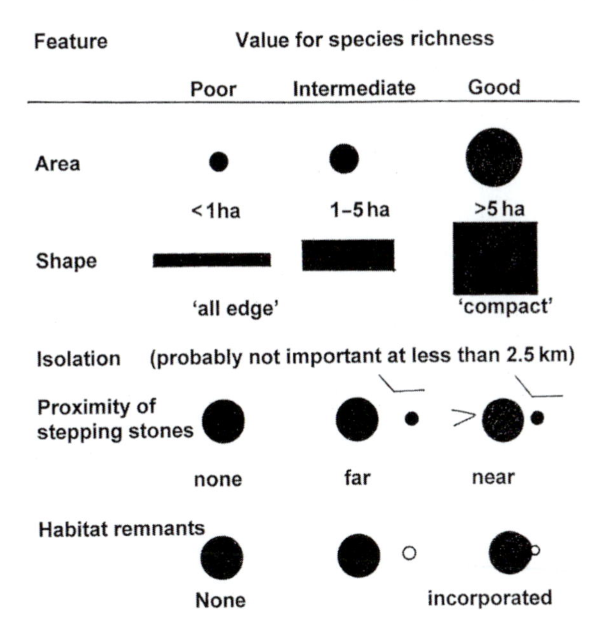

Figure 10.4. Parameters for favourable attributes of farm woodland patches for invertebrates, based on surveys of macrolepidoptera in Yorkshire, England (Usher & Keiller, 1997). Using species richness as an index, relative situations for poor, intermediate and good representation are summarised.

- Planting as compact blocks is preferable to planting in elongated strips, largely because of the increased chance of creating 'interior conditions' rather than habitats dominated by edge effects. This aspect of planning is likely to lead to support of more species characteristic of woodland habitats.
- Isolation may be an important consideration for species that disperse slowly or are reluctant to leave woodland environments. However, within distances of 2.5 km, isolation did not appear to be important for either plant or moth species richness.
- 'Stepping stones' of habitat remnants or field margins of woodland shrubs may assist colonisation of newly established woodland patches and may be one reason why isolation, as above, can seem unimportant.
- Very small remnants of woodlands can still support small (sometimes, depauperate) communities of woodland invertebrates. If new woodlands can be located near these remnants, then the remnants may constitute stepping stones. If physically incorporated into new woodlands, such remnants are immediate sources of species for colonisation of the

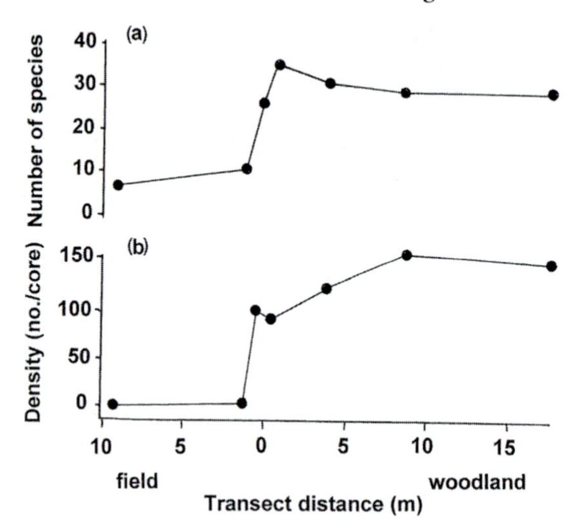

Figure 10.5. Cryptostigmata (moss mites) collected in the top 10 cm of soil (cores extracted) along a transect from arable field into farm woodland in Yorkshire, England (Sgardelkis & Usher, 1994): (a) number of species and (b) density (number/core) (0 = boundary point).

new habitats. However, as Usher & Keiller (1998) warned, it is also important that these remnants are indeed of woodland rather than of some other natural habitats whose species could be affected adversely by new woodland conditions.

Farm woodlands can clearly enhance wildlife, with strong belief that they will be richer in natural biota than the arable systems they replace. A perhaps more widespread pattern was found for soil-dwelling cryptostigmatid mites in Yorkshire. Fig. 10.5 shows the species richness of these moss mites and their density (number of individuals per soil core) within the top 10 cm of soil along transects from an arable field into a farm woodland. Richness and abundance both increased sharply at the woodland margin over the arable field levels. This discontinuity might reflect one or more of (1) the use of agrochemicals in the field, (2) repeated soil disturbances in the field with cultivation, (3) the greater amounts of organic material in the woodland soils, or other factors (Sgardelkis & Usher, 1994). In contrast to these mites, carabid beetles and spiders have been shown to be most diverse in the field near the woodland margin, where species from both major habitats can thrive, and considerably fewer species were found in the woodland than in either the margin or the field (Bedford & Usher, 1994). As Usher & Keiller (1998) noted,

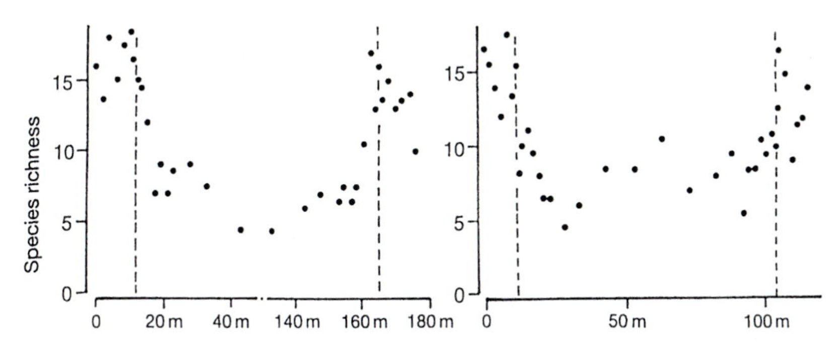

Figure 10.6. Species richness of Carabidae along transects of pitfall traps through two woodlands in Yorkshire, England, into arable fields each side of wood. Dashed lines mark woodland boundaries (Bedford & Usher, 1994).

the above studies indicated the nature of the large changes that can be expected in arthropod assemblages with establishment of farm woodlands and, although the trends may be relatively general, they can at present be formulated and predicted only rather tentatively. Thus:

- large changes in species composition are likely, but the timescales over which these will occur have not been defined;
- the extent of such changes will depend on the nature of the land-use change, including whether it is total or patchy;
- different types of land cover tend to be associated with different levels of species richness for agroforestry and nearby arable regions.

More work is needed on 'edge effects' in such contexts. For some Carabidae and spiders, these effects extended only about 5 m into woodlands (Bedford & Usher, 1994), and the influence of the farm woodland/field boundary on selected Carabidae is indicated in Fig. 10.6. *Pterostichus madidus* was widespread in woodland, with few individuals trapped in the adjacent field, whereas the reverse was true for *Bembidion tetracolum*. For such species, it seems that such habitat edges constitute an effective imposed barrier although, conversely, establishment of the woodland clearly increases overall richness of the local fauna by providing the resources needed by additional species that otherwise would not occur there. Bedford & Usher (1994) emphasised the potential for even small patches of woodland to support woodland-associated insects, with the minimal edge effects they found suggesting that patches greater than about 100 m^2 could contain effective interior habitat. Design parameters should seek to maximise this, namely the amount of habitat more than 5 m from the boundary.

Agroforestry has undergone considerable expansion in the past few decades as a facet of diversification of agricultural practices. It can be defined, broadly, as 'a land use that involves deliberate retention, introduction or mixture of trees or other woody perennials in crop/animal production fields to benefit from the resultant ecological and economic interactions' (MacDicken & Vergara, 1990). It is also generally compatible with the cultural patterns of local people, but it involves a distinct land-use system to increase total productivity and, because of the greater structural diversity over most conventional monoculture crops, can have direct conservation benefits in providing resources and linkages not generally available in field crops or pastures. Although newly fashionable, agroforestry itself is by no means new, with trees having been used in cropping systems for at least several thousand years. However, the primary motivations (in addition to economic return) have emphasised beneficial impacts on ecological services (such as nutrient recycling, soil fertility, reduction of climatic impacts, erosion control, and others) and in pest management (as barriers and as refuges for natural enemies) rather than wider benefits for biodiversity *per se*. Many agroforestry exercises are based on single tree species, but Huxley's (1990) suggestion of 'association ideotypes' (different species that collectively utilise a wider range of spatial and temporal features in a plantation and can coexist) merits serious support for increasing habitat diversity, particularly when combinations of local, native species, rather than exotic plantation trees, are used. A variety of spatial arrangements are employed, as in Fig. 10.7, such as planting in alternate rows or blocks and planting in more intimate patterns.

Agroforestry practices are thus very diverse, and many workers recognise three broad component-based categories:

- Agrisilviculture: crops and trees (including shrubs and/or vines).
- Silvopastoral: pasture and/or animals and trees.
- Agrosilvopastoral: crops, pastures and/or animals, and trees (Nair, 1990).

The first of these is perhaps the most widely used practice; Table 10.8 lists some of the main categories of agroforestry noted by MacDicken & Vergara (1990) to exemplify the diversity of approaches. However, another conservation consideration has also emerged from recent agroforestry exercises in southern Australia and is of much wider importance than to that region alone. In Victoria, one of the most popular agroforestry crop trees is the Tasmanian blue gum (*Eucalyptus globulus*), which

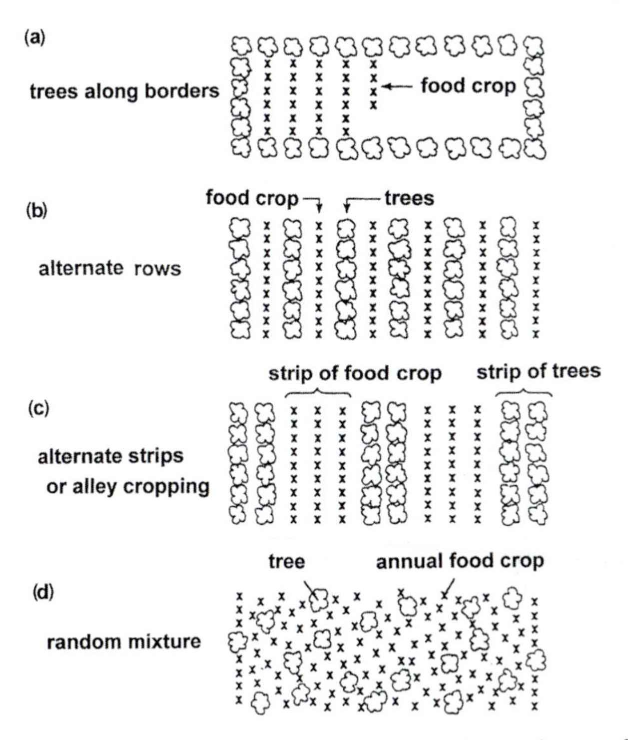

Figure 10.7. Some spatial aspects of planting for agroforestry: four examples of common planting patterns (after Nair, 1990).

is fast-growing and very well suited to Victorian conditions. It is, however, an exotic species in the region, and – whereas it may very adequately replenish habitat quality for many woodland vertebrates, in providing stepping stones and residential habitat – it may not constitute truly natural habitat for many ecologically specialised phytophagous insects native to Victoria and the communities they comprise. Wherever feasible to do so, it is likely that invertebrate conservation can be enhanced more by the use of local tree species in such operations.

Species translocations

Hand in hand with habitat enhancement, deliberate manipulation of species incidence by translocation (often involving individual taxa of conservation interest) or introduction (for example of biological control agents) may be contemplated, with the precise nature of habitat enhancement or restoration tailored to such focal needs. Conservation biological

Table 10.8. *Some categories of agroforestry practice (MacDicken & Vergara, 1990)*

Trees on croplands
Trees on croplands, simply planted or allowed to regenerate or persist in crop fields
Tree gardens, multispecies associations often with shade-tolerant crops, and may involve complex combinations of several tree species and perennial field crops
'Taungya' is the raising of a forest crop in combination with a temporary agricultural crop and was a common medieval practice in Europe; nowadays it is a popular form of agroforestry in parts of the tropics
Improved fallows is the replacement or enhancement of natural fallow vegetation by introduction of trees or shrubs
Plantation crop combinations involve deliberate combinations of perennial trees or shrubs with annual crops
Alley cropping is a practice in which arable crops are grown between hedgerows of planted shrubs or trees
Windbreaks and shelterbelts are strips of living trees maintained to provide shelter from sun, wind or precipitation
Trees with pastures or livestock
Trees on rangeland or pastures: primary use is often for provision of shade, but some species also provide fodder
Plantation crops with pastures: this emphasises plantation crop production, with grazing by stock as a secondary benefit
Live fences: used to enclose stock or to provide protection or privacy, but can also serve as windbreaks and sources of wood and foliage, so conferring multiple benefits

control implicitly considers the latter in many cases. The former option is rarer but has been used in a number of cases, particularly of butterflies, to help consolidate species within parts of their historical range from which they have been lost. This is a responsible exercise, and codes of conduct have been designed to ensure that all relevant aspects are considered (e.g. Butterfly Conservation, 1995; Sands & New, 2002) in increasing the number of populations (or metapopulation units) that can be supported within a landscape or region.

Implications of biotechnology

The multiple ramifications of 'biotechnology' in agriculture include several themes of considerable concern to conservationists, and some

Table 10.9. *Some possible disadvantages of development of biotechnology in agriculture (after Mannion, 1995)*

Ecological and environmental
Potential to create ecologically damaging organisms, such as invasive plants that become pests or plants whose engineered genes escape into wild relatives
Potential to create higher plants, bacteria or animals that are toxic to other organisms, including people
Potential to create organisms, perhaps particularly bacteria, that may adversely affect biogeochemical cycles or that could be used in biological cycles (or even biological warfare)
Engineered species with resistance to environmental stress may encourage spread of agriculture, rather than its intensification, so threatening additional natural ecosystems and their biota
Major difficulties in guaranteeing safety and establishing widely accepted protocols for risk assessment of transgenic organisms
Countries (predominantly less developed ones) without regulations for testing transgenic organisms may be exploited for field testing
Cultural and economic
Industry dominated by transnational companies based in the developed world, often with monopolies on transgenic seed production
Biotechnology in food industry reduces markets for some products produced traditionally in the developing world
Developing world resources (such as genes of wild species) are being exploited without recompense

have been noted earlier in other contexts (Table 10.9). General concerns include aspects of environmental contamination, increased uniformity and intensification of production processes, and facilitating spread of cropping into ever more marginal land hitherto unused because of its unsuitability. With accelerated development of crops capable of overcoming various environmental constraints, such unsuitability becomes less important and additional 'natural' habitat areas are taken over. As Janzen (1988) put it, in the context of genetically modified species in the tropics, 'with the production of crops and livestock that flourish in tropical rain forest habitats, millions of free-lance farmers will push into rain forest'.

Considerable impetus exists to using biotechnology to increase agricultural efficiency, with genetically modified crops promoted widely in many parts of the world to incorporate various desirable features of quality, uniformity and traits such as resistance to pests (see p. 105). Many

Table 10.10. *Features of genetically engineered insect resistance in crop plants (after Gatehouse* et al. *(1992) and Mannion (1995))*

Continuous protection of the crop, irrespective of weather, season and insect lifecycle variation
If seeds marketed at reasonable price, economically advantageous over chemical pesticides
No need for repeated applications, so saving labour costs
Costs of development to market availability less than those of developing a new chemical insecticide
Entire plant, including underground parts, is protected
Protection provided *in situ*, and contamination of wider environment cannot occur
Protection is target-specific, so (1) beneficial insects are not affected and (2) less disruption occurs to community interactions (food webs) than when pesticides are used
Active factor is biodegradable, with little (? no) chance of it being concentrated in the wider environment; but some possibility that resulting gene products could be toxic to animals/humans
Reduction of chemical pesticide use decreases input of fossil-fuel energy into agricultural systems
Engineered pesticide resistance overcomes problem of increasing resistance in insect pests to conventional pesticides; probably a short-lived advantage

of these traits relate only distantly to wider invertebrate conservation matters, but concerns over *Bt*-expressing cultivars have led to considerable debate over their compatability with biological control and such plants can be incorporated easily into ecologically based pest-management schemes or may have wider spill-over effects within the local ecological community. With substantial interest in promoting insect resistance in crop plants, this issue has tended to dominate the debate on development of 'biotechnology' and focused on the possible harmful effects – despite the major benefits that accrue (Table 10.10).

As one example, the enhanced uses of recombinant baculoviruses (recbv) against insects, with genetic modification possibly increasing persistence of the viruses in the local environment (Miller, 1995), may increase non-target effects. Thus, a field trial in which recbv of a noctuid moth (*Autographa californica*) incorporated toxin from a buthid scorpion led to suggestion that about 10% of all UK Lepidoptera could be placed at risk from similar exercises (Richards *et al.*, 1998). Considerable uncertainty persists, although one modified virus trialled did not adversely affect honeybees and some natural enemies (Heinz *et al.*, 1995). Hoy *et al.*

(1998) discussed the integration of genetically modified crops with conservation biological control, and raised two themes of wider relevance. First, it may be important that the 'new' plant continues to provide a suitable environment for natural enemies – such as maintaining supply of nectar and pollen and enabling seasonal effectiveness. It is undesirable for the traits that confer pest resistance (such as changes to physical features of the plant surface) to also have those effects on desirable natural enemies and hamper their search for pests. Second, the conditions in which the new crop is grown should still be suitable for the control agents. As Hoy *et al.* (1998) noted, modifications to crops may also be able to target ways to increase conservation of natural enemies and thereby become increasingly compatible with wider considerations of invertebrate conservation management.

Conservation of pests

Attempts to control pests are often formulated with the explicit purpose of strong suppression, if not eradication, and many agriculturists would support measures that would remove pests permanently from their working environments. However, as pests are components of 'biodiversity', such aims are commonly at odds with the ethical ideals of conservationists. Nevertheless, suggestions of 'conserving pests' are commonly met with incredulity. Conversely, the difficulties of eradicating pests are sometimes held as evidence that other invertebrates are not threatened by pest-management practices.

The most famous case of an insect pest having been rendered extinct within historical times is that of the Rocky Mountain locust, *Melanoplus spretus*, in North America. Formerly very widespread and abundant as a pest, this is the major (and, perhaps, only) case of an insect pest having been exterminated by – at least in part – agricultural activities. Lockwood & Debrey (1990) documented this species in some detail and pointed out that pesticides were not involved in extinction, but widespread habitat destruction followed by more localised and intensive influences of exotic species (including cattle destroying riparian vegetation by trampling of egg beds and overgrazing, and increases in introduced insectivorous birds such as fowls) led to its loss in ways that are not commonly considered as major threats in agricultural contexts.

Clearing of the habitat of tsetse flies (*Glossina* spp.) for agriculture, coupled with reduction of their wildlife hosts in Africa, is viewed as a possible threat to some of these important disease vectors. It could thus be viewed as beneficial to human interests, but it also helps to raise the important ethical problem of the extinction of invertebrate species inimicable to human priority interests. As Reid *et al.* (2000) noted, the tsetse fly is seen by some ecologists as 'the guardian of African ecosystems', because it prevents people and their livestock from overusing large parts of the continent. Extirpation could, indeed, provide short-term benefits through rural and human population expansion, but the longer-term environmental costs (such as loss of wildlife, natural habitats, ecological services and soil fertility) might be incalculable and severe.

Three major groups of tsetse fly species occupy largely distinct habitats and are categorised as 'savannah', 'forest' and 'riverine' species, respectively. The first two of these groups decline as habitat is cleared for agriculture, but some riverine species survive in densely populated villages and are not as dependent on more natural habitats; in essence, many of them can adapt to anthropogenic habitats. Although there is no perceived danger of tsetse flies becoming extinct – even though some savannah and forest species may, indeed, decline towards this state – riverine species will almost certainly not be affected by human population growth and its ramifications (Reid *et al.*, 2000).

In such cases, the identity of the insect species is reasonably assured. Samways (1997) emphasised the importance in some other cases of considering 'evolutionary significant units' (ESUs) in the balance between pest management and conservation, where – at times – the same apparent taxon may be present in both pans of the scales. It may then be important to recognise slightly different (and biologically distinct) taxa or 'races' in discussion, rather than simply to refer all to the apparent parent species. Samways noted the case of the gypsy moth (*Lymantria dispar*) for which the original British form has long been extinct, but recent incursions of the same species in Britain have led to its pest status and control in forests. Several workers (see Samways & Lockwood, 1998) have addressed this paradox for Orthoptera, in which closely related congenors may variously be pests (through abundance and wide distribution) or of serious conservation concern (through scarcity and narrow-range endemism). Regional agricultural differences may become highly important influence on the status of various species. The Moroccan locust (*Dociostaurus maroccanus*) is a polyphagous major pest on a variety of crops around the Mediterranean Sea, with (as for other pest locusts, see p. 113) enormous effort needed to control outbreaks. However, at the edges of its range, climatic conditions limit increase of the locust's numbers and grasslands have been reduced to increase cropping areas, leading to extirpations of the locust in regions where unique local genotypes may occur (Latchininsky, 1998).

Concluding comment

The approaches to invertebrate conservation through promoting more holistic management at both farm and landscape levels noted in this chapter are founded in the development of goodwill, increasing awareness and the expectation of mutual benefits. Continued progress towards environmentally sensitive and responsible IPM and promotion of agricultural practices that do not lead to further environmental degradation are clear central themes in such endeavour, but management must be founded in more than platitudes if it is to be persuasive. In this book I have tried to summarise some of the grounds for (1) considering agricultural systems as a vital arena for invertebrate conservation, (2) realising that invertebrates are functionally important in ecosystem maintenance and (3) appraising the wider problems of bringing invertebrate importance to much wider notice as a strategy for promoting need for their conservation. It may appear trite to repeatedly advocate the need for widespread education and consultation to communicate these themes, but this is indeed a central aspect of converting theory to conservation practice. It is perhaps of greater than usual relevance when the twin considerations of dealing with poorly understood biota with a substantial 'image problem' and large financial benefits and human livelihoods are involved. Ethical aspects may be even harder to communicate – perhaps particularly for points such as that noted last above, of pests themselves being conserved, a sentiment that may seem heretical in some quarters.

The more pragmatic issues involved in promoting holistic management (Cuperus *et al.*, 2000) include making substantial changes to agricultural practices and can have wider ramifications in the socioeconomic environment. Dilemmas occur over matters such as whether mandatory aspects of pest management (including 'enforcement' of area-wide management) should occur; the ramifications of biotechnology, such as increasing use of genetically modified crops and their management to overcome consumer resistance and to assure 'safety'; wider aspects of commodity safety and access to markets; and the overall environmental conscience of the populace. Much of this devolves on defining values to the parties involved – with, for example, IPM regarded as 'a business opportunity within the context of farming operations' (Hutchins, 1995), so that clear practical or economic advantage must be seen. Any increased costs (or lessened benefits) from measures promoted on conservation grounds are usually a deterrent to their adoption. Incentives must therefore be tangible. Although, for example, legislative measures to prohibit undesirable current practices with imposition of fines (or other penalties) for transgression are indeed possible, they are not as beneficial as the alternative of subsidies or other positive incentives for adopting more desirable practices (see Dent (1991) for discussion). For IPM adoption, much depends on the drive, goodwill and understanding of extension workers and other local agents who meet frequently, and on grounds of mutual respect, with primary producers and other land-management interests. Their roles in the broader conservation context noted here are at least equally significant.

The precise messages needed will vary somewhat in different contexts, different cropping systems and different environments, but – very broadly – encompass the topics listed below:

- Invertebrates can be very diverse in agricultural systems and collectively have vital ecological roles in all terrestrial ecosystems, including helping to maintain soil quality and to control pest taxa.
- Relatively few invertebrates are 'bad' (as damaging pests). Others have beneficial roles in agriculture (most tangibly as pollinators or as natural enemies of pests), and numerous others do not impinge directly on human interests. It is important to distinguish these broad functional categories and not to undertake pest-suppression methods that have marked non-target effects, either by harming other taxa or by impairing their ecological functions.
- Many measures undertaken historically to reduce populations and impacts of pests have had undesirable side effects, including causing mortality of non-target species and wider environmental impacts. Many of these are not obvious initially but may lead to considerably increased costs for crop protection by necessitating higher inputs of agricultural chemical and other intensive management.
- Development of IPM has sought more environmentally sensitive pest management, which has both economic and conservation benefits through increasing sustainability of production and market access, promoting effective crop/pasture/livestock protection based on sound ecological understanding, and reducing use of more damaging control methods.
- Cultural control methods are important in IPM and help to fulfil the universal needs of invertebrates by the protection, enhancement or

construction of habitats containing critical resources. Both field-margin quality and that of the crop or pasture itself can be modified for greater values to enhance effectiveness of pollinators and natural enemies of pests and to provide conditions resembling more closely those needed by invertebrates resident in natural communities.

- Likewise, moves towards less intensive agriculture can have wide environmental benefits through reducing impacts based on agricultural chemicals and fossil fuel usage.
- Although primary producers and conservationists emhasise different priorities (and will continue to do so), there is much common ground between them, particularly in mutual needs to sustain productivity, conserve invertebrate biodiversity and the services that accrue from this, and reduce any damaging effects of current practices by replacing these with more acceptable land-management strategies – if this can be achieved without economic loss. These topics can promote wider mutual understanding of common goals. Some of the parallels are recapitulated in Table 10.11.
- Increasing scales of management, planning at the landscape level to manage pests, provide essential and varied structural features and promote ecosystem functions, drawing on the practical expertise and skills of all parties to enhance 'holistic management', within which invertebrates are likely to be major beneficiaries.
- The basic principles of landscape ecology are highly relevant to invertebrate conservation, by seeking to promote connectivity and hospitality of both matrix and other features. Specific considerations for agricultural management can be incorporated into wider landscape planning to enhance conservation of a far wider range of organisms and environments than those usually considered in detail.

Fundamentally, conservation (including conservation of invertebrates) occurs within ecosystems, and ecosystems comprise landscapes, necessitating a variety of conservation approaches that transcend a range of temporal and spatial scales (Knight & Landres, 2002) in order to sustain invertebrate diversity as well as individual focal species and functional groups. Agroecosystems are major landscape elements, sometimes the predominant form present. The ideal duty of care for agroecosystem lands goes far beyond the most familiar purpose of producing food for humanity and encompasses responsibility for the wellbeing of the natural biota displaced by agricultural development in many parts of the world. In moving towards a less fragmented and piecemeal global approach to

Table 10.11. *Some parallels between agricultural pest management and conservation (after Gillespie & New, 1998)*

	Pest management	Conservation
Goals	To suppress species that cause economic harm or compromise outputs/quality of agricultural products Broadly, to increase resistance to increase in biotic diversity from invasive arthropods and plants and monitor these to counter them effectively and economically Pesticide use is integral to many operations.	To maintain natural diversity and to foster its wellbeing This is not necessarily linked with expectation of economic gain, though focused goals to to safeguard exploitable species incorporate such considerations implicitly
Knowledge of species	Many pests are well known and their ecology studied comprehensively	Most species in natural communities are poorly known, with biological knowledge for many almost non-existent
Exotic species	Often employed deliberately as natural enemies in biological control programmes	Viewed widely as one of greatest threats to natural the communities and common targets for eradication or suppression

invertebrate conservation than has prevailed until now, the principles discussed in this book have important roles to play. Factors such as their immense diversity that were viewed only a decade or so ago as impediments to invertebrate conservation are also opportunities for promoting conservation. Conservation on private land, with agricultural land being the largest component of this, includes pursuing greater compatability between pest management, wider crop and grazing management, and conservation management. McQuillan's (1999) comment on Tasmanian grassland geometrid moths that 'survival depends on appropriate management of the remaining grassland estate outside the formal reserve system' emphasises yet again the values of pastoral systems that so often pass unremarked. Sustainable agricultural production and sustainable invertebrate biodiversity are linked intimately and depend on intact ecological processes, and interest in preserving biodiversity is likely to increase as land, water and high-quality soils for use become ever scarcer with continued intensive agriculture.

An important conceptual change still to be adopted universally is to acknowledge the benefits to biodiversity that result from integrated farming systems and, within these, to incorporate environmental benefits directly as objectives rather than as by-products. Increasing knowledge and appreciation of the values of invertebrates in ecological sustainability is a crucial and pragmatic facet of this promotion and in helping to develop and focus more holistic management to sustain both the components (many of them invertebrates) and processes in agroecosystems. Rational incentives, and measures to deter disincentives, seem destined to play a part in promoting invertebrate conservation in agroecosystems and in the wider landscape.

References

Abbot, K. L. & Seymour, J. (1998). Conserving rainforest insect species while managing a pest species, *Bactrocera papayae* Drew and Hancock (Diptera: Tephritidae). In Zalucki, M. P., Drew, R. A. I. & White, G. G. (eds.). *Pest management: Future Challenges*, Vol. 2. Brisbane, Australia: University of Queensland, pp. 209–16.

A'Brook, J. (1968). The effect of plant spacing on the numbers of aphids trapped over the groundnut crop. *Annals of Applied Biology* **61**, 289–94.

Altieri, M. A. (1991). Increasing biodiversity to improve insect pest management in agroecosystems. In Hawksworth, D. L. (ed.). *The Biodiversity of Microorganisms and Invertebrates: Its Role in Sustainable Agriculture*. Wallingford, UK: CAB International, pp. 165–82.

Altieri, M. A. & Nicholls, C. I. (1999). Biodiversity, ecosystem function and pest management in agricultural systems. In Collins, W. W. & Qualset, C. O. (eds.). *Biodiversity in Agroecosystems*. Boca Raton, FL: CRC Press, pp. 69–84.

Andersen, A. N. (1990).The use of ant communities to evaluate changes in Australian terrestrial ecosystems: a review and a recipe. *Proceedings of the Ecological Society of Australia* **3**, 347–57.

Andersen, A. N. (1993). Ants as indicators of restoration success at a uranium mine in tropical Australia. *Restoration Ecology* **1**, 156–67.

Andersen, A. N. (1995). Measuring more of biodiversity: genus richness as a surrogate for species richness in Australian ant faunas. *Biological Conservation* **73**, 39–43.

Andersen, A. N. (1997). Using ants as bioindicators: multivariate issues in ant ecology. *Conservation Ecology* **1**, 8.

Andow, D. A. (1986). Plant diversification and insect population control in agroecosystems. In Pimentel, D. (ed.). *Some Aspects of Integrated Pest Management*. New York: Cornell University Press, pp. 277–348.

Andow, D. A. (1991). Vegetational diversity and arthropod population response. *Annual Review of Entomology* **36**, 561–86.

Andrezejevska, L. (1979). Herbivorous fauna and its role in the economics of grassland ecosystems. 1. Herbivores in natural and managed meadows. *Polish Ecological Studies* **5**, 5–44.

Asher, J., Warren, M., Fox, R., *et al.* (2001). *The Millenium Atlas of Butterflies in Britain and Ireland*. Oxford: Oxford University Press.

Asquith, A. & Miramontes, E. (2001). Alien parasitoids in native forests: the ichneumonid wasp community in a Hawaiian rainforest. In Lockwood, J. A., Howarth, F. G. & Purcell, M. R. (eds.). *Balancing Nature: Assessing the Impact of Importing*

Non-native Biological Control Agents (an International Perspective). Lanham, MD: Entomological Society of America, pp. 54–69.

Baldi, A. & Kisbenedek, T. (1997). Orthoptera assemblages as indicators of grassland naturalness in Hungary. *Agriculture, Ecosystems and Environment* **66**, 121–9.

Banaszak, J. (1992). Strategy for conservation of wild bees in an agricultural landscape. *Agriculture, Ecosystems and Environment* **40**, 179–92.

Barbosa, P. (ed.) (1998). *Conservation Biological Control*. San Diego, CA: Academic Press.

Barbosa, P. & Benrey, B. (1998). The influences of plants on insect parasitoids: implications for conservation biological control. In Barbosa, P. (ed.). *Conservation Biological Control*. San Diego, CA: Academic Press, pp. 55–82.

Barbosa, P. & Wratten, S. D. (1998). Influence of plants on invertebrate predators: implications to conservation biological control. In Barbosa, P. (ed.). *Conservation Biological Control*. San Diego, CA: Academic Press, pp. 83–100.

Barker, A. M., Brown, N. J. & Reynolds, C. J. M. (1999). Do host-plant requirements and mortality from soil cultivation determine the distribution of graminivorous sawflies on farmland? *Journal of Applied Ecology* **36**, 271–82.

Barnard, P. C. (1999). *Identifying British Insects and Arachnids: An Annotated Bibliography of Key Works*. Cambridge: Cambridge University Press.

Barratt, B. I. P., Evans, A. A., Ferguson, C. M., *et al.* (1998). Curculionoidea (Insecta: Coleoptera) of agricultural grassland and lucerne as potential nontarget hosts of the parasitoids *Microctonus aethiopoides* Loan and *Microctonus hyperodae* Loan (Hymenoptera: Braconidae). *New Zealand Journal of Zoology* **25**, 47–63.

Barratt, B. I. P., Ferguson, C. M., McNeill, M. R. & Goldson, S. L. (1999). Parasitoid host specificity testing to predict field host range. In Withers, T. M., Barton Browne, L. E. & Stanley, J. (eds.). *Host Specificity Testing in Australasia: Towards Improved Assays for Biological Control*. Brisbane, Australia: Department of Natural Resources, Queensland, pp. 77–83.

Bathon, H. (1996). Impact of entomopathogenic nematodes on non-target hosts. *Biocontrol Science and Technology* **6**, 421–34.

Bedding, R. A., Tyler, S. & Rochester, N. (1996). Legislation on the introduction of exotic entomopathogenic nematodes into Australia and New Zealand. *Biocontrol Science and Technology* **6**, 465–75.

Bedford, S. E. & Usher, M. B. (1994). Distribution of arthropod species across the margins of farm woodland. *Agriculture, Ecosystems and Environment* **48**, 295–305.

Behan-Pelletier, V. & Walter, D. E. (2000). Biodiversity of oribatid mites (Acari: Oribatida) in tree canopies and litter. In Coleman, D. C. & Hendrix, P. F. (eds.). *Invertebrates as Webmasters in Ecosystems*. Wallingford, UK: CAB International, pp. 187–202.

Bennet, F. D. & Habeck, D. H. (1995). *Cactoblastis cactorum*: a successful weed control agent in the Caribbean, now a pest in Florida? In Delfosse, E. S. & Scott, R. R. (eds.). *Proceedings of the Eighth International Symposium on Biological Control of Weeds*. Melbourne, Australia: CSIRO, pp. 21–6.

Bennett, A. F. (1999). *Linkages in the Landscape: The Role of Corridors and Connectivity in Wildlife Conservation*. Gland, Switzerland: IUCN.

Benton, T. G., Vickery, J. A. & Wilson, J. D. (2003). Farmland biodiversity: is habitat heterogeneity the key? *Trends in Ecology and Evolution* **18**, 182–8.

Berggren, A., Birath, B. & Kindvall, O. (2002). Effect of corridors and habitat edges on dispersal behavior, movement rate, and movement angles in Roesel's bush-cricket (*Metrioptera roeseli*). *Conservation Biology* **16**, 1562–9.

Berlocher, S. H. (1989). The complexities of host races and some suggestions for their identification by enzyme electrophoresis. In Loxdale, H. D. & Den Hollander, J. (eds.). *Electrophoretic Studies on Agricultural Pests*. Oxford: Oxford University Press, pp. 51–68.

Berryman, A. A. (1987). The theory and classification of outbreaks. In Barbosa, P. & Schulte, J. C. (eds.). *Insect Outbreaks*. San Diego, CA: Academic Press, pp. 3–29.

Blumberg, A. Y. & Crossley, D. A. (1982). Comparison of soil surface: arthropod populations in conventional tillage, no-tillage and old field systems. *Agro-Ecosystems* **8**, 247–53.

Boatman, N. D. (1994). *Field Margins: Integrating Agriculture and Conservation*. Brighton, UK: British Crop Protection Council.

Boatman, N. D., Dover, J. W., Wilson, P. J., Thomas, M. B. & Cowgill, S. E. (1989). Modification of farming practices at field margins to encourage wildlife and promote pest biocontrol. In Buckley, G. P. (ed.). *Biological Habitat Reconstruction*. London: Belhaven Press, pp. 299–311.

Boettner, G. H., Elkinton, J. S. & Boettner, C. J. (2000). Effects of a biological control introduction on three non-target native species of saturniid moths. *Conservation Biology* **14**, 1798–806.

Bohac, J. (1999). Staphylinid beetles as bioindicators. *Agriculture, Ecosystems and Environment* **74**, 357–72.

Bohac, J. & Pospisil, J. (1988). Accumulation of heavy metals in invertebrates and its ecological aspects. Presented at the International Conference on Heavy Metals in the Environment, Norwich, UK, 2, 354–357.

Bongers, T. (1990). The maturity index: an ecological measure of environmental disturbance based on nematode species composition. *Oecologia* **83**, 14–19.

Booij, C. J. H. & Noorlander, J. (1992). Farming systems and insect predators. *Agriculture, Ecosystems and Environment* **40**, 125–35.

Boswell, G. P., Britton, N. F. & Franks, N. R. (1998). Habitat fragmentation, percolation theory and the conservation of a keystone species. *Proceedings of the Royal Society of London. Series B. Biological Sciences* **265**, 1921–5.

Bottrell, D. G., Barbosa, P. & Gould, F. (1998). Manipulating natural enemies by plant variety selection and modification: a realistic study. *Annual Review of Entomology* **43**, 347–67.

Bouché, M. B. (1972). *Lombriciens de France: Ecologie et Systematique*. Paris: Institut Nationale de Recherche Agronomique.

Bouché, M. B. (1977). Strategies lombriciennes. *Ecological Bulletin, Stockholm* **25**, 122–32.

Bourchier, R. S. & McCarty, L. S. (1995). Risk assessment of biological control (predators and parasitoids). *Bulletin of the Entomological Society of Canada* **27**, 126–38.

Bourn, N. A. D., Pearman, G. S., Goodger, B., Thomas, J. A. & Warren, M. S. (2000). Changes in the status of two endangered butterflies over two decades and the influence of grazing management. In Rook, R. D. (ed.). *Grazing Management: The Principles and Practice of Grazing, For Profit and Environmental Gain Within Temperate Grassland Systems*. Reading, UK: British Grassland Society, pp. 141–6.

Briese, D. T. (1999). Open field host-specificity testing: is 'natural' good enough for risk assessment? In Withers, T. M., Barton Browne, L. E. & Stanley, J. (eds.). *Host Specificity Testing in Australasia: Towards Improved Assays for Biological Control*. Brisbane, Australia: Department of Natural Resources, Queensland, pp. 44–59.

Brown, R. A. (1989). Pesticides and non-target terrestrial invertebrates: an industrial approach. In Jepson, P. C. (ed.). *Pesticides and Non-target Invertebrates*. Wimborne, UK: Intercept, pp. 19–42.

Brunkhorst, D. J., Thackway, R., Coyne, P. & Cresswell, I. D. (1998). Australian protected areas: toward a representative system. *Natural Areas Journal* **18**, 255–60.

Brussaard, L., van Veen, J. A., Kooistra, M. J. & Lebbink, G. (1990). The Dutch programme on soil ecology of arable farming systems. I. Objectives, approach and prelimnary results. *Ecological Bulletin* **39**, 35–40.

Brust, G. E. (1994). Natural enemies in straw-mulch reduce Colorado potato beetle populations and damage in potatoes. *Biological Control* **4**, 163–9.

Buchmann, S. L. & Nabham, G. L. (1996). *The Forgotten Pollinators*. Washington, DC: Island Press.

Buckley, G. P. (1989). *Biological Habitat Reconstruction*. London: Belhaven Press.

Bugg, R. L. & Pickett, C. H. (1998). Enhancing biological control: habitat management to promote natural enemies of agricultural pests. In Pickett, C. H. & Bugg, R. L. (eds.). *Enhancing Biological Control*. Berkeley, CA: University of California Press, pp. 3–23.

Bugg, R. L. & Waddington, C. (1994). Using cover crops to manage arhropod pests of orchards: a review. *Agriculture, Ecosystems and Environment* **50**, 11–28.

Bugg, R. L., Anderson, J. H., Thomsen, G. D. & Chandler, J. (1998). Farmscaping in California: managing hedgerows, roadside and wetland planting, and wild plants for biointensive pest management. In Pickett, C. H. & Bugg, R. L. (eds.). *Enhancing Biological Control*. Berkeley, CA: University of California Press, pp. 339–74.

Bunce, R. G. H. & Hallam, C. J. (1993). The ecological significance of linear features. In Bunce, R. G. H., Ryzkowski, L. & Paoletti, M. G. (eds.). *Landscape Ecology and Agroecosystems*. Boca Raton, FL: Lewis Publishers, pp. 11–19.

Bunce, R. G. H. & Jongman, R. H. G. (1993). An introduction to landscape ecology. In Bunce, R. G. H., Ryzkowski, L. & Paoletti, M. G. (eds.). *Landscape Ecology and Agroecosystems*. Boca Raton, FL: Lewis Publishers, pp. 1–10.

Bunce, R. G. H., Barr, C. J., Howard, D. C. & Hallam, C. J. (1994). *The Current Status of Field Margins in the UK*. British Crop Protection Council Monograph 58, pp. 13–20.

Burel, F. (1989). Landscape structure effects on carabid beetles spatial patterns in western France. *Landscape Ecology* **2**, 215–26.

Burel, F. & Baudry, J. (1990). Hedgerow networks as habitats for colonisation of abandoned agricultural land. In Bunce, R. H. G. & Howard, D. C. (eds.). *Species Dispersal in Agricultural Environments*. Lymington, UK: Belhaven Press, pp. 238–55.

Burges, A. (1967). The decomposition of organic matter in the soil. In Burges, A. & Raw, F. (eds.). *Soil Biology*. New York: Academic Press, pp. 479–92.

Burn, A. J. (1988). Assessment of the impacts of pesticides on invertebrate predation in cereal crops. *Aspects of Applied Biology* **17**, 279–88.

Burn, A. J. (1989). Long-term effects of pesticides on natural enemies of cereal crop pests. In Jepson, P. C. (ed.). *Pesticides and Non-target Invertebrates*. Wimborne, UK: Intercept, pp. 177–93.

Butler, L., Zivkovich, C. & Sample, B. E. (1995). Richness and abundance of arthropods in the oak canopy of West Virginia's Eastern Ridge and Valley Section during a study of impact of *Bacillus thuringiensis* with emphasis on macrolepidoptera larvae. *Bulletin of the Agriculture and Forestry Experimental Station W. V. University* **711**, 1–19.

BUTT (Butterflies Under Threat Team) (1986). *The Management of Chalk Grassland for Butterflies*. Peterborough, UK: Nature Conservancy Council.

Butterfly Conservation (1995). Lepidoptera restoration: Butterfly Conservation's policy, code of practice and guidelines for action. *Butterfly Conservation News* **60**, 20–21.

Caldenasso, M. L., Pickett, S. T. A., Weathers, K. C. & Jones, C. G. (2003). A framework for a theory of ecological boundaries. *BioScience* **53**, 750–58.

Carlson, G. & Rodriguez, R. (1984). Farmer adjustment to mandatory pest control. In Conway, G. R. (ed.). *Pest and Pathogen Control: Structural, Tactical and Policy Models*. Chichester, UK: John Wiley & Sons, pp. 429–40.

Carriere, Y. & Roitberg, B. (1996). Behavioral ecology and qualitative genetics as alternatives for studying evolution in insect herbivores. *Evolutionary Ecology* **10**, 289–305.

Carroll, R., Augspurger, C., Dobson, A., *et al.* (1996). Strengthening the use of science in achieving the goals of the Endangered Species Act: an assessment by the Ecological Society of America. *Ecological Applications* **6**, 1–11.

Carruthers, R. L. & Onsager, J. A. (1993). Perspective on the use of exotic natural enemies for biological control of pest grasshoppers (Orthoptera: Acrididae). *Environmental Entomology* **22**, 885–93.

Carruthers, G. F., Hooper, G. H. S. & Walker, P. W. (1993). Impact of fenitrothion on the relative abundance and diversity of non-target organisms. In Corey, S. A., Dall, D. J. & Milne, W. M. (eds.). *Pest Control and Sustainable Agriculture*. Melbourne: CSIRO, pp. 136–9.

Carson, R. (1962). *Silent Spring*. Cambridge, MA: Riverside Press.

Chapman, J. & Sheail, J. (1994). *Field Margins: An Historical Perspective*. British Crop Protection Council Monograph no. 58, pp. 3–12.

Chapman, M. H. & Hoy, M. A. (1991). Relative toxicity of *Bacillus thuringiensis* var. *tenebrionis* to the two-spotted spider mite (*Tetranychus urticae* Koch) and its predator *Metaseiulus occidentalis* (Nesbitt) (Acari, Tetranychidae and Phytoseiidae). *Journal of Applied Entomology* **111**, 147–54.

Cherrett, J. M., Ford, J. B., Herbert, I. V. & Probert, A. J. (1971). *The Control of Injurious Animals*. London: English Universities Press.

Chiverton, P. A. (1984). Pitfall-trap catches of the carabid beetle *Pterostichus melanarius*, in relation to gut contents and prey densities, in insecticide treated and untreated spring barley. *Entomologia experimentalis et Applicata* **36**, 23–30.

Claridge, M. F. (1989). Electrophoresis in agricultural pest research: a technique of evolutionary biology. In Lonsdale, H. D. & den Hollander, J. (eds.). *Electrophoretic Studies on Agricultural Pests*. Oxford: Clarendon Press, pp. 1–6.

Clark, L., Geier, P. W., Hughes, R. D. & Morris, R. F. (1967). *The Ecology of Insect Populations in Theory and Practice*. London: Methuen.

Clark, M. S., Gage, S. H. & Spence, J. R. (1997). Habitats and management associated with common ground beetles (Coleoptera: Carabidae) in a Michigan agricultural landscape. *Environmental Entomology* **26**, 519–27.

Clarke, B., Murray, J. & Johnson, M. S. (1984). The extinction of endemic species by a program of biological control. *Pacific Science* **38**, 97–104.

Clarke, G. M. (2002). Inferring demography from genetics: a case study of the endangered golden sun moth, *Synemon plana*. In Young, A. G. & Clarke, G. M. (eds.). *Genetics, Demography and Viability of Fragmented Populations*. Cambridge: Cambridge University Press, pp. 213–25.

Clarke, G. M. & O'Dwyer, C. (1999). Genetic variability and population structure of the endangered golden sun moth. *Biological Conservation* **92**, 371–81.

Clement, S. L. & Cristofaro, M. (1995). Open-field tests in host-specificity determination of insects for biological control of weeds. *Biocontrol Science and Technology* **5**, 395–406.

CLM-IKC (1994). *Achtergronden van de Milieumeetlat voor Bestrijzdingsmiddelen*. [Backgrounds of the environmental yardstick for pesticides.] Ede, the Netherlands: IKC-AT, Kerngroep MJP-G.

Coll, M. (1998). Parasitoid activity and plant species composition in intercropped systems. In Pickett, C. H. & Bugg, R. L. (eds.). *Enhancing Biological Control: Habitat Management to Promote Natural Enemies of Agricultural Pests*. Berkeley, CA: University of California Press, pp. 85–119.

Coll, M. & Bottrell, D. G. (1996). Movement of an insect parasitoid in simple and diverse plant assemblages. *Ecological Entomology* **9**, 217–35.

Collins, W. W. & Qualset, C. O. (1999). *Biodiversity in Agroecosystems*. Boca Raton, FL: CRC Press.

Colloff, M., Fokstuen, G. & Boland, T. (2003). *Toward the Triple Bottom Line in Sustainable Agriculture: Biodiversity, Ecosystem Services and an Environmental Management System for Citrus Orchards in the Riverland of South Australia*. Canberra, Australia: CSIRO Entomology.

Coombes, D. S. & Sotherton, N. W. (1986). The dispersal and distribution of polyphagous predatory Coleoptera in cereals. *Annals of Applied Biology* **108**, 461–74.

Corbet, S. A. (1995). Insects, plants and succession: advantages of long-term set-aside. *Agriculture, Ecosystems, and Environment* **53**, 201–17.

Corbett, A. (1998). The importance of movement in response to natural enemies to habitat management. In Pickett, C. H. & Bugg, R. L. (eds.). *Enhancing Biological Control: Habitat Management to Promote Natural Enemies of Agricultural Pests*. Berkeley, CA: University of California Press, pp. 25–48.

Cortesero, A. M., Stapel, J. O. & Lewis, W. J. (2000). Understanding and manipulating plant attributes to enhance biological control. *Biological Control* **17**, 35–49.

Costanza, R., d'Arge, R., de Groot, P., *et al.* (1998). The value of the world's ecosystem services and natural capital. *Nature* **387**, 253–60.

Coulson, J. R., Soper, R. S. & Wiliams, D. W. (1991). *Biological Control Quarantine: Needs and Procedures*. Beltsville, MD: USDA-ARS.

Cowie, R. H. (1992). Evolution and extinction of Partulidae, endemic Pacific land snails. *Philosophical Transactions of the Royal Society of London. Series B. Biological Sciences* **335**, 167–91.

Crickmore, N., Zeigler, D. R., Feitelson, J., *et al.* (1998). Revision of the nomenclature for the *Bacillus thuringiensis* pesticidal crystal proteins. *Microbiology and Molecular Biology Reviews* **62**, 807–13.

Crickmore, N., Zeigler, D. R., Schnepf, E., *et al.* (2000). *Bacillus thuringiensis* toxin nomenclature. www.biols.susx.ac.uk/home/Neil_Crickmore/Bt/index.html.

Crossley, D. A., Coleman, D. C. & Hendrix, P. F. (1989). The importance of the fauna in agricultural soils: research approaches and perspectives. *Agriculture, Ecosystems, and Environment* **27**, 47–55.

Culver, J. J. (1919). *A Study of* Compsilura concinnata, *an Imported Tachinid Parasite of the Gipsy Moth and Brown-tail Moth.* Washington, DC: United States Department of Agriculture, bulletin no. 66.

Cuperus, G. W., Mulder, P. G. & Royer, T. A. (2000). Implementation of ecologically-based IPM. In Rechcigl, J. E. & Rechcigl, N. A. (eds.). *Insect Pest Management: Techniques for Environmental* Protection. Boca Raton, FL: Lewis Publishers, pp. 171–204.

Curry, J. P. (1994). *Grassland Invertebrates*. London: Chapman & Hall.

Dafni, A. & Shmida, A. (1996). The possible ecological implications of the invasion of *Bombus terrestris* (L.) (Apidae) at Mt. Carmel, Israel. In Matheson, A., Buchmann, S. L., O'Toole, C., Westrich, P. & Williams, I. H. (eds.). *The Conservation of Bees*. London: Academic Press, pp. 183–201.

Dahlsten, D. L., Garcia, R. & Lorraine, H. (1989). Eradication as a pest management tool: concepts and contexts. In Dahlsten, D. L. & Garcia, R. (eds.). *Eradication of Exotic Pests*. New Haven & London: Yale University Press, pp. 3–15.

Dash, M. C. (1990). Oligochaeta: Enchytraeidae. In Dindal, D. L. (ed.). *Soil Biology Guide*. New York: John Wiley & Sons, pp. 311–40.

Davies, K. F. & Margules, C. R. (1998). Effects of habitat fragmentation on carabid beetles: experimental evidence. *Journal of Animal Ecology* **67**, 460–71.

Davies, K. F. & Margules, C. R. (2000). The beetles at Wog Wog: a contribution of Coleoptera systematics to an ecological field experiment. *Invertebrate Taxonomy* **14**, 953–6.

Davis, B. N. K., Lakhani, K. H. & Yates, T. J. (1991). The hazards of insecticides to butterflies of field margins. *Agriculture, Ecosystems and Environment* **36**, 151–61.

Day, M. D. (1999). Continuation trials: their use in assessing the host range of a potential biological control agent. In Withers, T. M., Barton Browne, L. & Stanley, J. (eds.). *Host Specificity Testing in Australasia: Towards Improved Assays for Biological Control*. Brisbane, Australia: Department of Natural Resources, Queensland, pp. 11–20.

Day, W. H. (1996). Evaluation of biological control of the tarnished plant bug (Hemiptera: Miridae) in alfalfa by the introduced parasite *Peristenus digoneutis* (Hymenoptera: Braconidae). *Environmental Entomology* **25**, 512–18.

DeBach, P. (1974). *Biological Control by Natural Enemies*. New York: Cambridge University Press.

DeBach, P. & Hagen, K. S. (1964). Manipulation of entomophagous species. In DeBach, P. (ed.). *Biological Control of Insect Pests and Weeds*. New York: Reinhold Publishing, pp. 283–304.

De Jong, F. M. W. & van der Nagel, M. C. (1994). A field bioassay for wider effects of insecticides with larvae of the large white butterfly *Pieris brassicae* (L.). Announcement 59/2a. Ghent: Faculty of Agricultural Science, University of Ghent.

Dempster, J. P. (1991). Fragmentation, isolation and mobility of insect populations. In Collins, N. M. & Thomas, J. A. (eds.). *The Conservation of Insects and their Habitats*. London: Academic Press, pp. 143–53.

Dempster, J. P., King, M. L. & Lakhani, K. H. (1976). The status of the swallowtail butterfly in Britain. *Ecological Entomology* **1**, 71–84.

Dennis, P. & Fry, G. L. A. (1992). Field margins: can they enhance natural enemy population densities and general arthropod diversity on farmland? *Agriculture, Ecosystems and Environment* **29**, 471–504.

Dennis, R. L. H. (1986). Motorways and cross-movements: an insect's 'mental map' of the M56 in Cheshire. *Bulletin of the Amateur Entomologists' Society* **44**, 77–82.

Dent, D. R. (1991). *Insect Pest Management*, Wallingford, UK: CAB International.

Dent, D. R. (1995). *Integrated Pest Management*. London: Chapman & Hall.

De Ruiter, P. C., Bloem, J., Bouwman, L. A., *et al.* (1994). Simulation of dynamics of nitrogen mineralisation in below-ground food webs of two arable farming systems. *Agriculture, Ecosystems and Environment* **51**, 199–208.

De Snoo, G. R. & Chaney, K. (1999). Unsprayed field margins: what are we trying to achieve? *Aspects of Applied Biology* **54**, 1–12.

De Snoo, G. R. & De Wit, P. R. (1998). Buffer zones for reducing pesticide drift to ditches and risks to aquatic organisms. *Ecotoxicology and Environmental Safety* **41**, 112–18.

Didham, R. K. (1997). An overview of invertebrate responses to forest fragmentation. In Watt, A. D., Stork, N. E. & Hunter, M. D. (eds.). *Forests and Insects*. London: Chapman & Hall, pp. 303–20.

Diehl, S. R. & Bush, G. L. (1984). An evolutionary and applied perspective of insect biotypes. *Annual Review of Entomology* **29**, 471–504.

Di Giulio, M., Edwards, P. J. & Meister, E. (2001). Enhancing insect diversity in agricultural grasslands: the roles of management and landscape structure. *Journal of Applied Ecology* **38**, 310–19.

Dodd, A. P. (1940). *The Biological Campaign against Prickly Pear*. Brisbane, Australia: Government Printer.

Dohmen, G. P. (1998). Testing side-effects of pesticides on carabid beetles: a standardised method for testing ground-dwelling predators in the laboratory for regulation purposes. In Haskell, P. T. & McEwen, P. (eds.). *Ecotoxicology: Pesticides and Beneficial Organisms*. London: Kluwer Academic Publishers, pp. 98–106.

Doubleday, O. P., Clark, A. & McLaughlin, B. (1994). *Hedges: A Farmer's view*. British Crop Protection Council Monograph no. 58, pp. 377–84.

Douglas, F. (2000). Sun moths in the back paddock. In Barlow, T. & Thorburn, R. (eds.). *Balancing Conservation and Production in Grassy Landscapes*. Canberra, Australia: Environment Australia, pp. 56–57.

Douthwaite, R. J. & Fry, C. H. (1982). Food and feeding behaviour of the little bee-eater, *Merops pusillus*, in relation to tsetse fly control by insecticides. *Biological Conservation* **23**, 71–8.

Dover, J. W. (1990a). Butterflies and wildlife corridors. In *The Game Conservancy Review of 1989*. Fordingbridge, UK: The Game Conservancy, pp. 62–4.

Dover, J. W. (1990b). *Butterfly Ecology on Arable Farmland, with Special Reference to Agricultural practices*. Fordingbridge, UK: The Game Conservancy.

Dover, J. W. (1991). The conservation of insects on arable farmland. In Collins. N. M. & Thomas, J. A. (eds.). *The Conservation of Insects and their Habitats*. London: Academic Press, pp. 294–318.

Dover, J. W. (1994). Arable field margins: factors affecting butterfly distribution and abundance. *British Crop Protection Council Monograph no. 58*, pp. 59–66.

Dover, J. W. (1999). Butterflies and field margins. *Aspects of Applied Biology* **54**, 117–24.

Dover, J. W., Clarke, S. A. & Rew, L. (1992). Habitats and movement patterns of satyrid butterflies (Lepidoptera: Satyridae) on arable farmland. *Entomologist's Gazette* **43**, 29–44.

Dover, J. W., Sparks, T. H. & Greatorex-Davies, J. N. (1997). The importance of shelter for butterflies in open landscapes. *Journal of Insect Conservation* **1**, 89–97.

Dover, J., Sparks, T., Clarke, S., Gobbett, K. & Glossop, S. (2000). Linear features and butterflies: the importance of green lanes. *Agriculture, Ecosystems and Environment* **80**, 227–42.

Duan, J. J. & Messing, R. H. (2000). Evaluating nontarget effects of classical biological control: fruit fly parasitoids in Hawaii as a case study. In Follett, P. A. & Duan, J. J. (eds.). *Nontarget effects of Biological Control*. Boston, MA: Kluwer Academic Publishers, pp. 95–109.

Duelli, P. (1990). Population movements of arthropods between natural and cultivated areas. *Biological Conservation* **54**, 193–207.

Duelli, P. (1992). Mosaikkonzept und Inseltheorie in der Kulturlandschaft. *Verhandlungen der Gesellschaft fur Okologie* **21**, 379–83.

Duelli, P. & Obrist, M. K. (1998). In search of the best correlates for local organismal biodiversity in cultivated areas. *Biodiversity and Conservation* **7**, 297–309.

Duelli, P., Studer, M., Marchand, I. & Jakob, S. (1990). Population movements of arthropods between natural and cultivated areas. *Biological Conservation* **54**, 193–207.

Duelli, P., Obrist, M. K. & Schmatz, D. R. (1999). Biodiversity evaluation in agricultural landscapes: above ground insects. *Agriculture, Ecosystems and Environment* **74**, 33–64.

Edwards, C. A. (1991). The assessment of populations of soil-inhabiting invertebrates. *Agriculture, Ecosystems and Environment* **34**, 145–76.

Edwards, C. A. (2000a). Ecologically based use of insecticides. In Rechcigl, J. E. & Rechcigl, N. A. (eds.). *Insect Pest Management: Techniques for Environmental Protection*. Boca Raton, FL: Lewis Publishers, pp. 103–30.

Edwards, C. A. (2000b). Soil invertebrate controls and microbial interactions in nutrient and organic matter dynamics in natural and agroecosystems. In Coleman, D. C. & Hendrix, P. F. (eds.). *Invertebrates as Webmasters in Ecosystems*. Wallingford, UK: CAB International, pp. 141–59.

Edwards, C. A. & Bohlen, P. J. (1991). The effect of toxic chemicals on earthworms. *Review of Environmental Contamination and Toxicology* **125**, 23–99.

Edwards, C. A. & Lofty, J. R. (1977). *Biology of Earthworms.* London: Chapman & Hall.

Edwards, E. D. (1994). *Survey of Lowland Grassland Sites in the ACT for the Golden Sun Moth*, Synemon plana. Canberra, Australia: CSIRO Division of Entomology.

Edwards, P. J., Kollmann, J. & Wood, D. (1999). Determinants of agrobiodiversity in the agricultural landscape. In Wood, D. & Lenné, J. M. (eds.). *Agrobiodiversity.* Wallingford: CAB International, pp. 183–210.

Ehler, L. E. (1990). Environmental impact of introduced biological control agents: implications for agricultural biotechnology. In Marois, J. J. & Bruening, G. (eds.). *Risk Assessment in Agricultural Biotechnology*. University of California Division of Agriculture and Natural Resources, publication no. 1928, pp. 85–96.

Ehler, L. E. (1998). Conservation biological control: past, present and future. In Barbosa, P. (ed.). *Conservation Biological Control.* San Diego, CA: Academic Press, pp. 1–8.

Ehler, L. E. (1999). Critical isues related to nontarget effects in classical biological control of insects. In Follet, P. A. & Duan, J. J. (eds.). *Nontarget Effects of Biological control.* Boston, MA: Kluwer Academic Publishers, pp. 3–13.

Ehrlich, P. R. & Murphy, D. D. (1987). Conservation lessons from long-term studies of checkerspot butterflies. *Conservation Biology* **1**, 122–31.

Ellis, D. V. (1985). Quality control of biological surveys. *Marine Pollution Bulletin* **19**, 506–12.

Elzen, G. W. (1989). Sublethal effects of pesticides on beneficial parasitoids. In Jepson, P. C. (ed.). *Pesticides and Non-target Invertebrates.* Wimborne, UK: Intercept, pp. 129–50.

Erhardt, A. & Thomas, J. A. (1991). Lepidoptera as indicators of change in the semi-natural grasslands of lowland and upland Europe. In Collins, N. M. & Thomas, J. A. (eds.). *The Conservation of Insects and their Habitats.* London: Academic Press, pp. 213–37.

Erwin, T. L. (1982). Tropical forests, their richness in Coleoptera and other arthropod species. *Coleopterists' Bulletin* **36**, 74–5.

Erwin, T. L. (1983). Tropical forest canopies: the last biotic frontier. *Bulletin of the Entomological Society of America* **29**, 14–19.

Erwin, T. L. (1998). Biodiversity at its utmost: tropical forest beetles. In Reaka-Kudla, M. L., Wilson, D. E. & Wilson, E. O. (eds.). *Biodiversity II. Understanding and Protecting our Biological Resources.* Washington, DC: Joseph Henry Press, pp. 27–40.

Everts, J. W. (1990). Why are field trials necessary? In Somerville, L. & Walker, C. H. (eds.). *Pesticide Effects on Terrestrial Wildlife.* London: Taylor & Francis, pp. 5–10.

Everts, J. W. & Ba, L. (1997). Environmental effects of locust control: state of the art and perspectives. In Krall, S., Peveling, R. & Ba Diallo, D. (eds.). *New Strategies in Locust Control.* Basel, Switzerland: Birkhauser-Verlag, pp. 331–6.

Fauvel, G. (1999). Diversity of Heteroptera in agroecosystems: roles of sustainability and bioindication. *Agriculture, Ecosystems and Environment* **74**, 275–303.

Feber, R. E. & Smith, H. (1995). Butterfly conservation on arable farmland. In Pullin, A. S. (ed.). *Ecology and Conservation of Butterflies*. London: Chapman & Hall, pp. 84–97.

Feber, R. E., Smith, H. & Macdonald, D. W. (1999). The importance of spatially variable field margin management for two butterfly species. *Aspects of Applied Biology* **54**, 155–62.

Ferro, D. N. & McNeil, J. N. (1998). Habitat enhancement and conservation of natural enemies of insects. In Barbosa, P. (ed.). *Conservation Biological Control*. San Diego, CA: Academic Press, pp. 123–32.

Fielding, D. J. & Brusven, M. A. (1993). Grasshopper (Orthoptera, Acrididae) community composition and ecological disturbance on southern Idaho rangeland. *Environmental Entomology* **22**, 71–81.

Finck, P., Riecken, U. & Schroder, E. (2002). Pasture landscapes and nature conservation: new strategies for preservation of open landscapes in Europe. In Redecker, B., Finck, P., Hardtle, W., Reicken, U. & Schroder, E. (eds.). *Pasture Landscapes and Nature Conservatiuon*. Berlin: Springer-Verlag, pp. 1–13.

Flexner, J. L., Lighthart, B. & Croft, B. A. (1986). The effects of microbial pesticides on non-target beneficial arthropods. *Agriculture, Ecosystems and Environment* **16**, 203–54.

Follett, M. A., Johnson, M. T. & Jones, V. P. (1999). Parasitoid drift in Hawaiian pentatomids. In Follett, P. A. & Duan, J. J. (eds.). *Nontarget Effects of Biological Control*. Boston, MA: Kluwer Academic Publishers, pp. 77–93.

Forcella, F. Buhler, D. D. & McGiffen, M. E. (1994). Pest management and crop residues. In Hatfield, J. L. & Stewart, B. A. (eds.). *Crops Residue Management*. Boca Raton, FL: Lewis Publishers, pp. 173–89.

Forman, R. T. T. (1995). *Land Mosaics: The Ecology of Landscapes and Regions*. Cambridge: Cambridge University Press.

Forman, R. T. T. & Godron, M. (1986). *Landscape Ecology*. New York: John Wiley & Sons.

Fourcat, A. & Lecoq, M. 1998. Major threats to a protected grasshopper, *Prionotropis hystrix rhodanica* (Orthoptera: Pamphagidae: Akicerinae), endemic to southern France. *Journal of Insect Conservation* **2**, 187–93.

Fournier, E. & Loreau, M. 1999. Effects of newly planted hedges on ground-beetle diversity (Coleoptera, Carabidae) in an agricultural landscape. *Ecography* **22**, 87–97.

Frampton, G. K., Cilgi, T., Fry, G. L. A. & Wratten, S. D. (1995). Effects of grassy banks on the dispersal of some carabid beetles (Coleoptera: Carabidae) on farmland. *Biological Conservation* **71**, 347–55.

Freckman, D. W. & Baldwin, J. G. (1990). Nematoda. In Dindal, D. L. (ed.). *Soil Biology Guide*. New York: John Wiley & Sons, pp. 155–200.

Freitag, R. & Poulter, F. (1970). The effects of the insecticides Sumithion and Phosphamidon on populations of five carabid beetle and two species of lycosid spiders in northwestern Ontario. *Canadian Entomologist* **102**, 1307–11.

Fry, G. L. A. (1991). Conservation in agricultural ecosystems. In Spellerberg, I. F., Goldsmith, F. B. & Morris, M. G. (eds.). *The Scientific Management of Temperate Communities for Conservation*. Oxford: Blackwell, pp. 415–43.

Fry, G. L. A. (1994). The role of field margins in the landscape. *British Crop Protection Council Monograph no. 58*, pp. 31–40.

Fry, G. L. A. & Main, A. R. (1993). Restoring seemingly natural communities on agricultural land. In Saunders, D. A., Hobbs, R. J. & Ehrlich, P. R. (eds.). *Nature Conservation*, Vol. 3: Reconstruction of fragmented ecosystems. Chipping Norton: Surrey Beatty & Sons, pp. 225–41.

Fry, G. L. A. & Robson, W. J. (1994). The effects of field margins on butterfly movement. *British Crop Protection Council Monograph no. 58*, pp. 111–16.

Fry, G., Robson, W. & Banham, A. 1992. *Corridors and Barriers to Butterfly Movements in Contrasting Landscapes*. NINA research report. Trondheim, Norway: Norwegian Institute for Nature Preservation.

Funasaki, G. Y., Lai, P. Y., Nakahara, L. M., Beardsley, J. W. & Ota, A. K. (1988a). A review of biological control introductions in Hawaii, 1890–1985. *Proceedings of the Hawaiian Entomological Society* **28**, 105–60.

Funasaki, G. Y., Nakahara, L. M. & Kumashiro, B. R. (1988b). Introductions for biological control in Hawaii: 1985 and 1986. *Proceedings of the Hawaiian Entomological Society* **28**, 101–4.

Fye, R. E. & Carranza, R. L. (1972). Movement of insect predators from grain sorghum to cotton. *Environmental Entomology* **1**, 790–91.

Gagné, W. C. & Howarth, F. G. (1985). Conservation status of endemic Hawaiian Lepidoptera. In Heath, J. (ed.). *Proceedings of the 3rd Congress of European Lepidopterology, Cambridge, 1982*. Karlsruhe, Societas Europaea Lepidopterologica, pp. 74–84.

Gall, G. A. E. & Orians, G. H. (1992). Agriculture and biological conservation. *Agriculture, Ecosystems and Environment* **42**, 1–8.

Gatehouse, A. M. R., Boulter, D. & Hilder, V. A. (1992). Potential of plant-derived genes in the genetic manipulation of crops for insect resistance. In Gatehouse, A. M. R., Hilder, V. A. & Boulter, D. (eds.). *Plant Genetic Manipulation for Crop Protection*. Wallingford, UK: CAB International, pp. 155–81.

Gates, G. E. (1961). Ecology of some earthworms with special reference to seasonal activity. *American Midland Naturalist* **66**, 61–86.

Georgis, R., Kaya, H. K. & Gaugler, R. (1991). Effect of steinernematid and heterorhabditid nematodes (Rhabditida: Steinernematidae and Heterorhabditidae) on nontarget arthropods. *Environmental Entomology* **20**, 815–22.

Gerard, P. W. (1995). *Agricultural Practices, Farm Policy and the Conservation of Biological Diversity*. USDI Biological Sciences report no. 4. Washington, DC.

Gillespie, R. G. & New, T. R. (1998). Compatability of conservation and pest management strategies. In Zalucki, M. P., Drew, R. A. I. & White, G. G. (eds.). *Pest Management: Future Challenges*, Vol. 2. Brisbane, Australia: University of Queensland, pp. 195–208.

Glare, T. R. & O'Callaghan, M. (2000). Bacillus thuringiensis: *Biology, Ecology and Safety*. Chichester, UK: John Wiley & Sons.

Gliessman, S. R. (1990). *Agroecology: Researching the Ecological Basis for Sustainable Agriculture*. New York: Springer-Verlag.

Gliessman, S. R. (1997). *Agroecology: Ecological Processes in Sustainable Agriculture*. Chelsea, MI: Sleeping Bear Press.

Goodwin, S. & Steiner, M. (1997). Introduction of *Bombus terrestris* for pollination of horticultural crops in Australia: A submission to AQIS and Environment Australia. www.tmag.tas.gov.au/workshop/append2.html.

Gould, J. R., Elkinton, J. S. & Wallner, W. E. (1990). Density-dependent suppression of experimentally created gypsy moth, *Lymantria dispar* (Lepidoptera: Lymantriidae) populations by natural enemies. *Journal of Animal Ecology* **59**, 213–33.

Grant, I. F. (1989). Monitoring insecticide side-effects in large-scale treatment programmes: tsetse spraying in Africa. In Jepson, P. C. (ed.). *Pesticides and Non-target Invertebrates*. Wimborne, UK: Intercept, pp. 43–69.

Greaves, M. P. & Marshall, E. J. P. (1987). Field margins: definitions and statistics. *British Crop Protection Council Monograph no. 35*, pp. 85–94.

Greenslade, P. (1997). Are Collembola useful as indicators of the conservation value of native grasslands? *Pedobiologia* **41**, 215–20.

Greenslade, P. J. M. (1978). Ants. in Low, W. A. (ed.). *The Physical and Biological Features of Kunoth Paddock in Central Australia*. Technical paper no. 4. Adelaide, Australia: CSIRO Division of Land Resources, pp. 114–23.

Greig-Smith, P. W. (1992). Risk assessment approaches in the UK for the side-effects of pesticides on earthworms. In Greig-Smith, P. W., Becker, H., Edwards, P. J. & Heimbach, F. (eds.). *Ecotoxicology of Earthworms*. Andover, UK: Intercept, pp. 159–68.

Groenendijk, D., van Mannekes, M., Vaal, M. & van den Berg, M. (2002). Butterflies and insecticides: a review and risk analysis of modern Dutch practice. *Proceedings in Experimental and Applied Entomology* **13**, 29–34.

Groffman, P. M. & Jones, C. G. (2000). Soil processes and global change: will invertebrates make a difference? In Coleman, D. C. & Hendrix, P. F. (eds.). *Invertebrates as Webmasters in Ecosystems*. Wallingford, UK: CAB International, pp. 313–66.

Gross, H. R. (1987). Conservation and enhancement of entomophagous insects: a perspective. *Journal of Entomological Science* **22**, 97–105.

Gurr, G. & Wratten, S. (eds.) (2000). *Biological control: measures of success*. Dordrecht: Kluwer Academic Publishers.

Gurr, G. M., Balou, N. D., Memmott, J., Wratten, S. D. & Greathead, D. J. (2000). A history of methodological, theoretical and empirical approaches to biological control. In Gurr, G. & Wratten, S. (eds.). *Biological Control: Measures of Success*. Dordrecht: Kluwer Academic Publishers, pp. 3–37.

Hadfield, M. G., Miller, S. E. & Carwilo, A. H. (1993). The decimation of Hawai'ian rain tree snails by alien predators. *American Zoologist* **33**, 610–22.

Hamilton, J. G. (1998). Forecasting and locust management, or why some insects aren't locusts. In Zalucki, M. P., Drew, R. A. I. & White, G. G. (eds.). *Pest Management: Future Challenges*, Vol. 2. Brisbane, Australia: University of Queensland, pp. 82–6.

Hammond, P. E. (1992). Species inventory. In Groombridge, B. (ed.). *Global Biodiversity: Status of Earth's Living Resources*. London: Chapman & Hall, pp. 17–39.

Hammond, P. E. (1994). Practical approaches to the estimation of the extent of biodiversity in speciose groups. *Philosophical Transactions of the Royal Society B* **345**, 119–36.

Hance, T. (2002). Impact of cultivation and crop husbandry practices. In Holland, J. M. (ed.). *The Agroecology of Carabid Beetles*. Andover, UK: Intercept, pp. 231–49.

Hance, T., Gegoire-Wibo, C. & Lebrun, P. (1990). Agriculture and ground-beetle populations. *Pedobiologia* **34**, 337–46.

Hani, F. J., Boller, E. F. & Keller, S. (1998). Natural regulation at the farm level. In Pickett, C. H. & Bugg, R. L. (eds.). *Enhancing Biological Control*. Berkeley, CA: University of California Press, pp. 161–209.

Hansen, R. A. (2000). Diversity in the decomposing landscape. In Coleman, D. C. & Hendrix, D. F. (eds.). *Invertebrates as Webmasters in Ecosystems*. Wallingford, UK: CAB International, pp. 203–15.

Hanski, I. & Simberloff, D. (1997). The metapopulation approach, its history, conceptual domain and application to conservation. In Hanski, I. & Gilpin, M. E. (eds.). *Metapopulation Biology: Ecology, Genetics and Evolution*. San Diego, CA: Academic Press, pp. 5–26.

Harley, K. L. S. & Forno, I. W. (1992). *Biological Control of Weeds: A Handbook for Practitioners and Students*. Melbourne, Australia: Inkata Press.

Harrison, S., Murphy, D. D. & Ehrlich, P. R. (1988). Distribution of the Bay checkerspot butterfly, *Euphydryas editha bayensis*: evidence for a metapopulation model. *American Naturalist* **132**, 360–82.

Hart, B. J., Manley, W. J., Limb, T. M. & Davies, W. P. (1994). Habitat creation in large fields for natural pest regulation. *British Crop Protection Council Monograph no. 58*, pp. 319–22.

Harwood, R. W. J., Hickman, J. M., Macleod, A. & Sherratt, T. N. (1994). Managing field margins for hoverflies. *British Crop Protection Council Monograph no. 58*, pp. 147–52.

Hassall, M., Hawthorne, A., Maudsley, M., White, O. & Cardwell, C. (1992). Effects of headland management on invertebrate communities in cereal fields. *Agriculture, Ecosystems and Environment* **40**, 155–78.

Hassan, S. A. (1989). Testing methodology and the concept of IOBC/WPRS Working Group. In Jepson, P. C. (ed.). *Pesticides and Non-target Invertebrates*. Wimborne, UK: Intercept, pp. 1–18.

Haughton, A. J., Bell, J. R., Johnson, P. J., *et al.* (1999). Methods of increasing invertebrate abundance within field margins. *Aspects of Applied Biology* **54**, 163–70.

Hawkins, B. & Marino, P. (1997). The colonisation of native phytophagous insects in North America by exotic parasitoids. *Oecologia* **112**, 566–71.

Heatwole, H. & Lowman, M. (1986). *Dieback: Death of an Australian Landscape*. Sydney: Reed.

Heinz, K. M., McCutcheon, B. M., Herrmann, R., Parella, M. P. & Hammock, B. D. (1995). Direct effects of recombinant nuclear polyhedrosis virus on selected non-target organisms. *Journal of Economic Entomology* **88**, 259–64.

Helps, M. B. (1994). Field margins: an agricultural perspective. *British Crop Protection Council Monograph no. 58*, pp. 21–30.

Henry, C. S., Brooks, S. J., Thierry, D., Duelli, P. & Johnson, J. B. (2001). The common green lacewing (*Chrysoperla carnea* s.lat.) and the sibling species problem. In McEwen, P., New, T. R. & Whittington, A. E. (eds.). *Lacewings in the Crop Environment*. Cambridge: Cambridge University Press, pp. 29–42.

Henry, C. S., Brooks, S. J., Duelli, P. & Johnson, J. B. (2002). Discovering the true *Chrysoperla carnea* (Insecta: Neuroptera: Chrysopidae) using song analysis, morphology and ecology. *Annals of the Entomological Society of America* **95**, 172–91.

Herms, C. P., McCullough, D. G., Baue, L.S., *et al.* (1997). Susceptibility of the endangered Karner blue butterfly (Lepidoptera: Lycaenidae) to *Bacillus thuringiensis* var. *kurstaki* used for gypsy moth suppression in Michigan. *Great Lakes Entomologist* **30**, 125–41.

Higley, L. G. & Wintersteen, W. K. (1992). A novel approach to environmental risk assessment of pesticides as a basis for incorporating environmental costs into economic injury level. *American Entomologist* **38**, 34–9.

Hill, R. J. (1999). Minimising uncertainty: in support of no-choice tests. In Withers, T. M., Barton Browne, L. E. & Stanley, J. (eds.). *Host Specificity Testing in Australasia: Towards Improved Assays for Biological Control.* Brisbane, Australia: Department of Natural Resources, Queensland, pp. 1–10.

Hingston, A. B. & McQuillan, P. B. (1999). Displacement of Tasmanian native megachilid bees by the recently introduced bumblebee *Bombus terrestris* (Linnaeus, 1758) (Hymenoptera: Apidae). *Australian Journal of Zoology* **47**, 59–65.

Hingston, A. B., Marsden-Smedley, J., Driscoll, D. A., *et al.* (2002). Extent of invasion of Tasmanian native vegetation by the exotic bumblebee *Bombus terrestris* (Apoidea: Apidae). *Austral Ecology* **27**, 162–72.

Hirose, Y. (1998). Conservation biological control of mobile pests: problems and tactics. In Barbosa, P. (ed.). *Conservation Biological Control.* San Diego, CA: Academic Press, pp. 221–34.

Hobbs, R. J. & Saunders, D. A. (eds.) (1993). *Reintegrating Fragmented Landscapes: Towards Sustainable Production and Nature Conservation.* New York: Springer-Verlag.

Hochberg, M. E. (2000). What, conserve parasitoids? In Hochberg, M. E. & Ives, A. R. (eds.). *Parasitoid Population Biology.* Princeton, NJ: Princeton University Press, pp. 266–77.

Hockey, P. A. S. & Bosman, A. C. (1986). Man as an intertidal predator in Transkei: disturbance, community consequences and management of a natural food species. *Oikos* **46**, 3–14.

Hokkanen, H. M. T. (1991). Trap cropping in pest management. *Annual Review of Entomology* **36**, 119–38.

Hokkanen, H. M. T. (1997). Role of biological control and transgenic crops in reducing use of chemical pesticides for crop protection. In Pimentel, D. (ed.). *Techniques for Reducing Pesticide Use.* New York: John Wiley & Sons, pp. 103–27.

Hokkanen, H. M. T. & Pimentel, D. (1984). New approaches for selecting biological control agents. *Canadian Entomologist* **116**, 1109–21.

Hokkanen, H. M. T. & Pimentel, D. (1989). New associations in biological control: theory and practice. *Canadian Entomologist* **121**, 828–40.

Holland, J. M. (2002). Carabid beetles, their ecology, survival and use in agroecosystems. In Holland, J. M. (ed.). *The Agroecology of Carabid Beetles.* Andover, UK: Intercept, pp. 1–40.

Holland, J. M. & Luff, M. L. (2000). The effects of agricultural practices on Carabidae in temperate agroecosystems. *Integrated Pest Management Reviews* **5**, 109–29.

Holland, J. M., Thomas, S. R. & Courts, S. (1994). *Phacelia tanacetifolia* flower strips as a component of integrated farming. *British Crop Protection Council Monograph no. 58*, pp. 215–20.

Holland, J. M., Frampton, G. K. & Van der Brink, P. J. (2002). Carabids as indicators within temperate arable farming systems: implications from SCARAB and LINK integrated farming systems projects. In Holland, J. M. (ed.). *The Agroecology of Carabid Beetles*. Andover: Intercept, pp. 251–77.

Holt, R. D. & Hochberg, M. E. (2001). Indirect interactions, community modules and biological control: a theoretical perspective. In Wajnberg, E., Scott, J. K. & Quimby, P. C. (eds.). *Evaluating Indirect Effects of Biological Control*. Wallingford, UK: CAB International, pp. 13–38.

Holt, S. J. (1987). Categorisation of threats to and status of wild populations. In Fitter, R. & Fitter, M. (eds.). *The Road to Extinction*. Gland, Switzerland: IUCN/UNEP, pp. 19–30.

Hopper, K. R. (1998). Assessing and improving the safety of introductions for biological control. In Zalucki, M. P., Drew, R. A. I. & White, G. G. (eds.). *Pest Management: Future Challenges*, Vol. 1. Brisbane, Australia: University of Queensland, pp. 501–10.

Hopper, K. R. (2001). Research needs concerning non-target impacts of biological control introductions. In Wajnberg, E., Scott, J. K. & Quimby, P. C. (eds.). *Evaluating Indirect Effects of Biological Control*. Wallingford, UK: CAB International, pp. 39–56.

Hopper, K. R., Roush, R. T. & Powell, W. (1993). Management of genetics of biological control introductions. *Annual Review of Entomology* **38**, 27–51.

Hopper, K. R., Coutinot, D., Chen, K., *et al.* (1998). Exploration for natural enemies to control *Diuraphis noxia* (Homoptera: Aphididae) in the United States. In Quisenberry, S. S. & Peairs, F. B. (eds.). *Response Model for an Introduced Pest: The Russian Wheat Aphid*. Lanham, MD: Entomological Society of America, pp. 166–82.

Horn, D. J. (1991). Potential impact of *Coccinella septempunctata* on endangered Lycaenidae (Lepidoptera) in Northeastern Ohio. In: Polagar, L., Chambers, R. J., Dixon, A. F. G. & Hodek, I. (eds.). *Behaviour and Impact of Aphidophaga*. The Hague: Academic Publishing, pp. 159–62.

Horne, P. A. & Edward, C. L. (1997). Preliminary observations on awareness, management and impact of biodiversity in agricultural ecosystems. *Memoirs of Museum Victoria* **56**, 281–5.

Hornsby, A. C. (1992). Site-specific pesticide recommendations: the final step in environmental impact prevention. *Weed Technology* **6**, 736–42.

Houlding, B., Ridsdill-Smith, T. J. & Bailey, W. J. (1991). Injectable abamectin causes a delay in scarabaeine dung beetle egg-laying in cattle dung. *Australian Veterinary Journal* **68**, 185–6.

House, G. J. & All, J. N. (1981). Carabid beetles in soybean agroecosystems. *Environmental Entomology* **10**, 194–6.

Howarth, F. G. (1983). Classical biocontrol: panacea or Pandora's box? *Proceedings of the Hawaiian Entomological Society* **24**, 239–44.

Howarth, F. G. (1985). Impacts of alien land arthropods and molluscs on native plants and animals in Hawaii. In Stone, C. P. & Scott, J. W. (eds.). *Hawaii's Terrestrial Ecosystems: Preservation and Management*. Honolulu: University of Hawaii Press, pp. 149–79.

Howarth, F. G. (1991). Environmental impacts of classical biological control. *Annual Review of Entomology* **36**, 485–509.

Howarth, F. G. (2000). Non-target effects of biological control agents. In Gurr, G. & Wratten, S. (eds.). *Biological Control: Measures of Success*. Dordrecht: Kluwer Academic Publishers, pp. 369–403.

Howarth, F. G. (2001). Environmental issues concerning the importation of non-indigenous biological control agents. In Lockwood, J. A., Howarth, F. G. & Purcell, M. F. (eds.). *Balancing Nature: Assessing the Impact of Non-native Biological Control Agents (an International Perspective)*. Lanham, MD: Entomological Society of America, pp. 70–99.

Hoy, C. W., Feldman, J., Gould, F., *et al.* (1998). Naturally occurring biological controls in genetically engineered crops. In Barbosa, P. (ed.). *Conservation Biological Control*. San Diego, CA: Academic Press, pp. 185–205.

Hughes, R. D., Hughes, M. A., Aeschlimann, J.-P., Woolcock, L. T. & Carver, M. (1994). An attempt to anticipate biological control of *Diuraphis noxia* (Hom., Aphididae). *Entomophaga* **39**, 211–23.

Hunter, D. M., Milner, R. J. & Spurgin, P. A. (2001). Aerial treatment of the Australian plague locust, *Chortoicetes terminifera* (Orthoptera: Acrididae) with *Metarhizium anisopliae* (Deuteromycotina: Hyphomycetes). *Bulletin of Entomological Research* **91**, 93–9.

Hunter, M. L. (1996). *Fundamentals of Conservation Biology*. Cambridge, MA: Blackwell Science.

Hutcheson, J. (1990). Characterisation of terrestrial insect communities using quantified, Malaise-trapped Coleoptera. *Ecological Entomology* **15**, 143–51.

Hutchings, T. R. (1993). An integrated approach to pasture management. In Delfosse, E. S. (ed.). *Pests of pastures: Weed, Invertebrate and Disease Pests of Australian Sheep Pastures*. Melbourne, Australia: CSIRO, pp. 381–5.

Hutchins, S. (1995). Free enterprise: the only sustainable solution to IPM implementation. *Journal of Agricultural Entomology* **12**, 211–17.

Huxley, J. S. (1963). Preface. In Carson, R. *Silent Spring*. London: Hamish Hamilton, pp. xxi–xxii.

Huxley, P. A. (1990). Experimental agroforestry. In MacDicken, K. G. & Vergara, N. T. (eds.). *Agroforestry: Classification and Management*. New York: John Wiley & Sons, pp. 332–53.

Insley, H. (1988). *Farm Woodland Planning*. Forestry Commission Bulletin no. 80. London: HMSO.

IUCN (1993). *Guidelines for Protected Areas Management Categories*. Gland, Switzerland: IUCN.

IUCN (1994). *IUCN Red List Categories*. Gland, Switzerland: IUCN.

IUCN (2002). *The IUCN Red List of Threatened Animal Species*. Gland, Switzerland: IUCN.

IUCN *et al.* (1991). *Caring for the World*. Gland, Switzerland: IUCN.

Jackson, T. A., Pearson, J. F., O'Callaghan, M., *et al.* (1993). Development of Invade^{TM}, a bacterial product for control of grass grub (*Costelytra zealandica*) in New Zealand pastures. In Corey, S. A., Dall, D. J. & Milne, W. M. (eds.). *Pest Control and Sustainable Agriculture*. Melbourne, Australia: CSIRO, pp. 259–60.

Janzen, D. H. (1983). No park is an island: increase in interference from outside as park size decreases. *Oikos* **41**, 402–10.

Janzen, D. H. (1988). Tropical ecological and biocultural restoration. *Science* **239**, 243–4.

Jepson, P. C. (1982). Ecology of *Myzus persicae* and its predators in sugar beet. Ph. D. thesis. Cambridge, UK: University of Cambridge.

Jepson, P. C. (1989). The temporal and spatial dynamics of pesticide side-effects on non-target invertebrates. In Jepson, P. C. (ed.). *Pesticides and Non-target Invertebrates*. Andover, UK: Intercept, pp. 95–127.

Jepson, P. C. (1994). Field margins as habitats, refuges and barriers of variable permeability to Carabidae. *British Crop Protection Council Monograph no. 58*, pp. 67–76.

Jesse, L. C. H. & Obrycki, J. J. (2003). Occurrence of *Danaus plexippus* L. (Lepidoptera: Danaidae) on milkweed (*Asclepias syriaca*) in transgenic Bt corn agroecosystems. *Agriculture, Ecosystems and Environment* **97**, 225–33.

Johansen, C. A. (1977). Pesticides and pollination. *Annual Review of Entomology* **22**, 177–92.

Johnson, D. M. & Stiling, P. D. (1996). Host specificity of *Cactoblastis cactorum* (Lepidoptera: Pyralidae), an exotic *Opuntia*-feeding moth in Florida. *Environmental Entomology* **25**, 743–8.

Jones, D. T., Susilo, F. X., Bignell, D. E., *et al.* (2003). Termite assemblage collapse along a land-use intensification gradient in lowland central Sumatra, Indonesia. *Journal of Applied Ecology* **40**, 380–91.

Jones, R. E. & Kitching, R. L. (1981). Why an ecology of pests? In Kitching, R. L. & Jones, R. E. (eds.). *The Ecology of Pests: Some Australian Case Histories*. Melbourne, Australia: CSIRO, pp. 1–5.

Jones, S. (1991). Hedgerows. In Fry, R. & Lonsdale, D. (eds.). *Habitat Conservation for Insects: A Neglected Green Issue*. Middlesex, UK: Amateur Entomologists' Society, pp. 117–28.

Kalkhoven, J. T. R. (1993). Survival of populations and the scale of the fragmented agricultural landscape. In Bunce, R. G. H., Ryszkowski, L. & Paoletti, M. G. (eds.). *Landscape Ecology and Agroecosystems*. Boca Raton, FL: Lewis Publishers, pp. 83–90.

Kampf, H. (2002). Nature conservation in pastoral landscapes: challenges, choices and controls. In Redecker, B., Finck, P., Hardtle, W., Reicken, U. & Schroder, E. (eds.). *Pasture Landscapes and Nature Conservation*. Berlin: Springer-Verlag, pp. 15–38.

Kaule, G. & Krebs, S. (1989). Creating new habitats in intensively used agricultural land. In Buckley, G. P. (ed.). *Biological Habitat Reconstruction*. London: Belhaven Press, pp. 161–9.

Keesing, V. & Wratten, S. D. (1997). Integrating plant and insect conservation. In Maxted, N., Ford-Lloyd, B. V. & Hawkes, J. G. (eds.). *Plant Genetic Conservation*. London: Chapman & Hall, pp. 220–35.

Kells, A. R., Holland, J. M. & Goulson, D. (2001). The value of uncropped field margins for foraging bumblebees. *Journal of Insect Conservation* **5**, 283–91.

Kelly, G. C. & Park, G. N. (1986). *The New Zealand Protected Natural Areas Programme: A Scientific Focus*. New Zealand Biological Resources Centre Publication No. 4. Wellington, New Zealand: Department of Scientific and Industrial Research.

Kemp, J. C. & Barrett, G. W. (1989). Spatial patterning: impact of uncultivated corridors on arthropod populations within soybean agroecosystems. *Ecology* **70**, 114–28.

Kerry, B. (1995). The potential impact of natural enemies on the survival and efficacy of entomopathogenic nematodes. In Griffin, C. T., Gwynn, R. L. & Masson, J. P. (eds.). *Ecology and Transmission Strategies of Entomopathogenic Nematodes*. Luxembourg: European Commission, pp. 7–13.

Kevan, P. G. (1977). Blueberry crops in Nova Scotia and New Brunswick: pesticides and crop reductions. *Canadian Journal of Agricultural Economics* **25**, 64.

Key, K. H. L. (1978). *The Conservation Status of Australia's Insect Fauna*. Canberra, Australia: Australian National Parks and Wildlife Service.

King, E. G., Hopper, K. R. & Powell, J. E. (1985). Analysis of systems for biological control of crop arthropod pests in the United States by augmentation of predators and parasites. In Hoy, M. A. & Herzog, R. C. (eds.). *Biological Control in Agricultural IPM Systems*. New York: Academic Press, pp. 210–27.

Kirkham, F. W., Sherwood, A. J., Oakley, J. N. & Fielder, A. G. (1999). Botanical composition and invertebrate populations in sown grass and wildflower margins. *Aspects of Applied Biology* **54**, 291–8.

Kirkpatrick, J., McDougall, K., & Hyde, M. (1995). *Australia's most Threatened Ecosystems: The Southeastern Lowland Native Grasslands*. Chipping Norton, UK: Surrey Beatty & Sons.

Knight, R. L. & Landres, P. B. (2002). Central concepts and issues of biological conservation. In Gutzwiller, K. J. (ed.). *Applying Landscape Ecology in Biological Conservation*. New York: Springer-Verlag, pp. 22–3.

Kobel-Lamparski, A., Gack, C. & Lamparski, F. (1993). Influence of mulching treatment on epigeic spiders in vineyards of the Kaiserstuhl area (SW Germany). *Arachnological Mitteilungen* **5**, 15–32.

Kogan, M. (1982). Plant resistance in pest management. In Metcalf, R. L. & Luckmann, W. L. (eds.). *Introduction to Insect Pest Management*, 2nd edn. New York: John Wiley & Sons, pp. 93–134.

Kogan, M. (1988). Integrated pest management theory and practice. *Entomologia experimentalis et applicata* **49**, 59–70.

Kogan, M. (1998). Integrated pest management: historical perspective and categorising developments. *Annual Review of Entomology* **43**, 243–70.

Kovach, J., Petzoidt, C., Degni, J. & Tette, J. (1992). A method to measure the environmental impact of pesticides. *New York State Agricultural Experiment Station Food and Life Sciences Bulletin* **139**.

Kremen, C., Colwell, R. K., Erwin, T. L., *et al.* (1993). Terrestrial arthropod assemblages: their use in conservation planning. *Conservation Biology* **7**, 796–805.

Kreutzweiser, D. P., Holmes, S. B., Capell, S. S. & Eichenberg, D. C. (1992). Lethal and sublethal effects of *Bacillus thuringiensis* var. *kurstaki* on aquatic insects in laboratory bioassays and outdoor stream channels. *Bulletin of Environmental Contamination and Toxicology* **49**, 252–8.

Kreutzweiser, D. P., Capell, S. S. & Thomas, D. R. (1994) Aquatic insect responses *to Bacillus thuringiensis* var. *kurstaki* in a forest stream. *Canadian Journal of Forest Research* **24**, 2041–9.

Kromp, B. (1999). Carabid beetles in sustainable agriculture: a review on pest control efficiency, cultivation impacts and enhancement. *Agriculture, Ecosystems and Ecology* **74**, 187–228.

Kruess, A. & Tscharntke, T. (2000). Effects of habitat fragmentation on plant-insect communities. In Ekbom, B., Irwin, M. E. & Robert, Y. (eds.). *Interchanges of Insects between Agricultural and Surrounding Landscapes*. Dordrecht: Kluwer Academic, pp. 53–70.

Kruger, K. & Scholtz, C. H. (1997). Lethal and sublethal effects of ivermectin on the dung-breeding beetles *Euonoticellus intermedius* (Reiche) and *Onitis alexis* Klug (Coleoptera, Scarabaeidae). *Agriculture, Ecosystems and Environment* **61**, 123–31.

Kuhbauch, W. (1998). Loss of biodiversity in European agriculture during the twentieth century. In Barthlott, W. & Winiger, M. (eds.). *Biodiversity: A Challenge for Development Research and Policy*. Berlin: Springer-Verlag, pp. 145–55.

Lagerlof, J., Starck, J. & Svenson, B. (1992). Margins of agricultural fields as habitats for pollinating insects. *Agriculture, Ecosystems and Environment* **40**, 117–24.

Lambeck, R. J. (1999). *Landscape Planning for Biodiversity Conservation in Agricultural Regions: A Case Study from the Wheatbelt of Western Australia*. Biodiversity Technical Paper No. 2. Canberra, Australia: Environment Australia.

Landis, D. A. & Menalled, F. D. (1998). Ecological considerations in the conservation of effective parasitoid communities in agricultural systems. In Barbosa, P. (ed.). *Conservation Biological Control*. San Diego, CA: Academic Press, pp. 101–21.

Landis, D. A., Wratten, S. D. & Gurr, G. M. (2000). Habitat management to conserve natural enemies of arthropod pests in agriculture. *Annual Review of Entomology* **45**, 175–201.

Landy, J. (1993). Foreword. Pest control: the key to pasture productivity. In Delfosse, E. (ed.). *Pests of Pasture: Weed, Invertebrate and Disease Pests of Australian Sheep Pastures*. Melbourne, Australia: CSIRO, p. ix.

Latchininsky, A. V. (1998). Moroccan Locust (*Dociostaurus maroccanus*) (Thunberg 1815): a faunistic rarity or an important economic pest? *Journal of Insect Conservation* **2**, 167–78.

Launer, A. E. & Murphy, D. D. (1994). Umbrella species and the conservation of habitat fragments: a case of a threatened butterfly and a vanishing grassland ecosystem. *Biological Conservation* **69**, 145–53.

Laurance, W. F. & Yensen, E. (1991). Predicting the impacts of edge effects in fragmented habitats. *Biological Conservation* **55**, 77–92.

Lee, J. C., Menalled, F. D. & Landis, D. A. (2001). Refuge habitats modify impact of insecticide disturbance on carabid beetle communities. *Journal of Applied Ecology* **38**, 472–83.

Lee, M. S. Y. (2000). A worrying systematic decline. *Trends in Ecology and Evolution* **15**, 346.

Leong, K. L. H., Yoshimura, M. A. & Kaya, H. K. (1992). Low susceptibility of overwintering monarch butterflies to *Bacillus thuringiensis* Berliner. *Pan Pacific Entomologist* **68**, 66–8.

Letourneau, D. K. & Altieri, M. A. (1999). Environmental management to enhance biological control in agroecosystems. In Bellows, T. S. & Fisher, T. W. (eds.). *Handbook of Biological Control*. San Diego, CA: Academic Press, pp. 319–54.

Levins, R. 1970. Extinction. Some mathematical questions in biology. In Gerstenhaberr, M. (ed.). *Lectures on Mathematics in Life Sciences*. Providence, RI: American Mathematical Society, pp. 77–107.

Levitan, L., Merwin, I, & Kovach, J. (1995). Assessing the relative environmental impacts of agricultural practices: the quest for a holistic method. *Agriculture, Ecosystems and Environment* **55**, 153–68.

Lipton, J., Galbraith, H., Berger, J. & Wartenberg, D. (1996). A paradigm for ecological risk assessment. *Environmental Management* **17**, 1–5.

Lockwood, J. A. (1993a). Environmental issues involved in the biological control of rangeland grasshoppers (Orthoptera: Acrididae) with exotic agents. *Environmental Entomology* **22**, 503–18.

Lockwood, J. A. (1993b). The benefits and costs of controlling rangeland grasshoppers with exotic organisms: the search for a null hypothesis and regulatory compromise. *Environmental Entomology* **22**, 904–14.

Lockwood, J. A. (1996). The ethics of biological control: understanding the moral implications of our most powerful ecological technology. *Agriculture and Human Values* **13**, 2–19.

Lockwood, J. A. (1998). Management of orthopteran pests: a conservation perspective. *Journal of Insect Conservation* **2**, 253–61.

Lockwood, J. A. & DeBrey, L. D. (1990). A solution for the sudden and unexplained extinction of the Rocky Mountain grasshopper (Orthoptera. Acrididae). *Environmental Entomology* **19**, 1194–205.

Lockwood, J. A. & Ewen, A. B. (1997). Biological control of rangeland grasshoppers and locusts. In Gangwere, S. K., Muralirangan, M. C. & Muralirangam, M. (eds.). *Bionomics of Grasshoppers, Katydids and their Kin*. Wallingford, UK: CAB International, pp. 421–42.

Longley, M. & Sotherton, N. W. (1997). Factors determining the effects of pesticides upon butterflies inhabiting arable farmland. *Agriculture, Ecosystems and Environment* **61**, 1–12.

Lonsdale, W. M., Briese, D. T. & Cullen, J. M. (2001). Risk analysis and weed biological control. In Wajnberg, E., Scott, J. K. & Quimby, P. C. (eds.). *Indirect Effects of Biological Control*. Wallingford, UK: CAB International, pp. 185–210.

Loope, L. L., Medeiras, A. C. & Cole, F. R. (1988). Effects of the Argentine ant on the endemic biota of subalpine shrubland, Haleakala National Park. pp. 52–62. In Thomas, L. K. (ed.). *Management of exotic species in Natural Communities*. Fort Collins, CO: United States Fish and Wildlife Service.

Losey, J. E., Rayor, L. S. & Carter, M. E. (1999). Transgenic pollen harms monarch larvae. *Nature* **399**, 214.

Louda, S. M., Kendall, D., Connor, J. & Simberloff, D. (1997). Ecological effects of an insect introduced for the biological control of weeds. *Science* **277**, 1088–90.

Louda, S. M., Simberlof, D., Boettner, G., *et al.* (1998). Insights from data on the nontarget effects of the flowerhead weevil. *Biocontrol News and Information* **19**, 70–71N.

Louda, S. M., Arnett, A. E., Rand, T. A. & Rusell, F. L. (2003a). Invasiveness of some biological control insects and adequacy of their ecological risk assessment and regulation. *Conservation Biology* **17**, 73–82.

Louda, S. M., Pemberton, R. W., Johnson, M. T. & Follett, P. A. (2003b). Nontarget effects: the Achilles' heel of biological control? Retrospective analyses to reduce risk associated with biocontrol introductions. *Annual Review of Entomology* **48**, 365–96.

Loxdale, H. D. & Den Hollander, J. (1989). *Electrophoretic Studies on Agricultural Pests*. Oxford: Clarendon Press.

Lubbe, E. (1988). National report on environmental management in agriculture: W. Germany. In Park, J. R. (ed.). *Environmental Management and Agriculture: European Perspectives*. London: Belhaven Press, pp. 83–94.

Lumaret, J. P., Galante, E., Lumbreras, C., *et al.* (1993). Field effects of ivermectin residues on dung beetles. *Journal of Applied Ecology* **30**, 428–36.

Lynch, L. D., Ives, A. R., Waage, J. K., Hochberg, M. E. & Thomas, M. B. (2002). The risks of biocontrol: transient impacts and minimum non-target densities. *Ecological Applications* **12**, 1872–82.

Lys, J.-A. & Nentwig, W. (1991). Surface activity of carabid beetles inhabiting cereal fields. Seasonal phenology and the influence of farming operations on five abundant species. *Pedobiologia* **35**, 129–38.

Lys, J.-A. & Nentwig, W. (1992). Augmentation of beneficial arthropods by strip-management. 4. Surface activity, movements and activity density of abundant carabid beetles in a cereal field. *Oecologia* **92**, 373–82.

MacDicken, K. G. & Vergara, N. T. (1990). Introduction to agroforestry. In MacDicken, K. G. & Vergara, N. T. (eds.). *Agroforestry: Classification and Management*. New York: John Wiley & Sons, pp. 1–30.

Macdonald, D. W. & Smith, H. (1990). New perspectives on agroecology. In Firbank, L. G., Carter, N., Darbyshire, J. F. & Potts, G. R. (eds.). *The Ecology of Temperate Cereal Fields*. Oxford: Blackwell Scientific Publishers, pp. 413–48.

Mader, H. J. (1984). Animal habitat isolation by roads and agricultural fields. *Biological Conservation* **29**, 81–96.

Mader, H. J., Schell, C. & Kornacken, P. (1990). Linear barriers to arthropod movements in the landscape. *Biological Conservation* **54**, 209–22.

Madsen, M., Overgaard Nielsen, B., Holter, P., *et al.* (1990). Treating cattle with ivermectin: effects on the fauna and decomposition of dung pads. *Journal of Applied Ecology* **27**, 1–5.

Maelfait, J.-P. & De Keer, R. (1990). The border-zone of an intensively grazed pasture as a corridor for spiders Araneae. *Biological Conservation* **54**, 223–38.

Main, A. R. (1993). Landscape reintegration: problem definition. In Hobbs, R. J. & Saunders, D. A. (eds.). *Reintegrating Fragmented Landscapes: Towards Sustainable Production and Nature Conservation*. New York: Springer-Verlag, pp. 189–208.

Majer, J. D. & Beeston, G. (1996). The biodiversity integrity index: an illustration using ants in Western Australia. *Conservation Biology* **10**, 65–73.

Majer, J. D., Recher, H. F., Wellington, A. B., Woinarski, J. C. Z. & Yen, A. L. (1998). Invertebrates of eucalypt formations. In Williams, J. & Woinarski, J. (eds.). *Eucalypt Ecology: Individuals to Ecosystems*. Cambridge: Cambridge University Press, pp. 278–302.

Malden, W. J. (1899). Hedges and hedge-making. *Journal of the Royal Agricultural Society of England* **6**, 87–115.

Mannion, A. M. (1995). *Agriculture and Environmental Change. Temporal and Spatial Dimensions*. Chichester: John Wiley & Sons.

Marc, P., Canard, A. & Ysnel, F. (1999). Spiders (Araneae) useful for pest limitation and bioindication. *Agriculture, Ecosystems and Environment* **74**, 229–73.

Margules, C. R. (1992). The Wog Wog habitat fragmentation experiment. *Environmental Conservation* **19**, 316–25.

Marino, P. C. & Landis, D. A. (1996). Effects of landscape structure on parasitoid diversity and parasitism in agroecosystems. *Ecological Applications* **6**, 276–84.

Marohasy, J. J. (1998). The design and interpretation of host-specificity tests for weed biological control with particular reference to insect behaviour. *Biocontrol News and Information* **19**, 13–20N.

Marrs, R. H., Frost, A. J., Plant, R. A. & Lunnis, P. (1993). Determination of buffer zones to protect seedlings of non target plants from the effects of glyphosphate spray drift. *Agriculture, Ecosystems and Environment* **45**, 283–93.

Mauremooto, J. R., Wratten, S. D., Worner, S. P. & Fry, G. L. (1995). Permeability of hedgerows to predatory carabid beetles. *Agriculture, Ecosystems and Environment* **52**, 141–8.

McEvoy, P. B. & Coombs, E. M. (1999). Why things bite back: unintended consequences of biological weed control. In Follett, P. A. & Duan, J. J. (eds.). *Nontarget Effects of Biological Control*. Boston, MA: Kluwer, pp. 167–94.

McEvoy, P. B. & Coombs, E. M. (2000). A parsimonious approach to biological control of plant invaders. *Ecological Applications* **9**, 387–401.

McFadyen, R. E. C. (1998). Biological control of weeds. *Annual Review of Entomology* **43**, 369–93.

McGeoch, M. A. (1998). The selection, testing and application of terrestrial insects as bioindicators. *Biological Reviews* **73**, 181–201.

McIntyre, S., Barrett, G. W., Kitching, R. L. & Recher, H. F. (1992). Species triage: seeing beyond wounded rhinos. *Conservation Biology* **6**, 604–6.

McLaughlin, A. & Mineau, P. (1995). The impact of agricultural practices on biodiversity. *Agriculture, Ecosystems and Environment* **55**, 201–12.

McLeod, A. H., McGugar, B. M. & Coppel, H. C. (1962). *A Review of the Biological Control Attempts Against Insects and Weeds in Canada*. Technical communication of the Commonwealth Institute for Biological Control 2.

McNeely, J. A. (1988). *Economics and Biological Diversity*. Gland, Switzerland: IUCN.

McQuillan, P. B. (1999). The effect of changes in Tasmanian grasslands on the geometrid moth tribe Xanthorhoini (Geometridae: Larentiinae). In Ponder, W. & Lunney, D. (eds.). *The Other 99%: The conservation and biodiversity of invertebrates. Transactions of the Royal Zoological Society of New South Wales, Mosman*, 121–8.

Meinzingen, W. F. (1997). Overview and challenges of new control agents. In Krall, S., Peveling, R. & Ba Diallo, D. (eds.). *New Strategies in Locust Control*. Switzerland: Birkhauser-Verlag, Basel, pp. 105–15.

Messing, R. H. (2000). The impact of nontarget concerns on the practice of biological control. In Follett, P. A. & Duan, J. J. (eds.). *Nontarget Effects of Biological Control*. Boston, MA: Kluwer Academic publishers, pp. 45–55.

Metcalf, R. L. (1980). Changing roles of insecticides in crop protection. *Annual Review of Entomology* **25**, 219–56.

Miller, J. C. (1990). Field assessment of the effects of a microbial pest control agent on nontarget Lepidoptera. *Bulletin of the Entomological Society of America* **36**, 135–9.

Miller, L. K. (1995). Genetically engineered insect virus pesticides: present and future. *Journal of Invertebrate Pathology* **65**, 211–16.

Moen, J. & Jonsson, B. G. (2003). Edge effects on liverworts and lichens in forest patches in a mosaic of boreal forest and wetland. *Conservation Biology* **17**, 380–88.

Moldenke, A., Shaw, C. & Boyle, J. R. (1991). Computer-driven image-based soil fauna taxonomy. *Agriculture, Ecosystems and Environment* **34**, 177–85.

Moreby, S. J., Sotherton, N. W. & Jepson, P. W. (1997). The effects of pesticides on species of non-target Heteroptera inhabiting cereal fields in southern England. *Pesticide Science* **51**, 39–48.

Morris, M. G. (1969). Populations of invertebrate animals and the management of chalk grasslands in Britain. *Biological Conservation* **1**, 225–31.

Morris, M. G. (1971a). Differences between the invertebrate fauna of grazed and ungrazed chalk grassland. IV. Abundance and diversity of Homoptera – Auchenorhyncha. *Journal of Applied Ecology* **8**, 37–52.

Morris, M. G. (1971b). The management of grassland for the conservation of invertebrate animals. In Duffey, E, & Watt, A. S. (eds.). *The Scientific Management of Animal and Plant Communities for Conservation*. Oxford: Blackwell Scientific Publications, pp. 527–52.

Morris, M. G. & Webb, N. R. (1987). The importance of field margins for the conservation of insects. *British Crop Protection Council Monograph no. 33*, pp. 53–65.

Munguira, M. L. & Thomas, J. A. (1992). Use of road verges by butterfly and burnet populations, and the effects of roads on adult dispersal and mortality. *Journal of Applied Ecology* **29**, 316–29.

Murcia, C. (1995). Edge effects in fragmented forests: implications for conservation. *Trends in Ecology and Evolution* **10**, 58–62.

Murdoch, W. W., Chesson, J. & Chesson, P. L. (1995). Biological control in theory and practice. *American Naturalist* **125**, 344–66.

Murray, J., Murray, E., Johnson, M. S. & Clarke, B. (1988). The extinction of *Partula* on Moorea. *Pacific Science* **42**, 150–53.

Nafus, D. R. (1992). Impact of intentionally and accidentally introduced biological control agents on *Hypolimnas anomala* and *H. bolina* (Lepidoptera: Nymphalidae). *Pacific Science* **46**, 394–5.

Nafus, D. R. (1993). Movement of introduced biological control agents onto non-target butterflies, *Hypolimnas* spp. (Lepidoptera: Nymphalidae). *Environmental Entomology* **22**, 265–72.

Nafus, D. R. (1994). Extinction, biological control and insect conservation on islands. In Gaston, K. J., New, T. R. & Samways, M. J. (eds.). *Perspectives on Insect Conservation*. Andover, UK: Intercept, pp. 139–54.

Nafus, D. R. & Schreiner, I. (1986). Intercropped maize and sweet potatoes. Effects on parasitisation of *Ostrinia furnacalis* by *Trichogramma chilonis*. *Agriculture, Ecosystems and Environment* **15**, 189–200.

Nair, P. K. R. (1990). Classification of agroforestry systems. In MacDicken, K. G. & Vergara, N. T. (eds.). *Agroforestry: Classification and Management*. New York: John Wiley & Sons, pp. 31–57.

NAS (National Academy of Sciences) (1993). *Risk Assessment in the Federal Government: Managing the Process.* Washington, DC: National Academy Press.

Neale, C., Smith, D., Beattie, G. A. C. & Miles, M. (1995). Importation, host specificity testing, rearing and release of three parasitoids of *Phyllocnistis citrella* (Stainton) (Lepidoptera: Gracillariidae) in eastern Australia. *Journal of the Australian Entomological Society* **34**, 343–8.

Nentwig, W. (1995). Ackerkrautstreifen als Systemansatz fur eine umweltfreundliche Landwirtschaft. *Mitteilungen der Deutschen Gesellschaft für Allgemeine und Angewandte Entomologie* **9**, 679–83.

Nentwig, W., Frank, T. & Lethmayer, C. (1998). Sown weed strips: artificial ecological compensation areas as an important tool in conservation biological control. In Barbosa, P. (ed.). *Conservation Biological Control.* San Diego, CA: Academic Press, pp. 133–53.

Neville, P. J. & New, T. R. (1999). Ant genus to species ratios: a practical trial for surrogacy value in Victorian forests. In Ponder, W. & Lunney, D. (eds.). *The Other 99%: The Conservation and Biodiversity of Invertebrates. Transactions of the Royal Zoological Society of New South Wales. Mosman*, 133–7.

New, T. R. (1983). Systematics and ecology: reflections from the interface. In Highley, E. & Taylor, R. W. (eds.). *Australian Systematic Entomology: A Bicentenary Perspective.* Melbourne: CSIRO, pp. 50–79.

New, T. R. (1984). *Insect Conservation: An Australian Perspective.* The Hague: W. Junk.

New, T. R. (1991). *Insects as Predators.* Kensington, NSW: New South Wales University Press.

New, T. R. (1993). Angels on a pin: dimensions of the crisis in invertebrate conservation. *American Zoologist* **33**, 623–30.

New, T. R. (1994). *Exotic Insects in Australia.* Adelaide, Australia: Gleneagles Press.

New, T. R. (1995). *Introduction to Invertebrate Conservation Biology.* Oxford: Oxford University Press.

New, T. R. (1998). *Invertebrate Surveys for Conservation.* Oxford: Oxford University Press.

New, T. R. (1999a). Limits to species focusing in insect conservation. *Annals of the Entomological Society of America* **92**, 853–60.

New, T. R. (1999b). Entomology and nature conservation. *European Journal of Entomology* **96**, 11–17.

New, T. R. (2000a). The conservation of a discipline: traditional taxonomic skills in insect conservation. *Journal of Insect Conservation* **4**, 211–13.

New, T. R. (2000b). How to conserve the 'meek inheritors'. *Journal of Insect Conservation* **4**, 151–2.

New, T. R. (2002). *Insects and Pest Management in Australian Agriculture.* Melbourne, Australia: Oxford University Press.

Newsom, L. D. (1967). Consequences of insecticide use on non-target organisms. *Annual Review of Entomology* **12**, 257–86.

Noss, R. F. (1990). Indicators for monitoring biodiversity: a hierarchical approach. *Conservation Biology* **4**, 355–64.

NRC (National Research Council) (1989). Problems in U.S. agriculture. In *Alternative Agriculture.* Washington, DC: National Academy Press, pp. 89–134.

Oades, J. M. & Walters, L. J. (1994). Indicators for sustainable agriculture: policies to paddock. In Pankhurst, C. E., Doube, B. M., Gupta, V. V. S. R. & Grace, P. R. (eds.). *Soil Biota: Management in Sustainable Farming Systems*. Melbourne, Australia: CSIRO, pp. 219–23.

Obrycki, J. J., Elliott, N. C. & Giles, K. L. (1999). Coccinellid introductions: potential for and evaluation of nontarget effects. In Follet, P. A. & Duan, J. J. (eds.). *Nontarget Effects of Biological Control*. Boston, MA: Kluwer, pp. 127–45.

Odum, E. P. (1969). The strategy of ecosystem development. *Science* **164**, 262–70.

O'Dwyer, C. (1999). The g-habitat of the golden sun moth. In Ponder, W. & Lunney, D. (eds.). *The Conservation and Biodiversity of Invertebrates*. Mosman, Australia: Royal Zoological Society of New South Wales, pp. 322–40.

O'Dwyer, C. & Attiwill, P. M. (1999). A comparative study of habitats of the golden sun moth *Synemon plana* Walker (Lepidoptera: Castniidae): implications for conservation. *Biological Conservation* **89**, 131–41.

Oliver, I. & Beattie, A. J. (1993). A possible method for the rapid assessment of biodiversity. *Conservation Biology* **7**, 562–8.

Olkowski, W. & Zhang, A. (1998). Habitat management for biological control: examples from China. In Pickett, C. H. & Bugg, R. L. (eds.). *Enhancing Biological Control*. Berkeley, CA: University of California Press, pp. 255–70.

O'Neill, K. M., Olson, B. E., Rolston, M. G., *et al.* (2003). Effects of livestock grazing on rangeland grasshopper (Orthoptera: Acrididae) abundance. *Agriculture, Ecosystems and Environment* **97**, 51–64.

O'Neill, R. J., Giles, K. L., Obrycki, J. J., *et al.* (1998). Evaluation of the quality of four commercially available natural enemies. *Biological Control* **11**, 1–8.

Ovenden, G. N., Swash, A. R. H. & Smallshire, D. (1998). Agri-environment schemes and their contribution to conservation of biodiversity in England. *Journal of Applied Ecology* **35**, 955–60.

Paine, R. W. (1994). *Recollections of a Pacific entomologist*, 1925–1966. Canberra, Australia: Australian Centre for International Agricultural Research.

Pankhurst, C. E. & Lynch, J. M. (1994). The role of soil biota in sustainable agriculture. In Pankhurst, C. E., Doube, B. M., Gupta, V. V. S. R. & Grace, P. R. (eds.). *Soil Biota and Management in Sustainable Farming Systems*. Melbourne, Australia: CSIRO, pp. 3–9.

Paoletti, M. G. (1999). Invertebrate biodiversity bioindicators of sustainable landscapes. Practical use of invertebrates to assess sustainable land use. *Agriculture, Ecosystems and Environment* **74**, 1–444.

Paoletti, M. G. & Hassall, M. (1999). Woodlice (Isopoda: Oniscoidea): their potential for assessing sustainability and use as bioindicators. *Agriculture, Ecosystems and Environment* **74**, 157–65.

Parkman, J. P., Frank, J. H., Nguyen, K. B. & Smart, G. C., Jr (1994). Inoculative release of *Steinernema scapterisci* (Rhabditida: Steinernematidae) to suppress pest mole crickets (Orthoptera: Gryllotalpidae) on golf courses. *Environmental Entomology* **23**, 1331–7.

Paton, D. C. (1996). *Overview of the Impacts of Feral and Managed Honeybees in Australia*. Canberra, Australia: Australian Nature Conservation Agency.

Pearce, D. & Tinch, R. (1998). The base price of pesticides. In Vorley, W. & Keeney, D. (eds.). *Bugs in the System: Redesigning the Pesticide Industry for Sustainable Agriculture*. London: Earthscan, pp. 50–93.

Pearson, D. L. (1994). Selecting indicator taxa for the quantitative assessment of biodiversity. *Philosophical Transactions of the Royal Society of London. Series B: Biological Sciences* **345**, 75–9.

Pedigo, L. P. (1996). *Entomology and Pest Management*, 2nd edn. Upper Saddle River, NJ: Prentice Hall.

Pedigo, L. P. (1999). *Entomology and Pest Management*, 3rd edn. Upper Saddle River, NJ: Prentice Hall.

Pedigo, L. P., Hutchins, S. H. & Higley, L. G. (1986). Economic injury levels in theory and practice. *Annual Review of Entomology* **31**, 341–68.

Pekar, S. (1999). Effect of IPM practices and conventional spraying on spider population dynamics in an apple orchard. *Agriculture, Ecosystems and Environment* **73**, 155–66.

Penrose, L. J., Thwaite, W. G. & Bower, C. C. (1994). Rating index as a basis for decision making on pesticide use reduction and for accreditation of fruit produced under integrated orchard management. *Crop Protection* **13**, 146–52.

Perfecto, I. & Snelling, R. (1995). Biodiversity and the transformation of a tropical agroecosystem: ants in coffee plantations. *Ecological Applications* **5**, 1084–97.

Perkins, J. H. & Patterson, B. R. (1997). Pests, pesticides and the environment: a historical perspective on the prospects for pesticide reduction. In Pimentel, D. (ed.). *Techniques for Reducing Pesticide Use: Economic and Environmental Benefits*. New York: John Wiley & Sons, pp. 13–34.

Pesek, J. (1993). Historical perspective. In Hatfield, J. L. & Karlen, D. L. (eds.). *Sustainable Agricultural Systems*. Boca Raton, FL: Lewis Publishers, pp. 1–19.

Pickett, S. T. A. & White, P. (1985). *The Ecology of Natural Disturbances and Patch Dynamics*. Orlando, FL: Academic Press.

Pik, A. J., Oliver, I. & Beattie, A. J. (1999). Taxonomic sufficiency in ecological studies of terrestrial invertebrates. *Australian Journal of Ecology* **24**, 555–62.

Pimentel, D. (1995). Amounts of pesticides reaching target pests: environmental impacts and ethics. *Journal of Agricultural and Environmental Ethics* **8**, 17–29.

Poehling, H. M. (1996). Agricultural practices which enhance numbers of beneficial arthropods. *Acta Jutlandica* **71**, 269–75.

Pollard, E. (1968). The effect of the removal of the bottom flora of a hawthorn hedgerow on the Carabidae of the hedge bottom. *Journal of Applied Ecology* **5**, 125–39.

Pollard, E. & Yates, T. J. (1993). *Monitoring Butterflies for Ecology and Conservation*. London: Chapman & Hall.

Pollard, E., Hooper, M. D. & Moore, N. W. (1974). *Hedges*. London: Collins.

Powell, W. (1986). Enhancing parasitoid activity in crops. In Waage, J. & Greathead, D. (eds.). *Insect parasitoids*. London: Academic Press, pp. 319–40.

Pretty, J. N. (1998). *The Living Land: Agriculture, Food and Community Regeneration in Rural Europe*. London: Earthscan.

Prinsloo, G. L. & Neser, O. C. (1994). The *Aphytis* fauna of the Afrotropical region. In Rosen, D. (ed.). *Advances in the Study of* Aphytis. Andover, UK: Intercept, pp. 279–302.

Prinsloo, G. & Samways, M. J. (2001). Host specificity among introduced chalcidoid biological control agents in South Africa. In Lockwood, J. A., Howarth, F. G. & Purcell, M. F. (eds.). *Balancing Nature: Assessing the Impact of Importing Non-native Biological Control Agents (An International Perspective)*. Lanham, MD: Entomological Society of America, pp. 31–40.

Prokopy, R. J. (1994). Integration in orchard pest and habitat management: a review. *Agriculture, Ecosystems and Environment* **50**, 1–10.

Pschorn-Walcher, H. (1977). Biological control of forest insects. *Annual Review of Entomology* **22**, 1–22.

Qualset, C. O., McGuire, P. E. & Warburton, M. L. (1995). 'Agrobiodiversity': key to agricultural productivity. *California Agriculture* **49**, 45–9.

Quinn, M. A., Kepner, R. L., Walgenbach, D. D., *et al.* (1993). Grasshopper stages of development as indicators of nontarget arthropod activity: implications for grasshopper management programs on mixed-grass rangeland. *Environmental Entomology* **22**, 532–40.

Quisenberry, S. S. & Peairs, F. B. (1998). *Response Model for an Introduced Pest: The Russian Wheat Aphid*. Lanham, MD: Entomological Society of America.

Rabb, R. L. (1972). Principles and contexts of pest management. In *Implementing Practical Pest Management Studies*. Lafayette, IN: Purdue University, pp. 6–29.

Radford, B. J., Wilson-Rummenie, A. C., Simpson, G. B., Bell, K. L. & Ferguson, M. A. (2001). Compacted soil affects soil macrofauna populations in a semiarid environment in central Queensland. *Soil Biology & Biodiversity* **33**, 1869–72.

Rands, M. R. & Sotherton, N. W. (1986). Pesticide use on cereal crops and changes in the abundance of butterflies on arable farmland. *Biological Conservation* **36**, 71–82.

Ranney, J. W., Brunner, M. C. & Levenson, J. B. (1981). The importance of edge in the structure and dynamics of forest islands. In Burgess, R. L. & Sharpe, D. M. (eds.). *Forest Island Dynamics in Man-dominated Landscapes*. New York: Springer-Verlag, pp. 67–96.

Raymond, F. & Waltham, R. (1996). *Forage Conservation and Feeding*, 5th edn. Ipswich, UK: Farming Press.

Read, J. L. & Andersen, A. N. (2000). The value of ants as early warning indicators: responses to pulsed cattle grazing at an Australian arid zone locality. *Journal of Arid Environments* **45**, 231–51.

Reid, R. S., Kruska, R. L., Deichmann, U. & Thornton, P. K. (2000). Human population growth and the extinction of the tsetse fly. *Agriculture, Ecosystems and Environment* **77**, 227–36.

Resh, V. H. & McElravy, E. P. (1993). Contemporary quantitative approaches to biomonitoring using benthic macroinvertebrates. In Rosenberg, D. M. & Resh, V. H. (eds.). *Freshwater Biomonitoring and Benthic Macroinvertebrates*. New York: Chapman & Hall, pp. 159–94.

Reus, J. A. W. & Pak, G. A. (1993). An environmental yardstick for pesticides. *Mededelingen Faculteit Landbouwwetenschappen Rijksuniversiteit Gent* **58/2a**, 249–55.

Reynolds, H. T., Adkisson, P. L., Smith, R. F. & Fisbie, R. E. (1982). Cotton insect pest management. In Metcalf, R. L. & Luckmann, W. L. (eds.). *Introduction to Insect Pest Management*, 2nd edn. New York: John Wiley & Sons, pp. 375–441.

Richards, A., Matthews, M. & Christian, P. (1998). Ecological considerations for the impact evaluation of recombinant baculovirus insecticides. *Annual Review of Entomology* **43**, 493–517.

Riechert, S. E. (1998). The role of spiders and their conservation in the agroecosystem. In Pickett, C. H. & Bugg, R. L. (eds.). *Enhancing Biological Control*. Berkeley, CA: University of California Press, pp. 211–37.

Riechert, S. E. & Bishop, L. (1990). Prey control by an assemblage of generalist predators: spiders in garden test systems. *Ecology* **71**, 1441–50.

Riegert, P. W., Ewen, A. B. & Lockwood, J. A. (1997). A history of chemical control of grasshoppers and locusts 1940–1990. In Gangwere, S. K., Muralirangan, M. C. & Muralirangan, M. (eds.). *Bionomics of Grashoppers, Katydids and their Kin*. Wallingford, UK: CAB International, pp. 385–406.

Robinson, J., Tuden, D. & Pease, S. (1995). *Taxing Pesticides to Fund Environmental Protection and Integrated Pest Management*. University of California, Berkeley, CA: Center for Occupational and Environmental Health.

Robinson, R. A. & Sutherland, W. J. (2002). Post-war changes in arable farming and biodiversity in Great Britain. *Journal of Applied Ecology* **39**, 157–76.

Robson, W. J. (1992). Landscape permeability to butterfly movement at an abandoned meadow in southern Norway. B. Sc. thesis. Southampton, UK: University of Southampton.

Rochester, W. A. & Zalucki, M. P. (1998). Measuring the impacts of *Helicoverpa armigera* migration on pest management during summer and autumn. In Zalucki, M. P., Drew, R. A. I. & White, G. G. (eds.). *Pest Management: Future Challenges*, Vol. 2. Brisbane, Australia: University of Queensland, pp. 94–8.

Roitberg, B. D. (2000). Threats, flies and protocol gaps: can evolutionary biology save biological control? In Hochberg, M. E. & Ivey, A. R. (eds.). *Parasitoid Population Biology*. Princeton, MA: Princeton University Press, pp. 254–65.

Root, R. B. (1973). Organisation of a plant-arthropod association in simple and diverse habitats: the fauna of collards (*Brassica oleracea*). *Ecological Monographs* **43**, 95–124.

Rosen, D. & DeBach, P. (1979). *Species of* Aphytis *of the world*. The Hague: W. Junk.

Rosenheim, J. A. (2001). Source-sink dynamics for a generalist insect predator in habitats with strong higher-order predation. *Ecological Monographs* **71**, 93–116.

Rothschild, G. H. (1998). Applied entomology: prospects and challenges for the next millennium. In Zalucki, M. P., Drew, R. A. I. and White, G. G. (eds.). *Pest Management: Future Challenges*. Brisbane, Australia: University of Queensland, pp. 1–10.

Roush, R. T. & Tabashnik, B. E. (1990). *Pesticide Resistance in Arthropods*. London: Chapman & Hall.

Roy, H. E. & Pell, J. K. (2000). Interactions between entomopathogenic fungi and other natural enemies: implications for biological control. *Biocontrol Science and Technology* **10**, 737–52.

Ruberson, J. R., Nemeto, H, & Hirose, Y. (1998). Pesticides and conservation of natural enemies in pest management. In Barbosa, P. (ed.). *Conservation Biological Control*. San Diego, CA: Academic Press, pp. 207–33.

Rudd, R. L. (1964). *Pesticides in the Living Landscape*. London: Faber & Faber.

Russell, E. P. (1989). Enemies hypothesis: a review of the effect of vegetation diversity on predatory insects and parasitoids. *Environmental Entomology* **18**, 590–99.

Rykiel, E. J. (1985). Towards a definition of ecological disturbance. *Australian Journal of Ecology* **10**, 361–5.

Samways, M. J. (1994). *Insect Conservation Biology*. London: Chapman & Hall.

Samways, M. J. (1997). Classical biological control and biodiversity conservation: what risks are we prepared to accept? *Biodiversity and Conservation* **6**, 1309–16.

Samways, M. J. & Lockwood, J. A. (1998). Orthoptera conservation: pests and paradoxes. *Journal of Insect Conservation* **2**, 143–9.

Sands, D. P. A. (1997). The 'safety' of biological control agents: assessing their impact on beneficial and other non-target arthropods. *Memoirs of Museum Victoria* **56**, 611–16.

Sands, D. P. A. & New, T. R. (2002). *The Action Plan for Australian Butterflies*. Canberra, Australia: Environment Australia.

Sands, D. P. A. & van Driesche, R. G. (2000). Evaluating the host range of agents for biological control of arthropods: rationale, methodology and interpretation. In van Driesche, R., Heard, T., McClay, A. & Reardon, R. (eds.). *Host Specificity Testing of Exotic Arthropod Biological Control Agents: The Biological Basis for Improvement in Safety*. Morgantown, WV: USDA Forest Service, pp. 69–83.

Saunders, D. A. (2000). Biodiversity does matter. In Barlow, T. & Thorburn, R. (eds.). *Balancing Conservation and Production in Grassy Landscapes*. Canberra, Australia: Environment Australia, pp. 14–18.

Saville, N. M., Dramstad, W. E., Fry, G. L. A. & Corbet, S. A. (1997). Bumblebee movement in a fragmented agricultural landscape. *Agriculture, Ecosystems and Environment* **61**, 145–54.

Schellhorn, N. A., Harmon, J. H. & Andow, D. A. (2000). Using cultural practice to enhance insect pest control by natural enemies. In Rechcigl, J. (ed.). *Environmentally Sound Approaches to Insect Pest Management*. Chelsea, MI: Ann Arbor Press, pp. 147–70.

Schonrogge, K., Stone, G. N. & Crawley, M. J. (1996). Alien herbivores and native parasitoids: rapid developments and structure of the parasitoid and inquiline complex in an invading gall wasp, *Andricus quercuscalicis* (Hymenoptera: Cynipidae). *Ecological Entomology* **21**, 71–80.

Schwartz, M. W., Brigham, C. A., Hoeksema, J. P. *et al.* (2000). Linking biodiversity to ecosystem function: implications for conservation biology. *Oecologia* **122**, 297–305.

Seastedt, T. R. (2000). Soil fauna and control of carbon dynamics: comparisons of rangelands and forests across latitudinal gradients. In Coleman, D. C. & Hendrix, P. F. (eds.). *Invertebrates as Webmasters in ecosystems*. Wallingford, UK: CAB International, pp. 293–312.

Sewall, D. K. & Croft, B. A. (1987). Chemotherapeutic and non-target side effects of benomyl to orange tortrix, *Argyrotaenia citrana* (Lepidoptera: Tortricidae) and its braconid endoparasite *Apanteles aristoteliae* (Hymenoptera: Braconidae). *Environmental Entomology* **16**, 507–12.

Sgardelkis, S. P. & Usher, M. B. (1994). Responses of soil Cryptostigmata across the boundary between a farm woodland and an arable field. *Pedobiologia* **38**, 36–49.

Shattuck, S. O. (1999). *Australian Ants: Their Biology and Identification.* Melbourne, Australia: CSIRO.

Shaw, M. W. & Hochberg, M. E. (2001). The neglect of parasitic Hymenoptera in insect conservation strategies: the British fauna as a prime example. *Journal of Insect Conservation* **5**, 253–63.

Sheppard, A. W. (1999). Which test? A mini-review of test usage in host specificity testing. In Withers, T. M., Barton-Browne, L. & Stanley, J. (eds.). *Host Specificity Testing in Australasia: Towards Improved Assays for Biological Control.* Brisbane, Australia: Department of Natural Resources, Queensland, pp. 60–69.

Simberloff, D. (1992). Conservation of pristine habitats and unintended effects of biological control. In Kaufman, W. C. & Nechols, J. E. (eds.). *Thomas Say Proceedings of the Entomological Society of America*, Vol. 1. Lanham, MD: Entomological Society of America, pp. 103–17.

Simberloff, D. & Stiling, P. (1996). How risky is biological control? *Ecology* **77**, 1965–74.

Smallshire, D. & Cooke, A. I. (1999). Field margins in the UK agri-environment schemes. *Aspects of Applied Biology* **54**, 19–28.

Smith, R. F. (1969). The new and the old in pest control. *Proceedings, Academia Nazionale de Lincei (Rome)* **366**, 21–30.

Smith, S. M. (1996). Biological control with *Trichogramma*: advances, successes, and potential of their use. *Annual Review of Entomology* **41**, 375–406.

Sohlenius, B. (1980). Abundance, biomass and contribution to energy flow by soil nematodes in terrestial ecosystems. *Oikos* **34**, 186–94.

Sommaggio, D. (1999). Syrphidae: can they be used as environmental bioindicators? *Agriculture, Ecosystems and Environment* **74**, 343–56.

Sotherton, N. W. (1989). Farming practices to reduce the exposure of non-target invertebrates to pesticides. In Jepson, P. C. (ed.). *Pesticides and Non-target Invertebrates.* Wimborne, UK: Intercept, pp. 195–212.

Sotherton, N. W. (1990). The effects of six insecticides used in UK cereal fields on sawfly larvae (Hymenoptera, Tenthredinidae). In *Proceedings of the Brighton Crop Protection Conference: Pests and Diseases.* Alton, UK: British Crop Protection Council, pp. 999–1004.

Sotherton, N. W. (1995). Beetle banks: helping nature to control pests. *Pest Outlook* **6**, 13–17.

Sotherton, N. W., Boatman, N. D. & Rands, M. R. W. (1989). The conservation headlands experiment in cereal ecosystems. *Entomologist* **108**, 135–143.

Southwood, T. R. E. (1962). Migration of terrestrial arthropods in relation to habitat. *Biological Reviews* **37**, 171–214.

Southwood, T. R. E. & Henderson, P. A. (2000). *Ecological Methods*, 3rd edn. Oxford: Blackwell Science.

Southwood, T. R. E. & Way, M. J. (1970). Ecological background to pest management. In Rabb, R. L. & Guthrie, F. E. (eds.). *Concepts of Pest Management.* Raleigh, NC: North Carolina State University, pp. 6–29.

Speight, M. R. (1997). Forest insects in the tropics: current status and future threats. In Watt, A. D., Stork, N. E. & Hunter, M. D. (eds.). *Forests and Insects.* London: Chapman & Hall, pp. 207–27.

Spellerberg, I. F. (1993). *Monitoring Ecological Change.* Cambridge: Cambridge University Press.

Spencer, J. (2002). Managing wood pasture landscapes in England: the New Forest and other more recent examples. In Redecker, B., Finck, P., Hardtle, W., Reicken, U. & Schroder, E. (eds.). *Pasture Landscapes and Nature Conservation.* Berlin: Springer-Verlag, pp. 123–36.

Stanisic, J. (1999). Land snails and dry vine thickets in Queensland: using museum invertebrate collections in conservation. pp. 257–63. In Ponder, W. & Lunney, D. (eds.). *The Other 99%: The Conservation and biodiversity of invertebrates. Transactions of the Royal Zoological Society of New South Wales*, Mosman.

Stary, P. & Pike, K. S. (1999). Uses of beneficial insect diversity in agroecosystem management. In Collins, W. W. & Qualset, C. O. (eds.). *Biodiversity in Agroecosystems.* Boca Raton, FL: CRC Press, pp. 49–67.

Stavola, A. M. & Craven, H. (1992). Terrestrial ecological risk assessment of pesticides in the United States. In Greig-Smith, P. W., Becker, H., Edwards, P. J.& Heimbach, F. (eds.). *Ecotoxicology of Earthworms.* Andover, UK: Intercept, pp. 177–84.

Stern, V. M. (1981). Environmental control of insects using trap crops, sanitation, prevention and harvesting. In Pimentel, D. (ed.). *CRC Handbook for Pest Management in Agriculture*, Vol. 1. Boca Raton, FL: CRC Press, pp. 199–207.

Stiling, P. & Simberloff, D. (2000). The frequency and strength of nontarget effects of invertebrate biological control agents of plant pests and weeds. In Follett, P. A. & Duan, J. J. (eds.). *Nontarget Effects of Biological Control.* Boston, MA: Kluwer, pp. 31–43.

Stinner, G. E. & House, G. J. (1990). Arthropods and other invertebrates in conservation tillage agriculture. *Annual Review of Entomology* **35**, 299–318.

Stork, N. E. (1998). Insect diversity: facts, fiction and speculation. *Biological Journal of the Linnean Society* **35**, 321–37.

Story, P. & Cox, M. (2001). Review of the effects of organophosphorus and carbamate insecticides on vertebrates. Are there implications for locust management in Australia? *Wildlife Research* **28**, 179–93.

Strong, D. R. & Pemberton, R. W. (2001). Food webs, risks of alien enemies, and reform of biological control. In Wajnberg, E., Scott, J. K. & Quimby, P. C. (eds.). *Evaluating Indirect Effects of Biological Control.* Wallingford, UK: CAB International, pp. 57–79.

Strong, L. (1992a). Avermectins: a review of their impact on insects of cattle dung. *Bulletin of Entomological Research* **82**, 265–74.

Strong, L. (1992b). The use and abuse of feed-through compounds in cattle treatments. *Bulletin of Entomological Research* **82**, 1–4.

Strong, L. (1993). Overview: the impact of avermectins on pastureland ecology. *Veterinary Parasitology* **48**, 3–17.

Sunderland, K. D. & Chambers, R. J. (1983). Invertebrate polyphagous predators as pest control agents: some criteria and methods. In Cavalloro, R. (ed.). *Aphid Antagonists*. Rotterdam: A. A. Balkema, pp. 100–108.

Sunderland, K. D., Axelsen, J. A., Dromph, K., *et al.* (1997). Pest control by a community of natural enemies. *Acta Jutlandica* **72**, 271–326.

Swift, M. J. & Anderson, J. M. (1993). Biodiversity and ecosystem function in agroecosystems. In Schultze, E. & Mooney, H. A. (eds.). *Biodiversity and Ecosystem Function*. New York: Springer-Verlag, pp. 57–83.

Talekar, N. S. & Shelton, AM. (1993). Biology, ecology and management of diamondback moth. *Annual Review of Entomology* **38**, 275–301.

Taylor, R. W. (1983). Descriptive taxonomy: past, present and future. In Highley, E. & Taylor, R. W. (eds.). *Australian Systematic Entomology: A Bicentenary Perspective*. Melbourne, Australia: CSIRO, pp. 93–134.

Teetes, G. L. (1981). The environmental control of insects using planting times and plant spacing. In Pimentel, D. (ed.). *CRC Handbook of Pest Management in Agriculture*, Vol. 1. Boca Raton, FL: CRC Press, pp. 209–21.

Thiele, H. V. (1977). *Carabid Beetles in their Environment*. Berlin: Springer-Verlag.

Thomas, C. D. (1994). Ecology and conservation of butterfly metapopulations in the fragmented British landscape. In Pullin, A. S. (ed.). *Ecology and Conservation of Butterflies*. London: Chapman & Hall, pp. 46–63.

Thomas, C. D. & Hanski, I. (1997). Butterfly metapopulations. In Hanski, I. & Gilpin, M. E. (eds.). *Metapopulation Biology: Ecology, Genetics and Evolution*. San Diego, CA: Academic Press, pp. 359–86.

Thomas, C. D. & Harrison, S. (1992). Spatial dynamics of a patchily-distributed butterfly species. *Journal of Animal Ecology* **61**, 437–46.

Thomas, C. D. & Jones, T. M. (1993). Partial recovery of a skipper butterfly (*Hesperia comma*) from population refuges: lessons for conservation in a fragmented landscape. *Journal of Animal Ecology* **62**, 472–81.

Thomas, C. F. G., Parkinson, L., Griffiths, G. J. K., Fernandez Garcia, A. & Maskell, E. J. P. (2001). Aggregation and temporal stabilty of carabid beetle distributions in field and hedgerow habitats. *Journal of Applied Ecology* **38**, 100–116.

Thomas, C. F. G., Holland, J. M. & Brown, J. M. (2002). The spatial distribution of carabid beetles in agricultural landscapes. In Holland, J. M. (ed.). *The Agroecology of Carabid Beetles*. Andover, UK: Intercept, pp. 305–44.

Thomas, J. A. (1983a). The ecology and conservation of *Lysandra bellargus* (Lepidoptera: Lycaenidae) in Britain. *Journal of Applied Ecology* **20**, 59–83.

Thomas, J. A. (1983b). The ecology and status of *Thymelicus acteon* (Lepidoptera: Hesperiidae) in Britain. *Ecological Entomology* **8**, 427–35.

Thomas, J. A. (1984). The conservation of butterflies in temperate countries: past efforts and lessons for the future. In Vane-Wright, R. I. & Ackery, P. R. (eds.). *The Biology of Butterflies*. London: Academic Press, pp. 333–53.

Thomas, J. A., Thomas, C. D., Simcox, D. J. & Clarke, R. T. (1986). The ecology and declining status of the Silver-spotted Skipper butterfly (*Hesperia comma*) in Britain. *Journal of Applied Ecology* **23**, 365–380.

Thomas, M. B. & Willis, A. J. (1998). Biocontrol: risky but necessary? *Trends in Ecology and Evolution* **13**, 325–9.

Thomas, M. B., Wratten, S. D. & Sotherton, N. W. (1991). Creation of 'island' habitats in farmland to manipulate populations of beneficial arthropods: predator densities and emigration. *Journal of Applied Ecology* **28**, 906–17.

Thomas, M. B., Wratten, S. D. & Sotherton, N. W. (1992a). Creation of 'island' habitats in farmland to manipulate populations of beneficial arthropods: predator densities and species composition. *Journal of Applied Ecology* **29**, 524–31.

Thomas, M. B., Sotherton, N. W., Coombes, D. S. & Wratten, S. D. (1992b). Habitat factors influencing the distribution of polyphagous predatory insects between field boundaries. *Annals of Applied Biology* **120**, 197–202.

Tothill, J. D., Taylor, T. H. C.& Paine, R. W. (1930). *The Coconut Moth in Fiji: A History of its Control by Means of Parasites*. London: Imperial Institute of Entomology.

Tscharntke, T. 2000. Parasitoid populations in the agricultural landscape. In Hochberg, M. E. & Ives, A. R. (eds.). *Parasitoid Population Biology*. Princeton, NJ: Princeton University Press, pp. 235–53.

Tscharntke, T. & Kruess, A. (1999). Habitat fragmentation and biological control. In Hawkins, B. A. & Cornell, H. V. (eds.). *Theoretical Approaches to Biological Control*. Cambridge: Cambridge University Press, pp. 190–205.

Turner, C. E. (1985). Conflicting interests and biological control of weeds. In Delfosse, E. S. (ed.). *Proceedings of the Sixth International Symposium on Biological Control of Weeds, Vancouver*. Ottawa: Agriculture Canada.

UNCED (1992). *Agenda 21: The United Nations Programme of Action from Rio*. New York: United Nations.

Usher, M. B. (1986). Wildlife conservation evaluation: attributes, criteria and values. In Usher, M. B. (ed.). *Wildlife Conservation Evaluation*. London: Chapman & Hall, pp. 3–44.

Usher, M. B. (1995). A world of change: land-use patterns and arthropod communities. In Harrington, R. Q., Stork, N. E. (eds.). *Insects in a Changing Environment*. London: Academic Press, pp. 372–97.

Usher, M. B. & Keiller, S. W. J. (1998). The macrolepidoptera of farm woodlands: determinants of diversity and community structure. *Biodiversity and Conservation* **7**, 725–48.

Vandermeer, J. & Perfecto, I. 1995). *Breakfast of Biodiversity: The Truth about Rainforest Destruction*. Oakland: CA: Food First Books.

Vandermeer, J., van Noordwijk, M., Anderson, J., Ong, C. & Perfecto, I. (1998). Global change and multi-species agroecosystems: concepts and issues. *Agriculture, Ecosystems and Environment* **67**, 1–22.

van der Meijden, E. & van der Ven-van Wijk, C. A. M. (1997). Tritrophic metapopulation dynamics. A case study of ragwort, the cinnabar moth and the parasite *Cotesia popularis*. In Hanski, I. & Gilpin, M. E. (eds.). *Metapopulation Biology: Ecology, Genetics and Evolution*. San Diego, CA: Academic Press, pp. 387–405.

van der Werf, H. M. G. (1996). Assessing the impacts of pesticides in the environment. *Agriculture, Ecosystems and Environment* **60**, 81–96.

Vanderwoude, C., Andersen, A. N, & House, A. P. N. (1997). Ant communities as bio-indicators in relation to fire management of spotted gum (*Eucalyptus maculata* Hook.) forests in south east Queensland. *Memoirs of Museum Victoria* **56**, 671–5.

van Driesche, R. G. & Hoddle, M. (1997). Should arthropod parasitoids and predators be subject to host range testing when used as biological control agents? *Agriculture and Human Values* **14**, 211–26.

van Driesche, R. G. & Hoddle, M. S. (2000). Classical arthropod biological control: measuring success, step by step. In Gurr, G. & Wratten, S. (eds.). *Biological Control: Measures of Success*. Dordrecht: Kluwer, pp. 39–75.

van Driesche, R., Heard, T., McClay, A. & Reardon, R. (2000). *Host Specificity Testing of Exotic Arthropod Biological Control Agents: The Biological Basis for Improvement in Safety*. Morgantown, WV: USDA Forest Service.

van Emden, H. F. (1983). The anatomy of a pest management programme. In Cavalloro, R. (ed.). *Statistical and Mathematical Methods in Population Dynamics and Pest Control*. Rotterdam and Boston: A. A. Balkema, pp. 127–35.

van Emden, H. F. & Williams, G. F. (1974). Insect stability and diversity in agroecosystems. *Annual Review of Entomology* **19**, 455–75.

van Hook, T. (1994). The conservation challenge in agriculture and the role of entomologists. *Florida Entomologist* **77**, 42–73.

van Lenteren, J. C. (1986). Evaluation, mass production, quality control and release of entomophagous insects. In Franz, J. M. (ed.). *Biological Plant and Health Protection*. Stuttgart: Fischer-Verlag, pp. 31–56.

van Nouhuys, S. & Hanski, I. (2000). Apparent competition between parasitoids mediated by a shared hyperparasitoid. *Ecological Letters* **3**, 82–4.

van Swaay, C. & Warren, M. (1999). *Red Data Book of European Butterflies (Rhopalocera)*. Strasbourg: Council of Europe Publishing.

Vereijken, P.(1989). From integrated control and integrated farming, an experimental approach. *Agriculture, Ecosystems and Environment* **26**, 37–43.

Vereijken, P., Wijnand, F. & Stol, W. (1995). *Progress Report 2. Designing and Testing Prototypes. Progress Reports of the Research Network on Integrated and Ecological Arable Farming Systems for EU and Associated Countries*. Wageningen, the Netherlands: Research Institute for Agroecology and Soil Fertility.

VFRAC (Victorian Recreational Fishermen's Advisory Committee) (1987). *Spraying of Agricultural Chemicals from Aircraft and for Insecticidal and Other Purposes. Report to Minister for Conservation*. Melbourne, Australia: VFRAC.

Vickerman, G. P. & Sunderland, K. D. (1977). Some effects of dimethoate on arthropods in winter wheat. *Journal of Applied Ecology* **14**, 767–77.

Volkl, W., Zwolfer, H., Romstock-Volkl, M. & Schmelzer, C. (1993). Habitat management in calcareous grassland: effects on the insect community developing in flower heads of Cynareae. *Journal of Applied Ecology* **30**, 307–15.

Waage, J. 1989. The population ecology of pest-pesticide-natural enemy interactions. In Jepson, P. C. (ed.). *Pesticides and Non-target Invertebrates* Wimborne, UK: Intercept, pp. 81–93.

Waage, J. 1997. 'Yes, but does it work in the field?' The challenge of transferring biological control technology. *Entomophaga* **41**, 315–32.

Walker, B. H. (1992). Biodiversity and ecological redundancy. *Conservation Biology* **6**, 18–23.

Walker, G. P., Zarech, N., Bayoun, I. M. & Triapitsyn, S. V. (1997). Introduction of western Asian egg parasitoids into California for biological control of beet leafhoppers, *Circulifer tenellus*. *Pan-Pacific Entomologist* **73**, 236–42.

Walter, D. E. & Proctor, H. C. (1999). *Mites: Ecology, Evolution and Behaviour*. Wallingford, UK: CAB International.

Wapshere, A. J. (1974). A strategy for evaluating the safety of organisms for the biological control of weeds. *Annals of Applied Biology* **77**, 201–11.

Wapshere, A. J. (1989). A testing sequence for reducing rejection of potential biological control agents for weeds. *Annals of Applied Biology* **114**, 515–26.

Wapshere, A. J. (1992). Comparing methods for selecting effective biological agents for weeds. pp. 557–60. In *Proceedings of the First International Weed Control Congress, Melbourne*. Melbourne, Australia: Weed Science Society of Victoria.

Wardhaugh, K. G. & Rodriguez-Menendez, H. (1988). The effects of the antiparasitic drug, ivermectin, on the development and survival of the dung-breeding fly, *Orthelia cornicina* (F.) and the scarabaeine dung beetles, *Copris hispanus* L., *Bubas bubalus* (Olivier) and *Onitis belial* F. *Journal of Applied Entomology* **106**, 381–9.

Watt, A. D., Stork, N. E., Eggleton, P., *et al.* (1997). Impact of forest loss and regeneration on insect abundance and diversity. In Watt, A. D., Stork, N. E. & Hunter, M. D. (eds.). *Forests and Insects*. London: Chapman & Hall, pp. 273-86.

Wells, S. M., Pyle, R. M. & Collins, N. M. (1983). The *IUCN Invertebrate Red Data Book*. Gland, Switzerland: IUCN.

Western, D. (1989). Conservation without parks: wildlife in the rural landscape. In Western, D. & Pearl, M. C. (eds.). *Conservation for the Twenty First Century*. New York: Oxford University Press, pp. 158–65.

Whaley, W. H., Anhold, J. & Schaalje, G. B. (1998). Canyon drift and dispersion of *Bacillus thuringiensis* and its effects on select nontarget lepidopterans in Utah. *Environmental Entomology* **27**, 539–48.

Whitcomb, W. H. & Bell, K. (1964). Predaceous insects, spiders and mites of Arkansas cotton fields. *Bulletin of the Arkansas Agricultural Experiment Station* **690**, 1–84.

White, E. G. (2002). *New Zealand Tussock Grassland Moths*. Lincoln, New Zealand: Manaaki Whenua Press.

Whitford, W. G. (2000). Keystone arthropods as webmasters in desert ecosystems. In Coleman, D. C. & Hendrix, P. F. (eds.). *Invertebrates as Webmasters in Ecosystems*. Wallingford, UK: CAB International, pp. 25–41.

Wiens, J. A. (1997). Metapopulation dynamics and landscape ecology. In Hanski, I. & Gilpin, M. E. (eds.). *Metapopulation Biology: Ecology, Genetics and Evolution*. San Diego, CA: Academic Press, pp. 43–62.

Wiens, J. A. (2002). Central concepts and issues of landscape ecology. In Gutzwiller, K. J. (ed.). *Applying Landscape Ecology in Biological Conservation*. New York: Springer-Verlag, pp. 3–21.

Wilson, E. O. (1987). The little things that run the world (the importance and conservation of invertebrates). *Conservation Biology* **1**, 344–6.

Wise, D. H. (1993). *Spiders in Ecological Food Webs*. Cambridge: Cambridge University Press.

With, K. A. (2002). Using percolation theory to assess landscape connectivity and effects of habitat fragmentation. In Gutzwiller, K. J. (ed.). *Applying Landscape Ecology in Biological Conservation*. New York: Springer-Verlag, pp. 195–30.

Withers, T. M., Barton-Browne, L. & Stanley, J. (1999). *Host Specificity Testing in Australasia. Towards Improved Assays for Biological Control.* Brisbane, Australia: Department of Natural Resources Queensland.

Withers, T., Barton-Browne, L. & Stanley, S. (2000). How time-dependent processes can affect the outcome of assays. In van Driesche, R., Heard, T., McClay, A. & Reardon, R. (eds.). *Host Specificity Testing of Exotic Arthropod Biological Control Agents: The Biological Basis for Improvement in Safety.* Morgantown, WV: USDA Forest Service, pp. 27–41.

Wolt, J. D., Peterson, R. K. D., Bystrak, P. & Meade, T. (2003). A screening level approach for nontarget insect risk assessment: transgenic Bt corn pollen and the monarch butterfly (Lepidoptera: Danaidae). *Environmental Entomology* **32**, 237–46.

Wood, D. & Lenné, J. M. (1999). Why agrobiodiversity?. In Wood, D. & Lenné, J. M. (eds.). *Agrobiodiversity: Characterisation, Utilisation and Management.* Wallingford: CAB International, pp. 1–13.

Wratten, S. D. & Thomas, C. H. (1990). Farm-scale spatial dynamics of predators and parasitoids in agricultural landscapes. In Bunce, R. G. H. & Howard, D. C. (eds.). *Species Dispersal in Agricultural Habitats.* London: Belhaven Press, pp. 219–37.

Wratten, S. D., van Emden, H. F. & Thomas, M. B. (1998). Within-field and border refugia for the enhancement of natural enemies. In Pickett, C. H. & Bugg, R. L. (eds.). *Enhancing Biological Control.* Berkeley, CA: University of California Press, pp. 375–403.

Yeates, G. W. (1979). Soil nematodes in terrestrial ecosystems. *Journal of Nematology* **11**, 213–29.

Yeates, G. W. (1994). Modification and quantification of the nematode maturity index. *Pedobiologia* **38**, 97–101.

Yeates, G. W. & Bongers, T. (1999). Nematode diversity in agroecosystems. *Agriculture, Ecosystems and Environment* **74**, 113–35.

Yeates, G. W., Bongers, T., de Goede, R. G. M., Freckman, D. W. & Georgieou, S. S. (1993). Feeding habit in soil nematode families and genera: an outline for ecologists. *Journal of Nematology* **25**, 315–31.

Yen, A. L. & New, T. R. (eds.) (1997). Proceedings of the conference 'Invertebrate biodiversity and conservation'. *Memoirs of Museum Victoria* **56**, 261–675.

Young, O. P. & Edwards, J. P. (1990). Spiders in United States field crops, and their potential effect on crop pests. *Journal of Arachnology* **18**, 1–27.

Zalucki, M. P., Rochester, W. A., Norton, G. A., *et al.*(1998). IPM and Heliothis: what we have to do to make it work. In Zalucki, M. P., Drew, R. A. I. & White, G. G. (eds.). *Pest Management: Future Challenges*, Vol. 2. Brisbane, Australia: University of Queensland, pp. 107–14.

Index

VERMONT STATE COLLEGES

Hartness Library
Vermont Technical College
One Main St.
Randolph Center, VT
DISCARD